Temperature Behavior of Bridges

Bridges are subject to daily and seasonal temperature fluctuations. The temperature variations can affect bridge materials and structural integrity, often interacting with other loads and masking their effects. Understanding temperature behaviors is crucial for accurate load assessment and bridge performance evaluation.

This book comprehensively studies temperature behaviors of bridges, covering beam, arch, cable-stayed, and suspension bridges using analytical, numerical, and field monitoring approaches. For each type of bridge, it not only reports field monitoring results but also presents an integrated heat-transfer and structural analysis framework, significantly enhancing the efficiency of simulating bridge temperature behaviors. Moreover, this book derives simple and general analytical formulas for temperature-induced deformations of bridges that can be easily adopted by engineers. This standout feature has not been previously studied and reported within academic and engineering societies.

A unique feature of this book is the presentation of 25-year field monitoring data of the Tsing Ma Suspension Bridge, the most extensive field data available, showing the long-term behavior of the bridge. This invaluable data demonstrates the effects of global warming on infrastructure and necessitates the review of current design codes in the context of climate change. Other typical bridges, including the Hong Kong-Zhuhai-Macao Bridge and the Hong Kong Polytechnic University Footbridge, are also used as examples to enhance understanding.

Temperature Behavior of Bridges is an essential resource for postgraduate students, researchers, and engineers seeking to master the temperature behaviors affecting modern bridge infrastructure.

Temperature Behavior of Bridges

Yong Xia and Yi Zhou

CRC Press
Taylor & Francis Group
Boca Raton London New York

CRC Press is an imprint of the
Taylor & Francis Group, an **informa** business

Cover image: Yong Xia and Yi Zhou

First edition published 2025
by CRC Press
4 Park Square, Milton Park, Abingdon, Oxon, OX14 4RN

and by CRC Press
2385 NW Executive Center Drive, Suite 320, Boca Raton FL 33431

CRC Press is an imprint of Informa UK Limited

British Library Cataloguing-in-Publication Data
A catalogue record for this book is available from the British Library

ISBN: 978-1-032-76470-2 (hbk)
ISBN: 978-1-032-76802-1 (pbk)
ISBN: 978-1-003-48013-6 (ebk)

DOI: 10.1201/9781003480136

Typeset in Times
by codeMantra

Contents

About the Authors

Professor Yong Xia is a Professor in the Department of Civil and Environmental Engineering at the Hong Kong Polytechnic University. He obtained his Bachelor's and Master's degrees from Huazhong University of Science and Technology, China in 1992 and 1997, respectively. He also received his PhD from Nanyang Technological University, Singapore, in 2002. Later on, Prof. Xia worked at The University of Tokyo and The University of Western Australia as a Postdoctoral Fellow, before he joined the Hong Kong Polytechnic University as an Assistant Professor in 2006. His research interests include structural health monitoring (SHM), structural damage detection, and thermal behavior of structures. He has secured over 50 research grants as PI, which include health monitoring of the Tsing Ma Bridge (1377 m long main span), the Stonecutters Bridge (1018 m long main span), the world's longest Hong Kong–Zhuhai–Macao Bridge, the footbridge in the campus of the University, Canton Tower (600 m tall), and Shanghai Tower (632 m tall). He and his team members have comprehensively and systematically investigated the temperature behavior of beam, arch, cable-stayed, and suspension bridges via laboratory testing, analytical analysis, field monitoring, and numerical simulation. He has published four conference proceedings, two research books, over 20 design standards, more than 150 refereed journal papers, and 140 conference papers. He is the co-editor-in-chief of *Advances in Structural Engineering*, a council member of the International Society for Structural Health Monitoring of Intelligent Infrastructure, and a fellow member of The Hong Kong Institution of Engineers. He has obtained many prestigious awards, including State Technological Innovation Award by the State Council, China; Natural Science Award by the Ministry of Education, China; Changjiang Scholars by the Ministry of Education, China, and The Nishino Prize, among others.

Dr. Yi Zhou is an Associate Professor in the Department of Civil Engineering at the University of Science and Technology Beijing. He received his Ph.D. from Tongji University, China, in 2015. He was awarded a "Hong Kong Scholar" postdoctoral fellowship and worked together with Prof. Yong XIA at the Hong Kong Polytechnic University. Currently, he serves as a council member of the Institute of Diagnosis Treatment and Maintenance for Engineering–Architectural Society of China and a research fellow of the National Observation and Research Station of Material Corrosion and Structural Safety of Hong Kong-Zhuhai-Macao Bridge in Guangdong. His research interests include condition monitoring, and intelligent operation and maintenance for civil engineering structures. He has been the principal investigator of many national-level research projects sponsored by the National Natural Science Foundation of China and the National Key Research and Development Program. He has published dozens of academic papers, authored one monograph, and has been granted over 10 invention patents. The results of his research projects have been applied in the health monitoring practice of multiple large-scale bridges. In 2021, he was awarded the First Prize of the 12th Beijing Municipal Young College Teachers' Basic Teaching Skills Competition.

1 Introduction

NOTATIONS

A	Cross-sectional area
I_y, I_z	Moments of inertia of a cross section
$T(y, z)$	Cross-sectional temperature distribution
h_y, h_z	Width and height of the cross section
ΔT_u	Uniform temperature change
ΔT_{My}, ΔT_{Mz}	Transverse and vertical linear temperature difference
$\Delta T_E(y,z)$	Nonlinear residual temperature

1.1 DEVELOPMENT OF BRIDGES

Bridges are overhead structures spanning rivers, valleys, or other transportation routes, providing passage for pedestrians, vehicles, pipelines, etc. Besides their functional value, bridges also possess aesthetic value. Many bridges have become distinctive cultural landmarks that are visually pleasing and representative of their era, constituting an important part of the world's architectural art. Therefore, bridges are the combined achievement of engineering technologies and human art.

Bridges generally consist of four basic components: superstructure, substructure, bearings, and accessories. The superstructure is the main load-bearing structure that spans the obstacle when the transportation route is interrupted. It refers to the portion of the bridge above the level of bearings. The substructure is the parts below the bearing level, including piers, abutments, and foundations. Piers and abutments support the superstructure and transmit the permanent loads and live loads like vehicles to the foundations. The abutments are typically located at the ends of a bridge, whereas piers are situated in the middle of a span or at intervals along the bridge. The foundation bears all loads transmitted from piers and abutments. Bearings are placed on the top of piers and abutments to support the superstructure and allow for movements and displacements according to design requirements. Accessories of a bridge include the bridge decking, expansion joints, transition slabs at bridge ends, and conical slopes, among others (Shao, 2019).

The history of bridges is as ancient as human civilization. The earliest bridges, such as fallen tree trunks or natural stone formations, likely formed naturally. Before the 18th-century Industrial Revolution, bridges were generally constructed using natural materials like wood, stone, or woven vines. However, organic materials such as wood have limited strength, are prone to rotting, and are not fire-resistant. Consequently, wooden bridges have low load-bearing capacity and short service lives.

While stone possesses high compressive strength and excellent weather resistance, it is unable to withstand significant tensile stresses caused by bending. As a result,

DOI: 10.1201/9781003480136-1

most stone bridges adopt the form of arches to maximize the utilization of the stone's compressive strength. Both the Romans and the Chinese excelled in constructing arch bridges, as exemplified by structures like the Pont du Gard in Nîmes, France and the Anji Bridge (or Zhaozhou Bridge) in Shijiazhuang, China. Additionally, with their superior compressive properties, bricks could serve as a substitute for natural stone and had been widely used in bridge construction in ancient times.

Bridge engineering grows with the development of new materials and techniques. The first significant leap in bridge engineering occurred with the emergence of metal materials, represented by iron and steel, replacing natural materials as the primary construction materials for bridges. This transformation was particularly notable during the mid-19th century, with the large-scale production of steel and subsequently with the appearance of high-strength steel and high-strength steel wires. These advancements greatly increased the spanning capacity of bridges, marking a significant advancement in bridge engineering.

In the early 20th century, the widespread use of reinforced concrete and the emergence of prestressed concrete technology in the 1930s brought about a second leap in bridge engineering. These advancements provided bridges with cost-effective, durable, and high-strength building materials, enabling the development of structures with significant stiffness and load-bearing capacity.

From the 1950s onward, with the rapid development of computer technology and finite element (FE) analysis, engineers were able to perform large-scale structural analyses that were previously impossible. This facilitated the design of bridges with longer spans, higher load-carrying capacities, and more innovative and complex forms. These technologies marked the third leap in the development of bridge engineering.

Since the 21st century, cutting-edge technologies such as smart sensors, data analytics, and artificial intelligence have been continuously integrated into the field of bridge engineering, injecting new vitality into the design, construction, and maintenance of bridges. The integration of these technologies has transformed bridges from stationary structures to intelligent systems that are capable of active sensing and adaptation to environmental changes. The field of bridge engineering has entered a new era of "smart" structures.

In summary, the development of bridge engineering reflects the progress of human civilization and represents technological advancements and human creativity.

1.2 CLASSIFICATION OF BRIDGES

Bridges can be classified in various ways. According to the mechanical characteristics of their superstructures, bridges can be categorized into four types: beam bridges, arch bridges, cable-stayed bridges, and suspension bridges.

According to their function, bridges can be classified as highway bridges, railway bridges, road-cum-rail bridges, pedestrian bridges, aqueduct bridges, and pipeline bridges.

According to the length of bridges, they can be classified as super major bridges, major bridges, and medium/short-span bridges. Generally, bridges with a single span of less than 150 m are considered medium/short-span bridges, while those with spans longer than 150 m are classified as major bridges. The classification of super major bridges varies depending on the bridge type. For example, suspension bridges have a starting span of 1000 m, cable-stayed bridges and steel arch bridges have a starting span of 500 m, and other bridge types have a starting span of 300 m (Shao, 2019).

Bridges can also be classified based on the primary load-bearing material used, such as masonry bridges (including brick, stone, and plain concrete bridges), timber bridges, reinforced concrete bridges, prestressed concrete bridges, steel bridges, and steel-concrete composite bridges.

Classification based on the nature of obstacles crossed includes river-crossing bridges, sea-crossing bridges, valley-crossing bridges, flyover or overpass bridges, interchange bridges, and viaduct bridges.

Based on their plan layout, bridges can be classified as orthogonal bridges, skew bridges, and curved bridges.

Based on the position of the bridge floor relative to the superstructure, bridges can be classified as deck, through, half-through, and suspended bridges.

Based on the mobility, bridges can be classified as fixed bridges and movable bridges. The latter include bascule, lift, swing, and floating bridges.

The above classifications are not mutually exclusive. Here, the main features of beam bridges, arch bridges, cable-stayed bridges, and suspension bridges are introduced based on the classification of structural mechanical characteristics.

1.2.1 Beam Bridges

Beam bridges are a type of bridge in which beams are the main load-bearing elements. These bridges are the simplest and oldest type in civil engineering. They are widely used for various span requirements due to their stable structure and straightforward appearance.

There are generally two types of main beams used in beam bridges: solid-web girders and open-web (or truss) girders. Solid-web girders have a simple construction and are easy to manufacture, erect, and maintain. They are extensively used for medium/short-span bridges. However, they are not economically efficient in material utilization. Truss girders, on the other hand, utilize the members to resist axial forces, allowing for optimal material usage, lighter self-weight, and greater spanning capacity. They are commonly employed in constructing long-span bridges, such as the famous Quebec Bridge in Canada. According to the degree of redundancy, beam bridges can be further classified into statically determinate and indeterminate bridges.

Simply supported beam bridges are the most common type of statically determinate beam bridges, featuring main load-bearing beams supported on piers or abutments at both ends. The structural analysis of this bridge type is straightforward and not influenced by settlement or displacement of the supports, making it suitable for various geological conditions. Moreover, a simply supported beam bridge has a simple construction, facilitating the production of standardized and prefabricated components. Both the manufacturing and erection processes are relatively convenient, contributing to its widespread use. However, as the span increases, the bending moment at the mid-span of a simply supported beam bridge increases dramatically, rendering it less cost-effective for longer spans. Generally, the economical range for single-span prestressed reinforced concrete simply supported bridges is between 20 and 50 m (Xiang et al., 2011).

Another type of statically determinate beam bridge is the overhang beam bridge. An overhang beam bridge refers to a simply supported beam that extends freely outward from one or both ends, serving as the main load-bearing component of the superstructure. The superstructure typically includes suspended spans, which rest on the cantilever arms with a hinge connection. Being a statically determinate structure, the internal forces of an overhang beam bridge are not affected by foundation settlement, resulting in lower foundation requirements. Under permanent loads, overhang beam bridges have negative moments at the intermediate supports. Therefore, the maximum positive bending moment at the mid-span is smaller than that of simply supported beam bridges, making them suitable for longer span bridge designs. However, due to intermediate hinges on the beam, the construction of overhang beam bridges is more complex, and the ride comfort is reduced. Additionally, it's challenging to use prefabricated construction methods, often requiring expensive and slow on-site casting on costly scaffolding. As a result, they are less used nowadays.

A continuous beam bridge is a prime example of statically indeterminate beam bridges. Its superstructure comprises beams continuously spanning three or more supports, serving as the primary load-bearing structure. These bridges allow longer spans due to the unloading effect of negative moments at the intermediate supports under permanent loads. This significantly reduces the maximum positive bending moment at the mid-span, as compared to simply supported beam bridges. Typically, prestressed reinforced concrete continuous beam bridges are economically applicable for spans between 50 and 150 m. On the other hand, steel continuous beam bridges are economically feasible for spans ranging from 200 to 300 m (Xiang et al., 2011). Due to the participation of adjacent spans in resisting bending actions, continuous beam bridges have greater structural stiffness and smaller deformations compared to simply supported beam bridges. Moreover, these bridges offer improved ride comfort, thanks to the continuous passage of the main beams through the intermediate supports and a reduced number of expansion joints. This makes them favorable for high-speed traffic. However, continuous beam bridges are statically indeterminate structures, and any uneven settlements in the supports may cause significant changes in internal forces. Thus, the construction of this type of bridge requires good geological conditions.

Additionally, if the bearings between the main beams and piers of a beam bridge are eliminated and the superstructure and piers are connected as a single unit, it becomes a rigid frame bridge. In this design, the piers bear the bending action of the main beams, resulting in smaller bending moments in the main beams than that of continuous beam bridges. This allows for greater spanning capacity. The economically feasible single-span range for prestressed reinforced concrete continuous rigid frame bridges is typically between 150 and 300 m (Xiang et al., 2011). Currently, the Chongqing Yangtze River Bridge is the longest-span rigid frame bridge, with a main span of 330 m.

1.2.2 Arch Bridges

An arch bridge is a bridge that uses curved arch rings or arch ribs as the main load-bearing structure. The arch rings or arch ribs primarily bear axial compression and can be constructed using materials such as bricks, stones, or concrete with good compressive strength. Under vertical loads, the arch bridge generates not only vertical reactions at the abutments but also horizontal reactions. The presence of horizontal thrust reduces the bending moments and shear forces in the arch compared to a beam bridge of the same span. Arch bridges have a greater spanning capacity than beam bridges. Stone arch bridges and reinforced concrete arch bridges have good durability and require less maintenance and repair costs. Additionally, the rainbow-like appearance of arch bridges adds aesthetic value. Arch bridges are competitive options for spans up to 500 m. Generally, the economically applicable range of single-span stone arch bridges is 20–50 m, reinforced concrete arch bridges are 50–150 m, concrete-filled steel tube arch bridges are 150–400 m, and steel arch bridges are 200–700 m (Xiang et al., 2011). Currently, the largest-span arch bridge is the Tian'e Longtan Bridge in China with a main span of 600 m.

Single-span arch bridges can be grouped as hingeless (or fixed) arches, single-hinged arches, two-hinged arches, and three-hinged arches according to the number of hinges; all are statically indeterminate structures except for three-hinged arches. Although hinged arches involve three hinge arrangements, only two hinges or three hinges are frequently used. Arch bridges can also be classified as deck arch, through arch, and half-through arch bridges according to the position of the deck relative to the arch. In a deck arch bridge, the deck is often supported by a number of vertical columns or solid spandrels, while in a through arch bridge, the deck is generally suspended by hangers. Half-through arch bridges usually have both compressive columns and tensile hangers, both of which transfer the deck load to the arch ribs.

Arch bridges exert thrust (pushing inward) forces under vertically downward loads and require high-quality foundations. The substructure and foundation, especially the abutments, must withstand significant horizontal thrust. For multi-span continuous arch bridges, special measures or unidirectional thrust piers may be required to resist unbalanced thrust and prevent the failure of a single span from causing the progressive collapse of the entire bridge, thus increasing the project cost. In plain areas, the construction of arch bridges leads to increased height, resulting in challenges for approach connections and longitudinal slopes, which can be unfavorable for traffic. Additionally, unlike beam bridges, arch bridges cannot maintain balance by themselves before completion, making construction challenging.

In cases where the foundation conditions are unsuitable for constructing arch bridges with significant thrust forces, a tied-arch bridge can be built. In this type of bridge, the horizontal thrust is resisted by tension members (tie-rods), which replace the need for large horizontal thrust at the abutments.

1.2.3 Cable-Stayed Bridges

A cable-stayed bridge is a structural system where the main girder is directly pulled onto the bridge towers by multiple inclined stay cables. It is composed of compressive towers, tensile stay cables, and a main girder that is subjected to both compression and bending. From a mechanical point of view, the tensioned stay cables lift the main girder at multiple points, transferring the dead load and live load on the girder to the bridge towers and then transmitting the load to the foundations through the compression of the towers. The main girder of a cable-stayed bridge can be considered a continuous beam bridge with multiple elastic supports provided by the cables, resulting in reduced bending moments in the girder. Due to the horizontal forces from the inclined stay cables, the main girder also experiences compression and bending.

Cable-stayed bridges are highly statically indeterminate structures with the most complex load distribution among all bridge types. The magnitude of bending moments in the main girder is closely related to the initial tension in the stay cables. An optimal distribution of cable forces exists to minimize the bending moments in the girder.

Due to the elastic support provided by stay cables, the bending moment in the main girder of cable-stayed bridges is relatively small. Consequently, the girder dimension can be significantly reduced, typically with a main girder height of 1/50–1/200 of the span. This reduction in weight not only saves materials but also greatly improves the spanning capacity of cable-stayed bridges. Generally, cable-stayed bridges with prestressed reinforced concrete or composite girders are economically applicable within a main span range of 200–700 m, while those with steel girders have a reasonable span range from 700 to 1400 m (Xiang et al., 2011). Currently, the Russky Bridge in Russia holds the record for the longest main span cable-stayed bridge, with a main span of 1104 m. Additionally, the stable triangular configuration formed by the towers, cables, and main girder results in higher structural stiffness and better resistance to wind than suspension bridges. However, when the span becomes very large, the cantilever construction of cable-stayed bridges may face challenges such as excessive stress on the main girder due to the long cantilever length, increased tower height, and decreased cable stiffness due to the sag of super long cables. These issues require further investigation.

The stay cable arrangements, tower forms, and girder cross-sectional shapes of cable-stayed bridges are diverse, and they should be coordinated with each other. The commonly used cable-stayed bridge is a three-span double-tower structure, but the single-tower double-span form is also common. In contrast, multi-tower cable-stayed bridges with more than three towers are rare. The specific structural form and arrangement should be determined based on considerations such as river or terrain conditions, navigation requirements, and aesthetics.

1.2.4 Suspension Bridges

A suspension bridge is a type of bridge where the main girder is suspended by vertical hangers from suspension cables. Modern suspension bridges consist of several essential components, including the main cables, main girders, towers, hangers, and anchorages. The main cable serves as the primary load-bearing component in a suspension bridge. It is supported at the top of the bridge towers and fixed at the anchorages on both ends. The loads on the main girder are transferred to the main cable through tensioned hangers, causing the main cable to be under tension. The main cable then transfers this tension to the anchorages. The bridge towers lift the main cables to keep the main girder horizontal, resulting in compression in the bridge towers.

To withstand the significant tension from the main cables, the anchorages must be large (gravity-type anchorages) or rely on natural rock formations (tunnel-type anchorages). The tension forces transmitted to the anchorages from the main cables can be decomposed into vertical and horizontal components. Therefore, suspension bridges, similar to arch bridges, can generate horizontal reactions under vertical loads.

Modern suspension bridges widely utilize high-strength steel wire strands grouped together to form steel cables, fully utilizing their excellent tensile properties. Characterized by efficient material usage and lightweight, suspension bridges have the greatest spanning capacity among all bridge types. The economically feasible span range of traditional earth-anchored suspension bridges is typically between 1000 and 5000 m (Xiang et al., 2011). The 1915 Çanakkale Bridge in Turkey, with a main span of 2023 m, is the longest-span bridge in the world.

As a flexible structural system, suspension bridges have the smallest stiffness among all bridge types. They are susceptible to significant deflection and vibration under loads. The structural design of suspension bridges is generally controlled by wind-induced vibrations and stability issues. Moreover, suspension bridges are rarely used in railway bridges, compared with highway bridges.

According to the number of suspended spans, suspension bridges can be categorized into single-span, three-span, and multi-span suspension bridges. Single-span and three-span suspension bridges are the most common types, both featuring two bridge towers. The difference lies in whether the side span deck is suspended by hangers or supported by piers. If one side span of a three-span suspension bridge is suspended by hangers while the other side span is supported by piers, it creates an asymmetrical two-span suspension bridge, such as the Tsing Ma Bridge in Hong Kong. On the other hand, multi-span suspension bridges have more than three towers and are typically used for wide water surfaces with reefs or other suitable conditions for erecting central towers.

Suspension bridges can also be classified based on the anchoring method of the main cables, namely earth-anchored suspension bridges and self-anchored suspension bridges. The key difference between these two types is the presence or absence of anchor blocks. In self-anchored suspension bridges, the main cables are directly anchored onto the stiffening girders. In this configuration, the horizontal component of the cable force is resisted by the main girder, while the vertical component is balanced by counterweights at the ends of the girder. The majority of suspension bridges are earth-anchored suspension bridges, which retain the anchor blocks.

Compared to earth-anchored suspension bridges, self-anchored suspension bridges involve a higher construction risk as the main girders are typically constructed before installing the main cables. Additionally, self-anchored suspension bridges require a significant amount of steel in the main girders to meet stability and stress requirements due to the large axial compression forces generated by the main cables. Therefore, self-anchored suspension bridges are generally more suitable for shorter spans.

1.3 TEMPERATURE BEHAVIOR OF BRIDGES

As shown in Figure 1.1, any temperature distribution on a section can be decomposed into four components (European Committee for Standardization CEN, 2003): uniform temperature change ΔT_u, transverse linear temperature difference ΔT_{My}, vertical

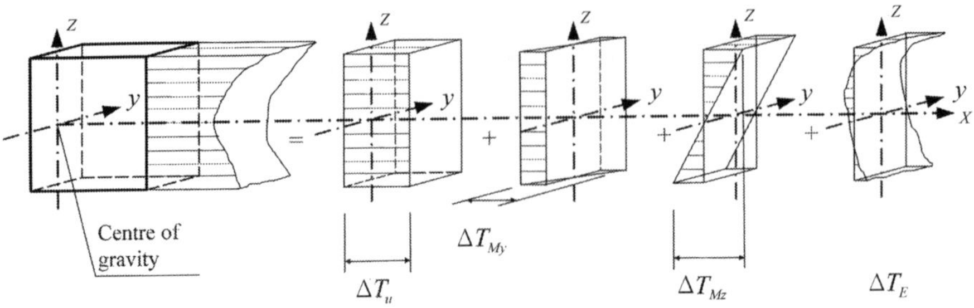

FIGURE 1.1 Temperature components of a cross section (European Committee for Standardization CEN, 2003).

linear temperature difference ΔT_{Mz}, and nonlinear residual temperature $\Delta T_E(y,z)$. These components can be calculated using the following formulas:

$$\Delta T_u = \frac{1}{A}\int_A T(y,z)\cdot dA \tag{1.1}$$

$$\Delta T_{My} = \frac{h_y}{I_z}\int_A T(y,z)y\cdot dA \tag{1.2}$$

$$\Delta T_{Mz} = \frac{h_z}{I_y}\int_A T(y,z)z\cdot dA \tag{1.3}$$

$$\Delta T_E(y,z) = T(y,z) - \Delta T_u - \Delta T_{My}\frac{y}{h_y} - \Delta T_{Mz}\frac{z}{h_z} \tag{1.4}$$

where $T(y,z)$ represents the cross-sectional temperature distribution, h_y and h_z denote the width and height of the cross section, A is the cross-sectional area, and I_y and I_z are the moments of inertia of the cross section. These four temperature components are illustrated in Figure 1.1.

For statically determinate bridges, uniform temperature changes ΔT_u and linear temperature differences ΔT_{My} and ΔT_{Mz} cause elongation and bending deformations without generating temperature-related internal forces. However, in statically indeterminate bridge structures, these temperature variations not only result in deformations but also induce internal forces, leading to thermal stresses.

The residual temperature $\Delta T_E(y,z)$ typically exhibits a nonlinear distribution across the section, but the structural deformation is assumed to satisfy the plane section condition. As a result, the elongation and contraction of structural fibers are constrained by adjacent fibers, creating self-restraint thermal stresses known as thermal self-stresses. Essentially, these self-stresses are a type of cross-sectional self-balanced stresses and are independent of the structural system's static determinacy. Under the plane section assumption, the thermal self-stresses have a zero resultant force and moment across the entire cross section, thereby not affecting the overall structural responses.

Bridge structures are constantly exposed to open-air environments and subjected to varying temperature conditions. Factors such as solar radiation and environmental temperature cause the temperature of bridges to fluctuate over time and exhibit non-uniform distribution patterns. These temperature changes can result in significant stress and deformation within the bridge structure. If these responses exceed certain limits, it can lead to structural cracking or alterations in alignment. In fact, field measurements have shown that temperature-induced stress and deformation can even surpass the effects of other live loads. For example, the monitoring system of the 480 m-long main span of Zhanjiang Bay cable-stayed bridge in Guangdong Province of China showed that the daily temperature-induced displacement of the bridge girder and towers was similar to those caused by a typhoon at the wind speed of 30 m/s, but larger than displacements caused by heavy traffic (Cao et al., 2011). Li and Ou (2016) concluded that the temperature-induced strain of a 648 m-long main span cable-stayed bridge was dominant in the total strain of the bridge.

To understand the impact of temperature variations on bridge structures, it's essential to obtain the structural temperature field. The temperature distribution across a bridge is influenced by external factors such as air temperature, solar radiation, wind, precipitation, snowfall, and other meteorological effects. Internal factors are related to the thermal physical properties of the structural materials, the bridge's orientation, and the geometry and dimensions of its components. The following major findings have been observed:

1. Longitudinal temperature distribution along the bridge tends to be relatively uniform; however, those in transverse and vertical directions often become non-uniform. Solar radiation and ambient temperature fluctuations predominantly affect the bridge's surface, subsequently conducting heat internally. Surface-near regions experience prompt temperature changes, and internal temperature adjustments lag due to the thermal inertia. In particular, concrete structures exhibit a more variable temperature discrepancy than steel ones due to the concrete's low thermal conductivity—roughly 1/25th that of steel. In complex bridge systems such as arch, cable-stayed, and suspension bridges, the use of distinct materials for different structural parts necessitates consideration of significant differential temperature among components.
2. Solar radiation is a primary driver of non-uniform temperature distribution within bridge cross sections. Bridges oriented north-south exhibit greater transverse temperature variations compared to east-west alignments; the intensity of solar radiation increases with lower latitudes. Reflective coatings, such as white paint, can mitigate temperature irregularities by absorbing less solar radiation. According to extensive data, the largest temperature unevenness in bridge sections typically occurs between 13:00 and 15:00, during the zenith of solar radiation. Steel bridges align more closely with these peak solar periods than concrete bridges, and the most uniform temperature distribution is observed in the hour preceding sunrise.

3. A thick asphalt concrete layer atop highway bridges can significantly cool the top plate. In contrast, a thin layer may have an adverse heating effect. The overhanging slab of box-girder bridges shades the web from solar radiation, which can enhance the overall vertical temperature gradient but moderate the local temperature variance of the web. The temperature distribution on vertical members like towers and piers varies with their orientation and the sun's position. High ground reflectivity, such as that of water or snow surfaces (0.6–0.8), provides a high reflected solar radiation underneath the main beam, exacerbating temperature unevenness. In contrast, typical ground reflectivity (around 0.2) has less influence. Additionally, open sections experience more uniform temperature distributions than closed ones due to the pronounced influence of wind speed.

The temperature distribution of bridges can be measured by installing thermometers on the bridges. With the advances in structural health monitoring (SHM) technology, the study of bridge temperature effects through field measurements has received increasing attention. Numerous researchers have investigated the behavior patterns of actual bridges under operational conditions.

However, the number of sensors is limited and may not capture the important thermal features. Numerical simulations, specifically FE analysis, are commonly employed to calculate the temperature behavior of bridges. This process typically involves two main steps: transient heat-transfer analysis to determine the temperature distribution and subsequent structural FE analysis to calculate the temperature-induced responses. In the case of large-scale structures like suspension bridges, establishing a detailed heat-transfer FE model can be time-consuming. Therefore, a divide-and-conquer approach is generally utilized by conducting a heat-transfer analysis on key components like towers and main girders to calculate the temperature distributions of the components, which are then manually assembled into the global structural model for response calculation (Xia et al., 2013). With the rapid development of computer technology, the global FE modeling of the entire bridge integrating heat-transfer and structural analyses has become increasingly possible (Li et al., 2023b).

SHM systems can provide the actual temperature data of bridges in real time, and numerical heat-transfer analyses can provide detailed temperature data. At the same time, the monitoring data can be used to verify the numerical model and results. Integrating field monitoring and numerical analysis enables a more comprehensive and accurate temperature behavior of the bridges.

Temperature effects are closely related to the composition of materials and the structural system. Li et al. (2023a) comprehensively reviewed the studies of the thermal behavior of beam bridges, arch bridges, cable-stayed bridges, and cable-suspension bridges. Here's a summary:

1. In beam bridges, the horizontal displacement at beam ends is primarily influenced by uniform temperature changes, while the vertical displacement at mid-span is controlled by vertical temperature differences in the main beam. Steel-concrete composite beams should account for the impact of different thermal conductivity properties between materials,

which can cause significant thermal stresses at the interface of steel and concrete. Additionally, frictional resistance at supports may affect the temperature-induced deformation of the superstructure and substructure.

2. In arch bridges, the uniform temperature changes in arch ribs can produce substantial overall deformation, and the temperature difference between structural components of arch bridges can also impact the vertical deck displacement. The elongation and contraction of arch bridge decks can generally be estimated using one-dimensional thermal expansion formula $\Delta L = \alpha \cdot L \cdot \Delta T$. Concrete-filled steel tubular arch bridges should also consider the influence of differing thermal conductivity properties of materials. As statically determinate structures, three-hinged arches do not develop internal forces or stresses due to temperature changes.
3. In cable-stayed bridges, the temperature effect analysis requires consideration of the uniform temperature changes in the tower, main girder, and stay cables, as well as the non-uniform temperature distribution within the cross sections of the tower and main girder. For the most commonly used double-tower symmetrical cable-stayed bridges, the vertical displacement at the mid-span of the main span is predominantly affected by the temperature changes in the main girder and stay cables, while the main girder length and the tower-top distance are primarily influenced by the main girder temperature.
4. In suspension bridges, temperature effects typically involve variations in the temperatures of the towers, main girders, main cables, and hangers. Specifically, the overall deformation of suspension bridges, including vertical displacement at the mid-span and horizontal displacement at the tower top, is dominated by changes in main cable temperature with a linear relationship between them. An increase in main cable temperature leads to a decrease in mid-span elevation in a double-tower suspension bridge and a reduction in the distance between the tower tops. Furthermore, changes in the length of the main girder are determined by the uniform temperature change within the main girder itself.

1.4 ORGANIZATION OF THE BOOK

In conclusion, comprehensive analysis and continuous monitoring are necessary to address the complex nature of bridge temperature behavior. The integrated application of SHM technologies can offer more precise data and facilitate a deeper understanding of bridge temperature behavior during operation.

This chapter overviews beam, arch, cable-stayed, and suspension bridges and their temperature behavior. Chapter 2 will overview SHM technologies for bridges. Chapter 3 details the heat-transfer numerical analysis and structural analysis. Chapters 4–7 investigate the thermal behavior of beam, arch, cable-stayed, and suspension bridges by integrating theoretical analysis, field monitoring, and numerical simulation. The long-term temperature behavior of bridges is reported in Chapter 8 using the Tsing Ma Bridge's 26-year field monitoring data. The temperature effects on the vibration properties of bridges are described in Chapter 9. Finally, the prospects of the temperature behavior of bridges are highlighted in Chapter 10.

REFERENCES

Cao, Y. H., Zhang, Y., Zhao, Y. and Wang, M. L., 2011, Distributed health monitoring system for Zhanjiang Bay Bridge. *Advances in Structural Engineering* **14(1)**: 1–12.

European Committee for Standardization CEN, 2003, *Eurocode 1: Actions on Structures, Part 1–5: General Actions — Thermal Actions.* EN1991-1-5: 2003E. Brussels, European Committee for Standardization CEN.

Li, H. and Ou, J., 2016, The state of the art in structural health monitoring of cable-stayed bridges. *Journal of Civil Structural Health Monitoring* **6(1)**: 43–67.

Li, L., Chen, B., Zhou, L., Xia, Q., Zhou, Y., Zhou, X. and Xia, Y., 2023a, Thermal behaviors of bridges: A literature review. *Advances in Structural Engineering* **26(6)**: 985–1010.

Li, L., Shan, Y., Jing, Q., Yan, Y., Xia, Z. and Xia, Y., 2023b, Global temperature behavior monitoring and analysis of a three-tower cable-stayed bridge. *Engineering Structures* **295**: 116855.

Shao, X. D., 2019, *Bridge Engineering*. Beijing, China Communications Press Co., Ltd.

Xia, Y., Chen, B., Zhou, X. Q. and Xu, Y. L., 2013, Field monitoring and numerical analysis of Tsing Ma Suspension Bridge temperature behavior. *Structural Control and Health Monitoring* **20(4)**: 560–575.

Xiang, H. F., Xiao, R. C., Xu, L. P., Shi, X. F., Ge, Y. J., Wei, H. Y. and Han, Z. Y., 2011, *Conceptual Design of Bridges*. Beijing, China Communications Press Co., Ltd.

2 Structural Health Monitoring of Bridges

2.1 SHM SYSTEMS

Bridges play a crucial role as nodes in transportation networks. Their safe, reliable, and cost-effective operation is vital for society's economy. Over their lifespan, which can stretch for several decades or even a century, bridges are exposed to various environmental and operational actions, such as cyclic temperature changes, traffic loads, strong winds, earthquakes, wave currents, scour of piers, foundation settlement, and material degradation (such as steel corrosion and concrete carbonation). These factors can impact a bridge's performance and, in severe cases, compromise its structural integrity, posing significant safety risks. Laboratory tests and numerical simulations have been widely adopted in bridge engineering to understand the bridge's behavior under different scenarios. However, the small-scale laboratory specimens and ideal numerical models differ from the real bridges on site. They do not fully represent the actual environments, loads, and responses of the bridges.

The recently developed SHM technology provides a real platform for the bridges concerned. The SHM technology adopts a comprehensive sensory system and a sophisticated data processing system implemented with advanced information technology and structural analysis algorithms. The main objectives of the SHM are to monitor the loading conditions of a structure, to assess its performance under various service loads, to verify or update the rules used in its design stage, to detect its damage or deterioration, and to guide its inspection and maintenance (Aktan et al., 2000; Xu and Xia, 2012). The development of SHM technology in bridge engineering has gradually gained widespread application and recognition during the past few decades.

2.1.1 Components of an SHM System

A modern SHM system for bridges typically integrates four key subsystems: the sensory subsystem, the data acquisition and transmission system, the data storage and management system, and the structural evaluation system. The first two subsystems are usually located on-site at the structure, while the latter two are typically housed within the control office of the bridge management department.

The sensory system acts as the eyes and ears of an SHM system, responsible for capturing the dynamic responses of structures under various environmental and load actions. This system comprises various types of sensors that are strategically placed at key locations of the bridge to monitor the external environments, loads,

DOI: 10.1201/9781003480136-2

and critical responses like stress, strain, displacement, and vibration. The selection and placement strategy of these sensors must consider the structural characteristics, monitoring objectives, and environmental conditions to ensure accurate capture of structural health indicators.

The data acquisition and transmission system serves as the nervous center of an SHM system, responsible for collecting data from the sensory system and transmitting it to the data processing center via wired or wireless networks. The design of this system must ensure real-time data acquisition and integrity, along with high reliability and anti-interference capability. In the event that the central server is unavailable, the system should also have temporary data storage capabilities to prevent data loss.

The data storage and management system acts as the memory repository of an SHM system, managing the collected data effectively. This includes preprocessing steps such as data validation, normalization, cleaning, fusion, and compression, as well as long-term storage and rapid retrieval of data. Using data warehouse platforms and database technologies, the data storage and management system ensures data security and accessibility, providing a solid foundation for structural health assessment and historical data analysis.

The structural evaluation system is the brain of an SHM system, responsible for in-depth analysis of the stored data to assess the structural health status. The structural evaluation system typically includes online and offline working modes. The online evaluation mode compares the measured data with design values, simulation results, and predetermined thresholds to quickly assess the structural condition. The offline evaluation mode utilizes model-based and data-driven algorithms to perform tasks such as load identification, modal identification, model updating, bridge rating, and damage diagnosis and prediction, providing decision support for structural maintenance and management.

Through the collaborative work of the above four subsystems, the SHM system can provide comprehensive monitoring and management of the health status of large structures, thereby enhancing their safety and durability. When designing an SHM system, engineers need to consider monitoring requirements, structural characteristics, environmental conditions, hardware performance, and economic factors to ensure the system's reliability, robustness, and maintainability. Additionally, the system's scalability and flexibility should be considered to adapt to future technological developments and changing monitoring needs. The design and implementation of such a system require interdisciplinary expertise, including civil engineering, information technology, data analysis, and engineering economics.

2.1.2 Sensory System and Sensors

The sensory system is the core of an SHM system, capturing the static and dynamic structural responses to various environmental and operational actions through multiple types of sensors. This system provides the essential data for subsequent data acquisition, processing, and structural health assessment. When designing the sensory system for a bridge, the monitoring requirements and items are first determined based on the monitoring objectives and structural characteristics. Next, the locations and number of measurement points are established according to the size and complexity of the structure. Careful selection of sensor types is essential to ensure their performance meets the monitoring requirements. The sensor selection should consider measurement range, sampling rate, sensitivity, resolution, linearity, stability, accuracy, repeatability, frequency response, durability, and economic indicators such as sensor cost and installation and maintenance expenses.

Monitoring items of a bridge sensory system can be broadly classified into four categories: environmental factors, external load actions, structural responses, and changes in structural characteristics.

2.1.2.1 Environmental Factors

Environmental factors monitoring includes measuring environmental temperature, humidity, rainfall, solar radiation, atmospheric pressure, and other related factors. Environmental temperature is typically measured using temperature sensors, which include thermocouples, thermistors, resistance temperature detectors (RTDs), semiconductor thermometers, and fiber optic thermometers based on different measurement principles. Environmental humidity is usually measured using lithium chloride hygrometers, capacitive humidity sensors, and electrochemical humidity sensors. Rainfall sensors typically include tipping bucket rain gauges, capacitance rain gauges, and infrared scattering rain sensors. Solar radiation is measured using solar radiometers, which can be further subdivided into pyranometers, pyrheliometers, net radiometers, spectroradiometers, infrared radiometers, and ultraviolet radiometers, depending on the type of radiation measured. Atmospheric pressure can be measured using electronic barometers and capacitive pressure sensors. The sensors measuring environmental factors (such as temperature, humidity, rainfall, pressure, and solar radiation) and external load actions (like wind speed and direction) can be integrated into a weather station. Solar radiation intensity, ambient temperature, and wind conditions are critical for determining the temperature distribution within a structure, which indirectly influences its deformation response.

2.1.2.2 External Load Actions

External load actions include structural temperature changes, traffic, winds, and seismic ground motions. Structural temperature, a significant factor causing deformation or internal force changes in bridge components, can be measured using the aforementioned temperature sensors or infrared thermometers. Traffic loads are typically measured using weigh-in-motion (WIM) systems, which can be categorized into strain-based, piezoelectric, capacitive, and fiber optic systems based on their working principles. The WIM systems measure the axle weights of passing vehicles to assess the total weight, speed, and axle distances, thus evaluating the traffic load on the bridge. Wind loads are generally assessed through wind speed, direction, and pressure data. Wind speed and direction are monitored using propeller anemometers or triaxial ultrasonic anemometers. Wind pressure in air flows can be measured using Pitot tubes, and the wind pressure on the surface of bridge components is often measured using ceramic or silicon piezoelectric differential pressure sensors. Seismic ground motions are usually measured by force-balance accelerometers and strong-motion seismometers.

2.1.2.3 Structural Responses

Structural responses include structural displacement, tilt, strain, vibration, cable forces, and other related responses. Structural displacement monitoring focuses on the movements of key components such as main girders, main cables, towers, bearings, and expansion joints. For long-span suspension and cable-stayed bridges, GNSS technology is typically employed to measure three-dimensional displacements of main girders and towers. Bearing and longitudinal expansion displacements at girder ends are usually tracked using draw-wire, magnetostrictive, or laser displacement sensors. Meanwhile, the tilt of bridge girders and towers is often monitored with micro-electromechanical system (MEMS) tilt sensors. Strain monitoring includes static and dynamic strain measurements. Static strain is commonly measured using fiber optic and vibrating wire strain sensors, whereas dynamic strain is monitored with fiber optic and resistance strain sensors. Structural vibrations are gauged through accelerometers, which can be categorized into force-balance, capacitive, and piezoelectric types. Cable force monitoring can be performed using either indirect or direct methods. Indirect methods often rely on frequency-based calculations derived from accelerometers or elasto-magneto-electric sensors, while direct methods adopt load cells.

2.1.2.4 Changes in Structural Characteristics

Changes in structural characteristics monitoring involves tracking corrosion, cracks, cable wire breakage, and bolt/nut fastening forces. Corrosion is typically assessed using electrochemical principles with linear polarization resistance and electrochemical noise sensors. Crack monitoring often employs vibrating wire, resistance, long-gauge fiber optic crack sensors, and high-definition cameras. Cable wire breakage is generally monitored using acoustic emission sensors. Bolt/nut fastening forces can be monitored using direct methods like vibrating wire or resistance load cells or indirect methods such as ultrasonic or strain gauge techniques.

A well-designed sensory system is crucial for effective SHM. The system must be reliable. With the continuous emergence of advanced sensor technologies, sensory systems are advancing toward greater intelligence, integration, stability, and reliability, providing solid hardware support for the safe operation and maintenance of bridges and large civil engineering structures.

2.1.3 Data Acquisition and Transmission System

In an SHM system, data acquisition and transmission system is the critical link between sensors and the data processing center. Its primary function is to gather signals from various sensors, convert them into digital data, and transmit them to a central server or processing unit, through wired or wireless means, for further analysis and storage. Unlike the straightforward data collection in a lab test, the data acquisition and transmission system in bridge SHM systems must handle massive amounts of data generated by numerous sensors of different types and ensure signal quality and integrity during long-distance transmission. The design of a data acquisition and transmission system must account for the complexity of large structures, different signal types, and sampling rates while ensuring reliability under diverse environmental and operational conditions.

Typically, a data acquisition and transmission system comprises a local wiring network, independent data acquisition units (DAUs) or substations, and a global wiring network. The local wiring network connects distributed sensors to their respective DAUs, whereas the global wiring network transmits signals from the DAUs to the central database server. To minimize signal loss and noise during the long-distance transmission, DAUs are strategically placed at key structural locations to collect signals from nearby sensors, condition these signals, and then transmit the digitized data to the central database server in a synchronized manner. Strategic deployment of DAUs significantly enhances the quality of monitoring data.

Selecting data acquisition hardware involves considering the type, range, compatibility, accuracy, and resolution of sensor output signals. For digital signals, distributed data acquisition devices using communication protocols and standards such as RS-485, CAN, Modbus TCP, or UDP are appropriate, with considerations for transmission distance, bandwidth, and rate. Charge signals need conditioning through charge amplifiers before acquisition. Analog electrical signals, like standard industrial signals (4–20 mA and −5 to 5 V), should be captured using centralized data acquisition equipment based on PCI or PXI technologies. Optical signal data should be acquired using specialized fiber optic demodulation equipment, while vibrating wire sensor signals require dedicated vibrating wire acquisition instruments. Resistance strain sensor signals need amplification via Wheatstone bridge conditioning before acquisition.

The data acquisition modes should be tailored to the specific parameters being monitored and the analysis requirements of the data. For continuous monitoring of parameters such as environmental and structural temperature, humidity, wind speed/direction, structural displacement, strain, and vibration, real-time continuous acquisition is typically employed. This method ensures a steady stream of data, capturing every subtle change. On the other hand, triggered acquisition is ideal for monitoring vehicle loads, seismic activities, and cable wire breakage, as it only collects and records data when certain thresholds are exceeded, making it more efficient for these intermittent events.

The sampling frequency must align with the data analysis requirements. For slow-changing parameters like environmental and structural temperature and humidity, a low sampling frequency of 1/60 Hz is usually sufficient. Wind speed, direction, and pressure generally require a sampling frequency of around 10 Hz. Seismic acceleration is best captured with a triggered sampling rate of 50 Hz. Dynamic response monitoring, such as dynamic displacement, strain, and vibration acceleration, typically necessitates a sampling frequency of 20 times the fundamental frequency of the bridge. For static response monitoring, such as quasi-static displacement and strain, the sampling frequency is much lower at 1 Hz or even 1/60 Hz. Acoustic emission sensors for monitoring cable wire breakage usually operate at a high sampling rate of 2 MHz following triggered acquisition.

A robust data transmission network is crucial for the seamless operation of a data acquisition and transmission system. Typically, wired communication technologies like Ethernet, RS-232, RS-485, and IEEE-1394 are employed to ensure reliable data transfer. In large structures, where the distance from DAUs to the central control office can extend several kilometers, optical fiber is the preferred choice. To enhance the reliability of data transmission, it is essential to implement anti-interference measures, including common-mode and differential-mode interference suppression, lightning protection grounding, and shielding techniques.

Data synchronization between sensors is critical in data processing, particularly for modal analysis using distributed acceleration data in various DAUs. Data synchronization can be achieved by regularly sending a synchronization signal from the central station to all DAUs so that all DAUs work simultaneously. Alternatively, GPS timer can provide an easy and accurate time. Using GPS time in different DAUs can realize signal synchronization at a high accuracy of approximately 100 nanoseconds.

2.1.4 Data Storage and Management System

In SHM systems, the data storage and management system plays a pivotal role in receiving and processing raw data collected on-site from bridges. As a core component for ensuring effective data management and utilization, the data storage and management system not only manages long-term data storage and rapid retrieval but also performs essential preprocessing steps. These steps include data validation, normalization, cleaning, and fusion. By utilizing data warehousing platforms and advanced database technologies, the data storage and management system ensures the integrity, accuracy, consistency, timeliness, and accessibility of data, thereby laying a robust foundation for SHM.

In a data storage and management system, data encoding is vital for effective management and utilization. The SHM system gathers both structured and unstructured data. Structured data encompasses basic bridge information, monitoring data, feature value data, and alarm data, whereas unstructured data includes images, videos, and text. Regardless of data type, comprehensive upload, retrieval, and export functionalities are necessary, alongside organized classified storage. To achieve this, the data storage and management system typically employs a data dictionary and assigns unique identifiers to various data types, enhancing the efficiency of data organization and retrieval.

The data storage and management system ensures reliable and consistent data entry by handling key preprocessing tasks. Data validation checks for anomalies and missing values in sensor data to ensure credibility; cleaning filters data based on feature selection requirements; and fusion integrates information from various sensors to enhance data usability. With technological advancements, modern data acquisition and transmission devices incorporate basic data preprocessing functions such as signal conditioning and filtering, reducing the burden on the data storage and management system and improving overall efficiency.

To boost data management efficiency, multi-tiered storage is typically adopted, comprising on-site acquisition stations, monitoring centers, and cloud storage. On-site acquisition stations use cyclic updating methods to ensure local storage of raw data, even during network outages. Servers at the monitoring center house critical data such as bridge information, real-time monitoring data, feature value data, alarm data, and historical evaluation reports. To manage these data effectively, the data storage and management system uses a modular architecture, layering and categorizing storage into various sub-databases. Furthermore, cloud storage is widely utilized, with data uploaded to cloud servers to facilitate remote access and backup.

Quick data access, retrieval, and security control are paramount in the data storage and management system design. To cater to diverse data access needs, data extraction from sources should support multiple modes, such as full extraction, incremental extraction, and log-based extraction, along with customizable formats and processes to enhance the accuracy, speed, and flexibility of data retrieval. To ensure data security, the data storage and management system implements stringent access controls and comprehensive data auditing functions, tracking data access and modifications to prevent unauthorized changes or damage during transmission, storage, and processing. Sensitive fields or business data within the monitoring system should be encrypted during storage and transmission to prevent data theft.

2.1.5 Structural Evaluation System

The structural evaluation system retrieves and analyzes data from the data storage and management system to assess structural health and offer critical decision support for their maintenance and management. A structural evaluation system typically adopts online and offline evaluation modes. Online evaluations focus on real-time data analysis to quickly identify potential structural issues, which is especially useful for rapid structural assessments in emergencies. In contrast, offline evaluations delve into historical data, providing a comprehensive structural health assessment to inform long-term maintenance and management strategies. The structural evaluation system applications encompass various areas, such as structural damage identification, structural health level evaluation, early warning, and emergency management.

Structural damage identification primarily targets to promptly detect and address performance degradation or potential failures. It is typically divided into four stages: first, damage detection to identify any damage; second, damage localization to pinpoint the exact location of the damage; third, damage quantification to estimate the severity; and finally, assessing the damage's impact on overall structural performance. A common approach in bridge damage identification is based on dynamic characteristics. This method identifies damage by analyzing changes in modal parameters (such as natural frequencies, damping ratios, and mode shapes) or their variants or vibration time series directly (Hou and Xia, 2021). Variations in natural frequency may indicate reduced structural stiffness, while changes in damping ratio can suggest alterations in energy dissipation properties, both linked to structural damage. Incorporating FE model updating techniques further aids in accurately locating damage and assessing its impact.

Structural health level evaluation differs from damage identification by focusing on the overall health of the structure and its components. It provides a holistic evaluation of the structure's or components' working conditions by analyzing monitoring data on environmental factors, external load actions, structural responses, and changes in structural characteristics. According to the Chinese technical specifications for structural monitoring of highway bridges JT/T 1037-2022 (Ministry of Transport of the People's Republic of China, 2022), structural health level evaluation includes evaluations of both overall structural health and individual component health, each classified into four levels: Level I (essentially intact), Level II (minor anomalies), Level III (moderate anomalies), and Level IV (severe anomalies). There are two primary methods for assessing structural health levels. One directly analyzes monitoring data and compares results with preset thresholds to determine the health status. The other adopts damage identification and model updating to establish a reliable FE model and then compares calculated structural responses against preset thresholds under normal operational loads. These methods enable a systematic and comprehensive evaluation of bridge health levels, providing a scientific basis for effective maintenance and management plans.

Early warning and emergency management concern extreme scenarios of the bridge. Bridge early warning indicators typically encompass parameters such as environmental factors, external load actions, structural responses, and changes in structural characteristics. Common indicators include overloadings, extreme temperatures and wind speeds, displacements, strains, and cracks at key sections. The structural evaluation system can predict normal and abnormal ranges for these indicators by analyzing long-term historical monitoring data. Combining simulation results, design specification allowances, expert experience, inspection guidelines, and emergency management needs, the structural evaluation system establishes reasonable over-limit alarm thresholds. Chinese standard classifies the alarm thresholds into three levels (Ministry of Transport of the People's Republic of China, 2022). These thresholds correlate with the risk levels posed by structural anomalies, with different thresholds triggering various warning levels and corresponding responses. Bridge management departments can then undertake targeted inspections based on the warning levels. In emergency cases, such as vortex-induced vibrations, strong winds, earthquakes, abnormal vibrations in hangers or stay cables, vehicle overloading, and ship collisions, the structural evaluation system can promptly assess structural status by analyzing monitoring data, aiding management departments in making emergency response decisions. Therefore, the SHM system plays a crucial role in emergencies, ensuring timely identification and repair of bridge damages after disasters.

2.1.6 Standard Specifications for SHM Systems

Standardization for SHM systems is essential for their efficient and consistent implementation. Standards and guidelines provide a framework for the design, installation, and data analysis of these systems, offering crucial technical references for engineers and fostering industry growth.

The International Organization for Standardization (ISO) has yet to establish specific standards for bridge SHM. However, ISO 13822:2010 (ISO Standards, 2010) and ISO 2394:2015 (ISO Standards, 2015), issued by Technical Committees 59 and 98, offer general frameworks related to structural design and performance evaluation. While these standards can guide SHM system design, they lack detailed technical implementation instructions.

China has rapidly advanced its bridge SHM standards, paralleling the swift expansion of its transportation infrastructure and significant achievements in long-span bridge construction. The country's multi-tiered standard system includes national, industry, regional, association, and enterprise standards, comprehensively covering SHM system design, implementation, acceptance, operation, maintenance, and data management. A prime example is the "Technical Specifications for Structural

Monitoring of Highway Bridges" (JT/T 1037-2022) (Ministry of Transport of the People's Republic of China, 2022), issued by the Ministry of Transport of the People's Republic of China. This specification stipulates the monitoring scope, including long-span bridges and other in-service bridges requiring monitoring. It specifies that long-span suspension bridges over 500 m, cable-stayed bridges over 300 m, arch bridges over 200 m, and beam bridges over 160 m should be monitored. The document details sensor placement for various bridge types and provides specific reference examples. Furthermore, it introduces a bridge condition indicator and structural health level and provides the indicator evaluation methods. Overall, this specification provides a clear and detailed technical framework for bridge monitoring.

China has also issued other national standards, such as the "Technical Code for Monitoring of Building and Bridge Structures" (GB 50982-2014) (Ministry of Housing and Urban-Rural Development of the People's Republic of China, 2014) and the "Specifications for Operational Monitoring of Urban Rail Transit Facilities—Part 2: Bridge" (GB/T 39559.2-2020) (China's State Administration for Market Regulation and China's Standardization Administration, 2020). Numerous technical standards and codes related to SHM have been published by organizations such as the China Civil Engineering Society and the China Association for Engineering Construction Standardization. These include:

- "Technical Regulation on Sensor Selection and Placement for Bridge Health Monitoring" (T/CCES 15-2020) (China Civil Engineering Society, 2020b)
- "Mass Data Processing Standard for Structural Health Monitoring" (T/CCES 16-2020) (China Civil Engineering Society, 2020a)
- "Early Warning Threshold Standard for Structural Health Monitoring System of Long-Span Bridges" (T/CECS 529-2018) (China Association for Engineering Construction Standardization, 2018)
- "Design Standard for Structural Health Monitoring System" (CECS 333-2012) (China Association for Engineering Construction Standardization, 2012)
- "Standard for Operational Maintenance and Management of Structural Health Monitoring Systems" (T/CECS 652-2019) (China Association for Engineering Construction Standardization, 2019)
- "Standard for Construction and Acceptance of Structural Health Monitoring Systems" (T/CECS 765-2020) (China Association for Engineering Construction Standardization, 2020)

Each of these standards has its own focus, collectively providing strong support for the effective implementation and efficient operation of bridge monitoring systems.

Guangdong-Hong Kong-Macao Greater Bay Area Standards Promotion Council of Guangdong Province also issued a data standardization system for intelligent operation and maintenance of bridge-island-tunnel crossings (Guangdong-Hong Kong-Macao Greater Bay Area Standards Promotion Council of Guangdong Province, 2021). It consists of more than 30 guidelines, 18 of which were published as of 2024.

Additionally, standardization organizations in India and Russia have released guidelines related to bridge monitoring. The Indian Roads Congress issued

"Guidelines for Corrosion Prevention, Monitoring, and Remedial Measures for Concrete Bridge Structures" (IRC:SP:80-2008) in 2008 (Indian Roads Congress, 2008), which introduces instruments and testing methods for monitoring steel corrosion and concrete degradation in bridges. Meanwhile, the Ministry of Construction, Housing and Utilities of the Russian Federation published "Bridges. Monitoring of technical condition (Structural health monitoring)" (SP 274.1325800.2016) in 2016 (Ministry of Construction and Housing and Utilities of the Russian Federation, 2016). This document specifies the technical requirements for bridge health monitoring software and hardware and details the applicable scenarios, implementation methods, and precautions for continuous monitoring, periodic monitoring, and special monitoring.

Mandatory enforcing standards may be challenging, particularly when the SHM was immature 20 years ago. There have been some informal guidelines on SHM systems. For instance, the Canadian Network of Centres of Excellence on Intelligent Sensing for Innovative Structures released a pioneering SHM guideline in 2001, "Guidelines for Structural Health Monitoring—Design Manual No. 2" (Mufti, 2001). This document delves into basic SHM concepts, system components, and provides technical guidance for conducting static and dynamic field tests and ten case studies of bridges and wharfs. Other significant references include: "F08b Guideline for Structural Health Monitoring" developed in 2006 by the European thematic network SAMCO (Rücker et al., 2006), "Monitoring and Safety Evaluation of Existing Concrete Structures (State-of-the-Art Report)" published in 2003 by the International Federation for Structural Concrete (Bergmeister et al., 2003), Drexel University's proposal to the U.S. Federal Highway Administration titled "Development of a Model Health Monitoring Guide for Major Bridges" (Aktan et al., 2002), and publications by the Japan Society of Civil Engineers, such as "Guidelines for the Utilization of Monitoring Technology" (Subcommittee on New Technology Application Promotion of the Infrastructure Maintenance General Committee of the Japan Society of Civil Engineers, 2022) and "Performance-based Maintenance and Monitoring of Steel Bridges" (Subcommittee on Performance-Checking Maintenance and Monitoring of Steel Bridges of the Committee on Steel Structures of the Japan Society of Civil Engineers, 2019). Although these documents are informal standards, they offer extensive technical insights and practical case studies of SHM.

Standardization plays a crucial role in the implementation of the SHM technology. In turn, the application will provide feedback and further promote the development of the technology. Meanwhile, standardization needs international cooperation and sharing, which is necessary because different countries and regions have different standardization procedures and legal systems.

2.1.7 Prospects of SHM

With the continuous progress of technology, bridge health monitoring systems are undergoing rapid development and revolution, gradually moving toward a more intelligent direction. The technology is increasingly playing a more critical and prominent role in the bridge engineering industry. In the near future, we can expect the continuous advancement of bridge SHM toward new heights in the following aspects.

1. The future bridge SHM systems will integrate new technologies, such as Internet of Things, robots, AI, and digital twins. The monitoring system will flexibly adapt to special structural requirements of different bridge types, providing customized monitoring services for each bridge. This customized service will make the monitoring system more efficient and more closely aligned with the actual needs of the bridges.
2. The SHM systems for bridges in a region may be integrated to form a group or population. The bridge population is under similar environmental conditions (temperature, humidity, and corrosive environment) and is subjected to similar and spatiotemporally correlated loads (traffic, wind, and ground motion). The deterioration and damage mechanisms of the bridge population are also similar and spatiotemporally correlated. The population-based SHM will result in more accurate models and a more comprehensive understanding of the bridge's behavior than the single bridge-based SHM. It will also result in a new SHM system design framework.
3. Climate change is a global threat. Increasing climate change is causing more extreme heatwaves, floods, and storms, aggravating the risks faced by bridges. However, the bridge engineering society is not ready to design and maintain bridge structures for adaptation to climate change. For example, the current bridge codes of temperature and wind loads in all countries/regions are based on past recorded data and do not consider the perceivable variations associated with climate change in recent decades. Although the increasing frequency and intensity of extreme weather events and disasters have been observed, real load and response data of bridges in the context of climate change are lacking. The long-term SHM systems will shorten this gap by providing first-hand data of the bridges and studying the bridges' behavior in the future. In this sense, the SHM system will improve the safety and resilience of bridges in a climate-altered future.

Overall, the prospects of SHM systems are promising. The SHM will not only contribute to the modernization and intelligence of bridge engineering but also inject new dynamics into sustainable social development.

2.2 MONITORING OF BRIDGE TEMPERATURE BEHAVIORS

Modern bridges typically feature long spans and complex configurations, making their load conditions and structural behavior intricate. Temperature variation is a significant environmental factor for bridges. Changes in solar radiation and ambient temperature cause fluctuations in the temperature of bridge components, which may affect material properties and structural behavior, such as deformation, stress, and modal parameters. If not properly addressed, these effects can compromise the safety and functionality of bridges. For example, in the 1970s, the collapse of the Carnavino Bridge in Italy was closely related to the excessive temperature stress during cantilever construction (Wittfoht, 1983); the bearings of the Cleddau Bridge in the UK experienced severe wear during operation due to transverse bending induced by temperature changes (Kromanis et al., 2015).

Temperature variations not only independently affect bridges but also couple with other loads, potentially masking the effects of these loads or structural damage. To accurately assess the influence of a specific load or damage level, it is essential to distinguish the effects of temperature from other load effects. Therefore, temperature monitoring should be an integral part of SHM systems. Typically, structural temperature is tracked along with environmental factors like ambient temperature, wind speed and direction, and solar radiation, in addition to other loads and structural responses. For instance, the comprehensive SHM system installed on Hong Kong's Tsing Ma Bridge in 1997, which remains operational, includes 276 sensors with 115 dedicated to temperature measurement (Xu and Xia, 2012).

Since the mid-20th century, researchers have delved into the temperature behavior of bridges, focusing primarily on two key areas: the temperature distribution across bridges and the effect of temperature on structural responses, known as temperature actions. Temperature fluctuations influence both the dynamic characteristics and the quasi-static behavior of bridges. The former has been extensively studied due to the widespread use of bridge dynamic characteristics in vibration-based damage identification. It is widely accepted that temperature mainly affects structural dynamic characteristics by altering the elastic modulus of materials. However, different types of bridges—such as beam bridges, arch bridges, cable-stayed bridges, and suspension bridges—exhibit significant variations in quasi-static thermal behavior. Therefore, this section will introduce temperature monitoring methods and review past research on the quasi-static thermal behavior of various bridge types.

2.2.1 Temperature Monitoring

In general, temperature monitoring includes structural temperature and ambient temperature monitoring. For the former, thermometers should be embedded within the bridge structure or attached to its surface. For the latter, sensors should be placed in the air surrounding the bridge components (for example, on the light poles or inside box girders). These two types of temperature measurements often show significant differences. Temperature sensors should be sheltered from direct sunlight appropriately to prevent substantial deviations between the sensor readings and the actual temperature.

Different types of bridges exhibit distinct structural temperature distribution patterns. To effectively place temperature measurement points, it's essential to consider the temperature variation characteristics of the entire structure and its critical components. Given the limited number of thermometers available, they should be strategically positioned in areas with significant temperature changes and abrupt gradients. For example, the top surface of the main girder is directly exposed to sunlight and thus exhibits the highest temperature. As a result, the girder may experience a pronounced vertical temperature gradient. Consequently, thermometers should be installed at various depths along the girder's cross-section to accurately capture this gradient. Similarly, the temperature of the main cables plays a crucial role in the overall deformation of suspension bridges, necessitating precise measurement of their temperature variations. Numerical heat-transfer analysis can be employed to predict potential temperature distribution within the bridge, which provides valuable insights for optimal placement of temperature sensors.

Common temperature sensors include thermocouples, thermistors, RTDs, and fiber optic sensors. Thermocouples are cost-effective and easy to install. They operate based on the Seebeck effect, where a voltage is generated at the junction of two different metals when there is a temperature difference. This voltage can be measured to determine the temperature at the junction. However, thermocouples are vulnerable to external factors like corrosion and oxidation, resulting in a shorter lifespan and reduced stability.

Thermistors and RTDs, on the other hand, measure temperature based on the principle that a material's electrical resistance increases with temperature. These sensors are generally more accurate and stable than thermocouples. The primary difference between thermistors and RTDs lies in the materials used: thermistors typically use ceramics or polymers, whereas RTDs are usually made of platinum. For example, the widely used PT100 sensor, a type of RTD, has a nominal resistance of 100 Ω at 0°C. While thermistors have a smaller temperature range (−90°C to 130°C), they usually offer higher accuracy than RTDs.

Fiber optic sensors measure temperature changes using principles of light reflection and diffraction. These sensors provide higher accuracy and sensitivity over a broader range than other types. Their unique features include long-distance coverage, multi-point measurement capability, and immunity to corrosion and electromagnetic interference. However, fiber optic sensors are delicate, necessitating careful installation and maintenance. In addition, the data acquisition device of fiber optic sensors (or interrogator) is expensive.

The temperature distribution of bridges is typically uneven and varies over time. In SHM systems, however, thermometers are usually placed at only a few specific points. These sparsely distributed thermometers may not accurately capture the entire temperature profile of the bridge. To address this, numerical heat-transfer analysis can be employed to calculate extensive temperature in regions without thermometers. Solar radiation intensity and wind speed are influential boundary condition parameters in heat-transfer simulations of bridges. Therefore, sensors for these key parameters should be included in a temperature monitoring system. Unfortunately, almost all bridge monitoring systems do not measure solar radiation. In practice, researchers have to use solar radiation data from regional meteorological stations or empirical data.

2.2.2 Temperature Behaviors of Beam Bridges

Beam bridges are widely used for small to medium spans due to their simple structure, ease of construction, and low maintenance costs. The temperature distribution and behavior of these bridges have been extensively investigated.

Researchers typically use numerical simulations or deploy thermometers in laboratory models or real bridges to study temperature distribution. However, the limited number of thermometers on bridges makes it challenging to fully capture the temperature distribution. Thus, combining transient heat-transfer analysis with measurement data has proven to be an effective approach.

Hoffman et al. (1980) conducted on-site monitoring of a segmental box girder bridge. Although the bridge was curved in the north-south direction, no significant longitudinal or transverse temperature differences were observed. They found that the key vertical temperature distribution could be approximated with a fifth-order polynomial, consistent with New Zealand standards. The temperature distribution on the top slab of the box girder followed a linear pattern.

Song et al. (2012) examined the vertical and transverse temperature gradients and their statistical properties in a high-performance concrete box girder and compared these results with Chinese and international design standards. They found that existing codes overestimated the temperature effects and that the transverse temperature difference was non-negligible. They proposed a method to determine the temperature effects based on a binomial probability model.

Besides field monitoring, some researchers conducted scaled model tests. Zuk (1965) comprehensively analyzed the temperature behavior of composite bridges and determined the vertical temperature distribution by combining steady-state solutions with empirical formulas for the top and bottom surface temperatures. They found that an insulation layer significantly reduced the temperature gradient.

Priestley (1978) reviewed design standards for thermal loads in prestressed and reinforced concrete bridges, and discovered that a fifth-order polynomial temperature distribution model could accurately predict critical conditions for all sections. He developed a general analysis method for predicting temperature-induced stresses in concrete bridges, which was validated with measurement data from a quarter-scale box girder model.

Berwanger (1983) performed numerical and experimental studies on the temperature distribution and temperature-induced effects in composite plate girder bridges. He cooled a 0.354:1 scale model of a three-span continuous bridge by placing ice on the deck and measured strain, temperature, and displacement during a 77-minute cooling period. The correlation between numerical and measured temperatures showed high accuracy with a chi-squared probability of 0.99. The measured deflection was less than the calculated value, possibly due to concrete expansion caused by water penetration.

Temperature changes can cause structural deformation and strain/stress in bridges, sometimes even masking the effects of other loads. Temperature effects on bridges have also been extensively studied. Field measurement data shows that the temperature-induced vertical deflection at the mid-span of the Tsing Ma Bridge reached 2313 mm within a year, while the longitudinal displacement of the girder reached 903 mm (Xu et al., 2010), which is much larger than the responses caused by daily traffic loads.

Fu and DeWolf (2004) examined the impact of temperature gradients on the overall behavior of a three-span continuous curved post-tensioned concrete bridge. They collected tilt data to monitor transverse rotation and found that the transverse rotations due to the temperature gradient of the bridge were a cause of torsion cracks in the box girders. Similarly, Kromanis et al. (2015) reported severe wear of the roller bearings on the Cleddau Bridge caused by temperature-induced plan twist.

Catbas et al. (2008) developed a reliability model for the main truss components and the entire structural system of a long-span cantilever truss bridge. They reported that the peak-to-valley strain difference was about 400 $\mu\varepsilon$, ten times higher than the maximum strain caused by traffic loads based on one year of monitoring data. The temperature had a significant effect on the reliability analysis.

Hedegaard et al. (2013) investigated the effects of thermal gradients on longitudinal stresses and curvature of the I-35W St. Anthony Falls Bridge. They found that the stresses and deformations calculated using the bilinear design gradient model specified by AASHTO were considerably lower than those derived from the measured results. They concluded that the design thermal gradient might not be conservative for structures in all regions.

Studies on the temperature behaviors of beam bridges are summarized in Table 2.1.

TABLE 2.1
Studies on the Temperature Behaviors of Beam Bridges

Bridge	Focus	Study Approach	References
Composite box girders	Extreme temperature gradients in the composite box girder bridges of different configurations	Numerical parametric study	Dilger et al. (1982)
Concrete box girders	Extreme temperature gradients in the concrete bridges of different configurations	Numerical	Elbadry and Ghali (1983)
Concrete members	Surface temperature of concrete considering full boundary conditions	Numerical	Riding et al. (2007)
Composite bridges	Temperature behaviors under different insulation conditions	Experimental and numerical	Zuk (1965)
Composite slab-beam bridge	Temperature distribution and temperature-induced moment, stress, and strain	Experimental and numerical	Berwanger (1983)
Concrete box and π beam bridge	Vertical distribution of the thermally induced stress in bridges	Theoretical and experimental	Priestley (1978)
Curved prestressed concrete box girder	Temperature distribution and deflection in a curved box girder bridge	Field monitoring	Hoffman et al. (1980)
Steel I beam and box girder bridge	Thermal behavior of steel bridges in the tropical region	Numerical and statistical	Au et al. (2002)
Various bridges	Temperature distributions in the bridges in different locations	Statistical and numerical	Potgieter and Gamble (1989)
Steel I beam and box girder bridges with concrete decks	Thermal behavior of steel bridges in the tropical region	Numerical and statistical	Tong et al. (2002)
Concrete box girder with deep box	Vertical and transverse temperature gradients and their statistical characteristics	Numerical, field monitoring, and statistical	Song et al. (2012)

(Continued)

TABLE 2.1 (*Continued*)
Studies on the Temperature Behaviors of Beam Bridges

Bridge	Focus	Study Approach	References
Kishwaukee River Bridge	Structural responses to the nonlinear temperature distributions	Numerical and field monitoring	Potgieter and Gamble (1983)
Concrete bridges	Temperature distribution and inspection of the code specifications	Numerical parametric study	Branco and Mendes (1993)
Casilina bridge	Temperature distribution	Numerical and field monitoring	Froli et al. (1996)
Concrete box girder bridges	Temperature distribution	Numerical and field monitoring	Mirambell and Aguado (1990)
Concrete box girder	Temperature distributions and variations in concrete bridges under environmental thermal loads	Numerical and experiment	Abid (2015)
Various highway bridges	Statistical model of extreme thermal loadings in highway bridges	Numerical and statistical	Ho and Liu (1989)
Steel box and π girder bridges	Temperature distribution	Numerical and experimental	Tong et al. (2001)
Various bridges	Statistics of temperatures in various bridges	Statistical	Roeder (2003)
High-speed railway bridge	Extreme value model for uniform temperatures in bridge piers	Field monitoring and statistical	Dai et al. (2023)
Composite concrete slab on steel I beam bridge	Simplification of temperature distribution along the depth	Mathematical simplification	Kennedy and Soliman (1987)
Composite bridges	Numerical parametric study of temperature distribution and induced stresses	Numerical and parametric	Fu et al. (1990)
Concrete box girder bridge	Temperature gradients and correlations between ambient and structural temperatures	Field monitoring, correlation analysis	Wollman et al. (2002)
Cleddau Bridge	Quasi-static temperature effects on the bearing movements	Numerical and field monitoring	Kromanis et al. (2015)
Xiaoqing River No. 3 bridge	Effect of encased concrete on the temperature gradient	Field monitoring and statistical	Zhang et al. (2021a)
Beam bridges with different configurations	Deflections of beam bridges due to the vertical temperature difference	Theoretical	Zhou et al. (2021a)
Prestressed concrete beam	Prestress and deflection change due to temperature gradient	Theoretical	Liu and Zuk (1963)

(*Continued*)

TABLE 2.1 (*Continued*)
Studies on the Temperature Behaviors of Beam Bridges

Bridge	Focus	Study Approach	References
Steel box bridges with concrete decks	Temperature distribution and temperature-induced deformation	Numerical and field monitoring	Moorty and Roeder (1992)
Curved post-tensioned concrete bridge	Temperature distribution and induced displacement	Numerical and field monitoring	Fu and DeWolf (2004)
Long-span truss bridge	Temperature-induced strain; reliability assessment	Statistical and numerical	Catbas et al. (2008)
Three-span post-tensioned concrete dual-box bridge	Temperature distribution and induced stress	Numerical and field monitoring	Hedegaard et al. (2013)
Lieshihe Bridge	Temperature-induced strain and correlation between strain and temperature	Statistical and field monitoring	Yang et al. (2019)

2.2.3 Temperature Behaviors of Arch Bridges

Compared to beam bridges, arch bridges have greater spanning capacity by leveraging the compressive strength of materials and the geometric stiffness of the arch. Due to their statically indeterminate nature, temperature variations can significantly affect the deformation and stress of arch bridges. This has prompted numerous studies on their thermal behavior using field data and numerical simulations.

Xia et al. (2022) investigated the temperature behavior of an arch footbridge by integrating one-year field data with numerical simulations. They found that the maximum temperature difference of the bridge was about 6°C on the hottest day and 5°C on the coldest day. The root mean square error between simulated and measured temperatures was less than 1.2°C, and the maximum stress difference between FE analysis and monitoring data was under 1.5 MPa, confirming the numerical model's accuracy.

Zhou et al. (2017b) developed an extreme value model for thermal gradients in the Jiubao Bridge's prestressed concrete girder using the generalized Pareto distribution and Bayesian estimation. They concluded that traditional normal and logistic distributions failed to capture the characteristics of the temperature gradients, advocating for the generalized Pareto distribution to model these extremes.

Zhu et al. (2021) performed a statistical analysis on the effective temperature and gradients in the truss arch and deck of the Dashengguan Yangtze River Bridge. They discovered that the effective temperature followed a normal distribution, while the temperature gradients were best represented by a weighted sum of two lognormal distributions. Notably, the extreme values of effective temperature and gradients did not occur simultaneously, with solar radiation causing significant variations in exposed components.

Wang et al. (2016) conducted a numerical analysis of thermal distribution and stress in a concrete box arch bridge. They noted non-uniform temperature distribution along the arch axis due to varying azimuth and inclination angles. The vertical gradient in the middle web was modeled as a negative exponential function, while the end web followed a cubic parabolic function. Horizontal temperature gradients were also significant near the arch foot where the width was large.

Larsson and Karoumi (2011) explored the temperature distribution inside the hollow concrete arch of the New Svinesund Bridge through simulations and found significant vertical gradients compared to horizontal ones.

As for the thermal responses of arch bridges, Zhao et al. (2019) studied the effect of temperature on the deflection of the Dashengguan Yangtze River Bridge deck using field data. They applied the wavelet transform to separate temperature-induced deflection from train-induced deflection and used principal component analysis (PCA) to confirm temperature sensitivity. They found that the train-induced deflection increased with temperature. They proposed a real-time warning method using girder deflection to indicate track irregularities or deteriorating components, with an alarm trigger within 10 seconds.

Zhou et al. (2017a) reported preliminary findings from the Jiubao Bridge's health monitoring system, covering structural temperatures, girder deformations, and their nonlinear relationships. Later, Zhou et al. (2018) formulated these relationships as a weighted sum of key temperature variables using clustering analysis. They identified the effective temperature of the main arch rib and the vertical temperature gradient in the girder as critical factors and found that the mid-span vertical deflection could be predicted based on these variables with an error of less than 2 mm.

Duan et al. (2011) analyzed strain-temperature correlations in a tied arch bridge to distinguish between damage-induced and environmental responses. Using one-year SHM data, they observed significant hysteresis between strain and temperature. After removing linear and nonlinear temperature effects from the strain measurement, the processed data was used for novel detection and overload alarms.

Yarnold et al. (2012) developed a temperature-based structural identification method for the Tacony-Palmyra Bridge, using local temperature data as the input and strain and displacement data as the output. The correlation model established long-term performance criteria for bearings, expansion joints, deck or substructure cracking, member/connection overstresses, and structural linearity. This method was later compared to conventional vibration-based methods, showing greater sensitivity in detecting damage scenarios.

Ding et al. (2017) used an adaptive finite impulse response filter to analyze strain influence lines and dynamic load factors in the Dashengguan Yangtze River Bridge. They found that temperature effects on the influence line could be modeled with third-order polynomials, while the dynamic load factors of most members were temperature-independent.

The studies about the temperature distribution and effects on arch bridges are summarized in Table 2.2.

TABLE 2.2
Studies on the Temperature Behaviors of Arch Bridges

Bridge	Focus	Study Approach	References
New Svinesund Bridge	Temperature distribution in the hollow concrete arch	Numerical and field monitoring	Larsson and Karoumi (2011)
Concrete box arch bridge	Temperature distribution and the thermal stress	Numerical	Wang et al. (2016)
Steel arch bridge with hangers	Temperature distribution and temperature-induced stress and deflection	Numerical and field monitoring	Xia et al. (2022)
Jiubao Bridge	Extreme value model for the thermal gradients using the Bayesian estimation	Statistical and field monitoring	Zhou et al. (2017b)
Dashengguan Yangtze River Bridge (Railway)	Probability distribution of effective temperature and temperature gradients	Statistical and field monitoring	Zhu et al. (2021)
Non-prismatic solid structure	Temperature distribution and design suggestions	Numerical	Yuksel (2009)
Dashengguan Yangtze River Bridge (Railway)	Temperature effects on the deflection	Field monitoring and PCA	Zhao et al. (2019)
Jiubao Bridge	SHM system and preliminary monitoring results	Field monitoring	Zhou et al. (2017a)
Jiubao Bridge	Correlation of temperature and structural deformation	Statistical	Zhou et al. (2018)
Tied arch bridge	Correlation of temperature and induced strain and structural health monitoring	Statistical	Duan et al. (2011)
Tacony-Palmyra Bridge	Correlation of temperature and strain/deformation	Numerical and field monitoring	Yarnold et al. (2012); Yarnold and Moon (2015); (2015)
Dashengguan Yangtze River Bridge (Railway)	Correlation of temperature and strain influence line and dynamic load factors	Statistical and field monitoring	Ding et al. (2017)

2.2.4 Temperature Behaviors of Cable-Stayed Bridges

Cable-stayed bridges are popular in long-span bridge construction due to their efficient material use and aesthetic design. The main girder of a cable-stayed bridge is supported by cables attached to towers. Although the stay cables replace piers and allow for longer continuous spans, they indeed increase the redundancy and mechanical complexity of the bridge, making its temperature behavior more intricate.

Many researchers have studied the temperature characteristics of cable-stayed bridges using data from on-site SHM systems. Wang et al. (2019) investigated the temperature field of the prestressed concrete girder of the Puxi Cable-Stayed Bridge and its thermal effects on cable forces and identified a three-piece vertical temperature distribution pattern. On sunny days, the vertical temperature gradient between the top surface and the web could reach 15.4°C.

Cao et al. (2010) reported the monitoring results of the Zhanjiang Bay Cable-Stayed Bridge and found that the structural concrete temperature lagged behind the ambient temperature by 5–6 hours. Similarly, Yang et al. (2018a) reported that the structural temperature of the Huanggang Yangtze River Bridge lagged the ambient temperature by 3.2 hours, while the main girder deflection lagged the structural temperature by 2.2 hours.

Li et al. (2019) conducted a detailed heat-transfer analysis of the concrete tower of the Qingzhou Navigation Bridge (one of the three cable-stayed bridges of the Hong Kong-Zhuhai-Macau Bridge) using ABAQUS. They found a large temperature gradient along the thickness direction due to the large cross-sectional area and low thermal conductivity of the towers.

Regarding the thermal response of cable-stayed bridges, Xu et al. (2019) studied the thermal effects on the main girder deflection of the Nanjing Dashengguan Yangtze River Highway Bridge using a multivariable linear model. The weight coefficients, defined as the normalized responses to each type of temperature action (uniform temperature, horizontal and vertical temperature gradients, and the temperature gradient of the tower), were obtained from the numerical models and updated with the measured temperature and displacement data collected during the bridge closure. They also reported that the actual girder temperature profile agreed well with that recommended by the British standard, while the transverse temperature difference of the deck was non-negligible.

Wang et al. (2022) studied the temperature effects on the tower deformation of a cable-stayed bridge using SHM data. A strong linear relationship between the air temperature and the tower deformation was observed. The two towers departed from each other as the temperature rose with correlation coefficients of 0.89 and −0.92.

Wang et al. (2019) reported that the thermal effects on cable forces of the Puxi Bridge during extreme weather events were more adverse than those in normal weather. Field monitoring data showed that cable tensions at several locations (near the side pier, in the center of the side span, and near the towers) were most seriously affected by temperature.

Sousa Tomé et al. (2018) conducted a two-dimensional (2D) FE heat-transfer analysis and used the extracted temperatures as thermal loads in a simplified 3D model, assuming a constant longitudinal temperature distribution. They found that long-wave radiation heat transfer was crucial for ensuring analysis accuracy. The uniform temperature changes had the most significant impact on structural responses (bearing displacement, girder deflection, tower rotation, and cable force), accounting for about 90%. The temperature gradients had a minor effect on the longitudinal bearing displacement. The optimal deployment of thermometers was also discussed to more accurately estimate bridge responses.

Zhang et al. (2021b) numerically studied the temperature effects on a cable-stayed bridge with a separated side-box steel-concrete composite girder. The bridge was separately simulated by two models with different element compositions. The beam model consisted only of beam elements, thus considering only the longitudinal temperature gradients. By contrast, the hybrid model consisted of various element types according to bridge configurations. The maximum thermal stress of the bridge deck in each section of the main girder produced by the hybrid model was generally 20% larger than that by the beam model. They found that direct solar radiation had a significant effect on the temperature stress of the steel box girders, with a maximum stress of 31 MPa in summer. Moreover, the timing of the maximum temperature stress did not coincide with the maximum temperature gradient, and using only the maximum temperature gradient for structural thermal analysis would underestimate the temperature stress.

The studies about the temperature distribution and its effects on cable-stayed bridges are summarized in Table 2.3.

TABLE 2.3
Studies on the Temperature Behaviors of Cable-Stayed Bridges

Bridge	Focus	Study Approach	References
Qingzhou Bridge	Temperature behavior of the concrete tower	Numerical	Li et al. (2019)
Puxi Cable-stayed Bridge	Temperature field in the prestressed concrete girder	Numerical and field monitoring	Wang et al. (2019)
Zhanjiang Bay Cable-Stayed Bridge	Temperature temporal and spatial behaviors and structural deformations	Field monitoring and numerical	Cao et al. (2010)
Huanggang Yangtze River Bridge	Time lag of the structural temperature and girder deflection	Field monitoring	Yang et al. (2018a)
Dashengguan Yangtze River Bridge (Highway)	Site-specific extreme wind and structural temperature loads	Statistical and field monitoring	Xu et al. (2021)
Shanghai Yangtze River Bridge	Temperature-induced deformations of the bridges	Theoretical and field monitoring	Zhou and Sun (2019a,b); Zhou et al. (2020b)
Corgo Bridge	Temperature distribution and its influences on bridge responses	Numerical	Sousa Tomé et al. (2018)
A cable-stayed bridge	Temperature distribution and induced stresses	Numerical	Zhang et al. (2021b)
Dashengguan Yangtze River Bridge (Highway)	Thermal effects on the girder deflection using a multivariate linear-based model	Field monitoring and mathematical	Xu et al. (2019)
A cable-stayed bridge	Temperature effects on the tower deformation	Field monitoring and statistical	Wang et al. (2022)

(Continued)

TABLE 2.3 (*Continued*)
Studies on the Temperature Behaviors of Cable-Stayed Bridges

Bridge	Focus	Study Approach	References
Huanggang Yangtze River Bridge	Correlation of structural temperature and girder deflection	Statistical and field monitoring	Yang et al. (2018a)
Anqing Yangtze River Bridge	Correlation of structural temperature and the tower deformation	Statistical and field monitoring	Yang et al. (2018b)
Liuzhuang Bridge	Temperature distribution and induced deformation; computation method	Numerical	Zhu and Meng (2017)
Zhijiang Bridge	Cable damage detection through abnormal variation of the girder deflection-temperature correlation envelope curve	Numerical and statistical	Wang and Ye (2019)
Can Tho Bridge	Condition evaluation using Mahalanobis distance	Numerical and field monitoring	Le and Nishio (2015, 2019)
Zhaobaoshan Bridge	Damage detection of expansion joints	Statistical	Huang et al. (2018)

2.2.5 Temperature Behaviors of Suspension Bridges

Suspension bridges, known for their most spanning capacity, consist of main cables, towers, stiffening girders/trusses, suspenders, and anchorages. The temperature behavior of suspension bridges is significant due to the long span.

Wang et al. (2018) examined the temperature behavior of the Aizhai Suspension Bridge using FE simulations and SHM data. They found that the lateral and longitudinal temperature gradients and the temperature-induced tensile stress in the concrete deck exceeded the corresponding values specified in Chinese standards.

Ding et al. (2012) analyzed the thermal characteristics of the Runyang Suspension Bridge using long-term SHM data. They developed a probability function for the vertical and horizontal temperature gradients in the steel box girder's top and bottom plates. They found that the predicted extreme temperature gradients for a 100-year return period were consistent with British standards.

Zhang et al. (2021c) introduced a copula-based probabilistic model for the temperature gradient of the Nanxi Yangtze River Suspension Bridge's steel box girder. Using the Monte Carlo method, they extrapolated the temperature gradient over the bridge's service life. The results showed that the vertical temperature gradient was 96% of the corresponding value in the design standard.

Zhou et al. (2015) combined numerical simulations and field monitoring data to study the Humber Bridge's temperature behavior. They simulated convection using the conduction of air elements and modeled radiation using the radiation matrix method in ANSYS. The numerical analysis closely matched the measurements. The vertical and transverse temperature gradients were larger than 14°C and 18°C, respectively.

Xia et al. (2020a) analyzed the temperature distribution of the Jiangyin Suspension Bridge using a time-varying solar radiation model, which was adjusted for the temperature difference between the north and south anchorages. They reported significant temperature gradients between the inside and outside of the bridge towers, which could induce considerable thermal stress and should be considered in engineering practice.

In studies on temperature effects, Xu et al. (2010) reported the temperature variation statistics of the Tsing Ma Bridge and their correlations with the bridge displacement in three directions. The effective deck temperature was linearly correlated with the longitudinal displacements of the towers and the longitudinal and vertical displacements of the main girder and main cables. The temperature did not significantly influence the lateral deformations of the bridge.

Xia et al. (2013) studied the thermal effects on the Tsing Ma Bridge through numerical analysis. They modeled the girder, tower, and cross-frame segments separately and conducted the heat-transfer analysis to calculate their temperatures. The numerical results closely matched the measured ones. These temperatures were then input into a 3D FEM of the entire bridge to analyze the temperature-induced responses. The calculated displacements agreed well with the measured data.

Kim et al. (2005) presented the monitoring results of the Yeongjong Suspension Bridge. The field measurements showed that the average hanger tension, the horizontal displacement of expansion joints, and the vertical and lateral displacement of the girder were primarily governed by yearly and daily temperature variations. An ARX model was constructed to predict the displacements with high accuracy. The difference between the predicted and measured displacements was assumed to follow a normal distribution and was used to establish thresholds for structural damage warnings.

Xia et al. (2017b) analyzed one-year monitoring data to study the correlation between temperature and structural responses (strain, displacement, and suspender frequency/force) of the Jiangyin Yangtze River Bridge. The temperature changes caused a 5% variation in the suspender forces. The longitudinal temperature distribution was non-uniform, possibly due to varying airflow conditions along the main span.

Westgate et al. (2015) studied the temperature-induced behavior of the Tamar Bridge using FE simulation and field monitoring data. They found that the sway of the Plymouth Tower was related to the thermal expansion of the bridge deck, pulling it toward the mid-span as the bridge heated up. In contrast, the Saltash Tower, near the main span expansion joint, was almost unaffected by the girder's daily thermal behavior.

The studies on the temperature distribution and its effects on suspension bridges are summarized in Table 2.4.

TABLE 2.4
Studies on the Temperature Behaviors of Suspension Bridges

Bridge	Focus	Study Approach	References
Humber Bridge	Temperature distribution	Numerical and field monitoring	Zhou et al. (2015)
Jiangyin Yangtze River Suspension Bridge	Temperature distribution with a modified solar radiation model	Numerical and field monitoring	Xia et al. (2020a)
Jiangyin Yangtze River Suspension Bridge	Temperature distribution in the longitudinal direction	Statistical and field monitoring	Xia et al. (2017b)
Aizhai Bridge	Temperature distribution in the bridge	Statistical and field monitoring	Wang et al. (2018)
Runyang Suspension Bridge	Probability functions of the temperature gradients and extreme values	Statistical and field monitoring	Ding et al. (2012)
Nanxi Yangtze River Suspension Bridge	Probabilistic model for temperature gradients	Statistical and field monitoring	Zhang et al. (2021c)
Nanxi Suspension Bridge	Extreme value analysis and temperature distribution	Statistical and field monitoring	Deng et al. (2018)
Sutong Bridge	Temperature distribution and variation pattern	Field monitoring and statistical	Mao et al. (2018)
Various suspension bridges	Temperature-induced bridge deformations	Theoretical	Zhou et al.(2020c; 2021b; 2022)
Jiangyin Suspension Bridge	Relationship between the longitudinal boundary stiffness and the structural temperature	Mathematical	Xia et al. (2020b)
Jiangyin Suspension Bridge	Analytical method of calculating/separating thermal stress from the monitoring data	Theoretical	Xia et al. (2018)
Tamar bridge	Temperature distribution and temperature-induced behavior	Numerical	Westgate et al. (2015)
Tamar Bridge	Probabilistic structural identification	Mathematical	Jesus et al. (2019)
Tsing Ma Bridge	Temperature-induced stress	Numerical and field monitoring	Xia et al. (2013)
Great Belt Bridge	Regression analysis of temperature and structural response	Statistical	Farreras-Alcover et al. (2015)
Tamar Bridge	Thermal effects on structural deformation	Field monitoring	Koo et al. (2013)
Tsing Ma Bridge	Temperature behavior	Statistical and field monitoring	Xu et al. (2010)

(Continued)

TABLE 2.4 (*Continued*)
Studies on the Temperature Behaviors of Suspension Bridges

Bridge	Focus	Study Approach	References
Xihoumen Suspension Bridge	Temperature-induced and traffic-induced mid-span deflection and damage detection	Mathematical and field monitoring	Xu et al. (2020a,b)
Yeongjong Suspension Bridge	Correlation of structural responses and temperature and damage warning	Field monitoring and mathematical	Kim et al. (2005)
Jiangyin Suspension Bridge	Temperature effects on suspender forces	Statistical and field monitoring	Xia et al. (2017b)
Jiangyin Suspension Bridge	Damage detection using temperature and temperature-induced strain	Statistical and field monitoring	Xia et al. (2017a)
Humber bridge	Temperature-induced static response	Numerical and field monitoring	Zhou et al. (2020a)
Runyang Suspension Bridge	Seasonal correlation of frequency-temperature and expansion-temperature; damage detection	Mathematical	Deng et al. (2010)
Three long-span steel bridges	Expansion joint displacement and wear	Mathematical	Guo et al. (2015)
Tamar Bridge	Correlation between the deck extension and temperature measurement	Correlation and field monitoring	Battista et al. (2015)
Aizhai Suspension Bridge	linear relations between the structural deformations and the temperature	Mathematical and field monitoring	Yu and Ou (2017)
Nanxi Suspension Bridge	Deflection–vehicle load–temperature correlation model using deep learning	Mathematical and field monitoring	Deng et al. (2022)

2.2.6 Summary

Temperature actions exist throughout a bridge's entire lifespan, making monitoring temperature changes and their effects vital. One key purpose of temperature monitoring is to provide data supporting the development and updating of structural design standards. This leads to more optimized designs that mitigate the impact of temperature on bridge safety and serviceability. For example, determining the vertical and horizontal temperature gradients of girders relies on long-term field monitoring data. Additionally, temperature monitoring enables the elimination of temperature effects in bridge condition assessment. When evaluating the effect of traffic or wind loads, it is essential to remove temperature-induced displacements and strains from the raw measurements to obtain accurate results. Furthermore, the quantitative effect of temperature on structural responses can serve as an indicator of potential damage. For instance, if a bridge's main girder is restricted in its longitudinal movement, the resulting thermal displacement will differ significantly, highlighting areas of concern.

This section reviews past studies on temperature distribution and its effects on beam, arch, cable-stayed, and suspension bridges. Understanding temperature distribution is the first step in studying thermal effects. For large-span bridges, temperature usually exhibits non-uniform distribution due to their complex structures. Current studies often combine field monitoring with numerical simulations to investigate the temporal and spatial distribution of temperature data.

The non-uniform temperature distribution causes differential expansion and contraction in various parts of the bridge, resulting in complex thermal deformations and internal stresses. By developing mathematical models that correlate field-monitored temperature data with structural responses, researchers can reveal and quantify the impacts of temperature variations on the structure. This comprehensive understanding of temperature response mechanisms helps evaluate the bridge's operational state.

REFERENCES

Abid, S., 2015, Temperature distributions and variations in concrete box-girder bridges: Experimental and finite element parametric studies. *Advances in Structural Engineering* **18**: 469–486.

Aktan, A., Catbas, F. N., Grimmelsman, K. and Tsikos, C., 2000, Issues in infrastructure health monitoring for management. *Journal of Engineering Mechanics* **126(7)**: 711–724.

Aktan, A. E., Catbas, F. N., Grimmelsman, K. A. and Pervizpour, M., 2002, *Development of a Model Health Monitoring Guide for Major Bridges*. Technical Report No. DTFH61-01-P-00347. Philadelphia, PA, Drexel Intelligent Infrastructure and Transportation Safety Institute.

Au, F. T. K., Cheung, S. K. and Tham, L. G., 2002, Design thermal loading for composite bridges in tropical region. *International Journal of Steel and Composite Structures* **2(6)**: 441–460.

Battista, N. D., Brownjohn, J. M., Tan, H. P. and Koo, K.-Y., 2015, Measuring and modelling the thermal performance of the Tamar Suspension Bridge using a wireless sensor network. *Structure and Infrastructure Engineering* **11(2)**: 176–193.

Bergmeister, K., Aktan, A. E., Bucher, C., Dorfmann, L., Fehling, E., Frey, R. P., Geier, R., Huth, O., Inaudi, D., Maier, J. E., Santa, U., Schwesinger, P., Slowik, V. and Wenzel, H., 2003, *Monitoring and Safety Evaluation of Existing Concrete Structures (State-of-the-Art Report). fib Bulletin No. 22. fib Task Group 5.1.*

Berwanger, C., 1983, Transient thermal behavior of composite bridges. *Journal of Structural Engineering* **109(10)**: 2325–2339.

Branco, F. A., Mendes, P. A., 1993, Thermal actions for concrete bridge design. *Journal of Structural Engineering* **119(8)**: 2313–2331.

Cao, Y. H., Yim, J., Zhao, Y. and Wang, M. L., 2010, Temperature effects on cable stayed bridge using health monitoring system: a case study. *Structural Health Monitoring* **10(5)**: 523–537.

Catbas, F. N., Susoy, M. and Frangopol, D. M., 2008, Structural health monitoring and reliability estimation: Long span truss bridge application with environmental monitoring data. *Engineering Structures* **30(9)**: 2347–2359.

China Association for Engineering Construction Standardization, 2012, *Design Standard for Structural Health Monitoring System.* CECS 333-2012. Beijing, China Architecture Publishing & Media Co., Ltd.

China Association for Engineering Construction Standardization, 2018, *Early Warning Threshold Standard for Structural Health Monitoring System of Long-Span Bridges.* T/CECS 529-2018. Beijing, China Architecture Publishing & Media Co., Ltd.

China Association for Engineering Construction Standardization, 2019, *Standard for Operational Maintenance and Management of Structural Health Monitoring Systems.* T/CECS 652-2019. Beijing, China Architecture Publishing & Media Co., Ltd.

China Association for Engineering Construction Standardization, 2020, *Standard for Construction and Acceptance of Structural Health Monitoring Systems.* T/CECS 765-2020. Beijing, China Architecture Publishing & Media Co., Ltd.

China Civil Engineering Society, 2020a, *Mass Data Processing Standard for Structural Health Monitoring.* T/CCES 16-2020. Beijing, China Architecture Publishing & Media Co., Ltd.

China Civil Engineering Society, 2020b, *Technical Regulation on Sensor Selection and Placement for Bridge Health Monitoring.* T/CCES 15-2020. Beijing, China Architecture Publishing & Media Co., Ltd.

China's State Administration for Market Regulation and China's Standardization Administration, 2020, *Specifications for Operational Monitoring of Urban Rail Transit Facilities—Part 2: Bridge.* GB/T 39559.2-2020. Beijing, China Quality and Standards Publishing & Media Co., Ltd.

Dai, G., Wang, F., Chen, Y. F., Ge, H. and Rao, H., 2023, Modelling of extreme uniform temperature for high-speed railway bridge piers using maximum entropy and field monitoring. *Advances in Structural Engineering* **26(2)**: 302–315.

Deng, Y., Ding, Y. L. and Li, A. Q., 2010, Structural condition assessment of long-span suspension bridges using long-term monitoring data. *Earthquake Engineering and Engineering Vibration* **9(1)**: 123–131.

Deng, Y., Ju, H. W., Zhai, W. Q., Li, A. Q. and Ding, Y. L., 2022, Correlation model of deflection, vehicle load, and temperature for in-service bridge using deep learning and structural health monitoring. *Structural Control and Health Monitoring* **29(12)**: e**3113**.

Deng, Y., Li, A. Q., Liu, Y. and Chen, S. R., 2018, Investigation of temperature actions on flat steel box girders of long-span bridges with temperature monitoring data. *Advances in Structural Engineering* **21(14)**: 2099–2113.

Dilger, W. H., Ghali, A., Chan, M., Cheung, M. S. and Maes, M. A., 1982, Temperature stresses in composite box girder bridges. *Journal of Structural Engineering* **109(6)**: 1460–1478.

Ding, Y. L., Zhao, H. W. and Li, A. Q., 2017, Temperature effects on strain influence lines and dynamic load factors in a steel-truss arch railway bridge using adaptive FIR filtering. *Journal of Performance of Constructed Facilities* **31(4)**: 04017024.

Ding, Y. L., Zhou, G. D., Li, A. Q. and Wang, G. X., 2012, Thermal field characteristic analysis of steel box girder based on long-term measurement data. *International Journal of Steel Structures* **12(2)**: 219–232.

Duan, Y. F., Li, Y. and Qiang, X. Y., 2011, Strain-temperature correlation analysis of a tied arch bridge using monitoring data. In *2011 International Conference on Multimedia Technology*, Hangzhou, China, pp. 6025–6028.

Elbadry, M. M. and Ghali, A., 1983, Temperature variations in concrete bridges. *Journal of Structural Engineering* **109(10)**: 2355–2374.

Farreras-Alcover, I., Chryssanthopoulos, M. K. and Andersen, J. E., 2015, Regression models for structural health monitoring of welded bridge joints based on temperature, traffic and strain measurements. *Structural Health Monitoring* **14(6)**: 648–662.

Froli, M., Hariga, N., Nati, G. and Orlandini, M., 1996, Longitudinal thermal behaviour of a concrete box girder bridge. *Structural Engineering International* **6**(4): 237–242.

Fu, H. C., Ng, S. F. and Cheung, M. S., 1990, Thermal behavior of composite bridges. *Journal of Structural Engineering* **116(12)**: 3302–3323.

Fu, Y. D. and DeWolf, J. T., 2004, Effect of differential temperature on a curved post-tensioned concrete bridge. *Advances in Structural Engineering* **7(5)**: 385–397.

Guangdong-Hong Kong-Macao Greater Bay Area Standards Promotion Council of Guangdong Province, 2021, *Data Standard System for Intelligent Operation and Maintenance of Bridge-island-tunnel crossings — Guidelines on the Establishment of Standards.* T/GBAS 0001-2021.

Guo, T., Liu, J., Zhang, Y. F. and Pan, S. J., 2015, Displacement monitoring and analysis of expansion joints of long-span steel bridges with viscous dampers. *Journal of Bridge Engineering* **20(9)**: 04014099.

Hedegaard, B. D., French, C. E. W. and Shield, C. K., 2013, Investigation of thermal gradient effects in the I-35W St. Anthony Falls Bridge. *Journal of Bridge Engineering* **18(9)**: 890–900.

Ho, D., Liu and C. H., 1989, Extreme thermal loadings in highway bridges. *Journal of Structural Engineering* **115(7)**: 1681–1696.

Hoffman, P. C., McClure, R. M. and West, H. H., 1980, Temperature studies for an experimental segmental bridge. *PCI Journal* **28(2)**: 78–97.

Hou, R. and Xia, Y., 2021, Review on the new development of vibration-based damage identification for civil engineering structures: 2010–2019. *Journal of Sound and Vibration* **491**: 115741.

Huang, H. B., Yi, T. H., Li, H. N. and Liu, H., 2018, New representative temperature for performance alarming of bridge expansion joints through temperature-displacement relationship. *Journal of Bridge Engineering* **23(7)**: 04018043.

Indian Roads Congress, 2008, *Guidelines for Corrosion Prevention, Monitoring and Remedial Measures for Concrete Bridge Structures.* IRC:SP: 80-2008. New Delhi, Indian Road Congress.

ISO Standards, 2010, *Bases for Design of Structures: Assessment of Existing Structures.* ISO **13822**:2010. Geneva, ISO.

ISO Standards, 2015, *General Principles on Reliability for Structures.* ISO **2394**:2015. Geneva, ISO.

Jesus, A., Brommer, P., Westgate, R., Koo, K., Brownjohn, J. and Laory, I., 2019, Bayesian structural identification of a long suspension bridge considering temperature and traffic load effects. *Structural Health Monitoring* **18(4)**: 1310–1323.

Kennedy, J. B. and Soliman, M. H., 1987, Temperature distribution in composite bridges. *Journal of Structural Engineering* **113(3)**: 475–482.

Kim, S., Kim, C. Y. and Lee, J., 2005, Monitoring results of a self-anchored suspension bridge. In Ansari, F. (ed.), *Sensing Issues in Civil Structural Health Monitoring*, Springer, Dordrecht, pp. 475–484.

Koo, K. Y., Brownjohn, J. M. W., List, D. I. and Cole, R., 2013, Structural health monitoring of the Tamar suspension bridge. *Structural Control and Health Monitoring* **20(4)**: 609–625.

Kromanis, R., Kripakaran, P. and Harvey, B., 2015, Long-term structural health monitoring of the Cleddau bridge: evaluation of quasi-static temperature effects on bearing movements. *Structure and Infrastructure Engineering* **12(10)**: 1342–1355.

Larsson, O. and Karoumi, R., 2011, Modelling of climatic thermal actions in hollow concrete box cross-sections. *Structural Engineering International* **21(1)**: 74–79.

Le, H. V. and Nishio, M., 2015, Time-series analysis of GPS monitoring data from a long-span bridge considering the global deformation due to air temperature changes. *Journal of Civil Structural Health Monitoring* **5(4)**: 415–425.

Le, H. V. and Nishio, M., 2019, Structural change monitoring of a cable-stayed bridge by time-series modeling of the global thermal deformation acquired by GPS. *Journal of Civil Structural Health Monitoring* **9(5)**: 689–701.

Li, Y., He, S. H. and Liu, P., 2019, Effect of solar temperature field on a sea-crossing cable-stayed bridge tower. *Advances in Structural Engineering* **22(8)**: 1867–1877.

Liu, Y. N. and Zuk, W., 1963, Thermoelastic effects in prestressed flexural members. *PCI Journal* **8(3)**: 64–85.

Mao, J. X., Wang, H., Feng, D. M., Tao, T. Y. and Zheng, W. Z., 2018, Investigation of dynamic properties of long-span cable-stayed bridges based on one-year monitoring data under normal operating condition. Structural Control and Health Monitoring 25(**5**): e2146.

Ministry of Construction, Housing and Utilities of the Russian Federation, 2016, *Bridges. Monitoring of Technical Condition (Structural Health Monitoring).* SP 274.1325800.2016. Moscow, FGUP Standartinform.

Ministry of Housing and Urban-Rural Development of the People's Republic of China, 2014, *Technical Code for Monitoring of Buildings and Bridge Structures.* GB50982-2014. Beijing, China Construction Industry Press.

Ministry of Transport of the People's Republic of China, 2022, *Technical Specifications for Structural Monitoring of Highway Bridges.* JT/T 1037-2022. Beijing, China Communications Press Co., Ltd.

Mirambell, E. and Aguado, A., 1990, Temperature and stress distributions in concrete box girder bridges. *Journal of Structural Engineering* **116(9)**: 2388–2409.

Moorty, S. and Roeder, C. W., 1992, Temperature-dependent bridge movements. *Journal of Structural Engineering* **118(4)**: 1090–1105.

Mufti, A., 2001, *Guidelines for Structural Health Monitoring: Design Manual No.* 2. Winnipeg, MN, ISIS Canada Corporation.

Potgieter, I. C. and Gamble, W. L., 1983, *Response of Highway Bridges to Nonlinear Temperature Distributions.* Structural Research Series No. 505. Champaign, IL, University of Illinois at Urbana-Champaign.

Potgieter, I. C. and Gamble, W. L., 1989, Nonlinear temperature distributions in bridges at different locations in the United States. *PCI Journal* **34(4)**: 80–103.

Priestley, M. J. N., 1978, Design of concrete bridges for temperature gradients. *ACI Journal* **75(5)**: 209–217.

Riding, K. A., Poole, J. L., Schindler, A. K., Juenger, M. C. G. and Folliard, K. J., 2007, Temperature boundary condition models for concrete bridge members. *ACI Materials Journal* **104(4)**: 379–387.

Roeder, C. W., 2003, Proposed design method for thermal bridge movements. *Journal of Bridge Engineering* **8(1)**: 12–19.

Rücker, W., Hille, F. and Rohrmann, R., 2006, *F08b: Guideline for Structural Health Monitoring, Federal Institute of Materials Research and Testing (BAM)*. Berlin, SAMCO.

Song, Z. W., Xiao, J. Z. and Shen, L. M., 2012, On temperature gradients in high-performance concrete box girder under solar radiation. *Advances in Structural Engineering* **15(3)**: 399–415.

Sousa Tomé, E., Pimentel, M. and Figueiras, J., 2018, Structural response of a concrete cable-stayed bridge under thermal loads. *Engineering Structures* **176**: 652–672.

Subcommittee on New Technology Application Promotion of the Infrastructure Maintenance General Committee of the Japan Society of Civil Engineers, 2022, *Guidelines for the Utilization of Monitoring Technology*. Tokyo, Japan Society of Civil Engineers and Maruzen Publishing Co., Ltd.

Subcommittee on Performance-Checking Maintenance and Monitoring of Steel Bridges of the Committee on Steel Structures of the Japan Society of Civil Engineers, 2019, *Performance based Maintenance and Monitoring of Steel Bridges*. Tokyo, Japan Society of Civil Engineers and Maruzen Publishing Co. Ltd.

Tong, M., Tham, L. G. and Au, F. T. K., 2002, Extreme thermal loading on steel bridges in tropical region. *Journal of Bridge Engineering* **7(6)**: 357–366.

Tong, M., Tham, L. G., Au, F. T. K. and Lee, P. K. K., 2001, Numerical modelling for temperature distribution in steel bridges. *Computers and Structures* **79(6)**: 583–593.

Wang, D., Liu, Y. M. and Liu, Y., 2018, 3D temperature gradient effect on a steel–concrete composite deck in a suspension bridge with field monitoring data. *Structural Control and Health Monitoring* **25(7)**: e2179.

Wang, G. X. and Ye, J. H., 2019, Localization and quantification of partial cable damage in the long-span cable-stayed bridge using the abnormal variation of temperature-induced girder deflection. *Structural Control and Health Monitoring* **26(1)**: e2281.

Wang, J. F., Zhang, J. T., Xu, R. Q. and Yang, Z. X., 2019, Evaluation of thermal effects on cable forces of a long-span prestressed concrete cable–stayed bridge. *Journal of Performance of Constructed Facilities* **33(6)**: 04019072.

Wang, Y., Yang, D.-H., Zhou, Y.-Z. and Yi, T.-H., 2022, Cable effective length model error-based bridge performance warning method under thermal action. *Advances in Structural Engineering* **25(2)**: 306–320.

Wang, Y. B., Zhan, Y. L. and Zhao, R. D., 2016, Analysis of thermal behavior on concrete box-girder arch bridges under convection and solar radiation. *Advances in Structural Engineering* **19(7)**: 1043–1059.

Westgate, R., Koo, K. Y. and Brownjohn, J., 2015, Effect of solar radiation on suspension bridge performance. *Journal of Bridge Engineering* **20(5)**: 04014077.

Wittfoht, H., 1983, Causes of the partial collapse of the "Viadotto Cannavino" near Agro di Celico. *Beton- und Stahlbetonbau,* **78(2)**: 29–36.

Wollman, C. L. R., Breen, J. E. and Cawrse, J., 2002, Measurements of thermal gradients and their effects on segmental concrete bridge. *Journal of Bridge Engineering* **7(3)**: 166–174.

Xia, Q., Cheng, Y. Y., Zhang, J. and Zhu, F., 2017a, In-service condition assessment of a long-span suspension bridge using temperature-induced strain data. *Journal of Bridge Engineering* **22(3)**: 04016124.

Xia, Q., Liu, S. L. and Zhang, J., 2020a, Temperature analysis of a long-span suspension bridge based on a time-varying solar radiation model. *International Journal of Smart Structures and Systems* **25(1)**: 23–35.

Xia, Q., Wu, W. L., Li, F. N., Zhou, X. Q., Xu, Y. L. and Xia, Y., 2022, Temperature behaviors of an arch bridge through integration of field monitoring and unified numerical simulation. *Advances in Structural Engineering* **25(16)**: 3492–3509.

Xia, Q., Xia, Y., Wan, H. P., Zhang, J. and Ren, W. X., 2020b, Condition analysis of expansion joints of a long-span suspension bridge through metamodel-based model updating considering thermal effect. *Structural Control and Health Monitoring* **27(5)**: e2521.

Xia, Q., Zhang, J., Tian, Y. D. and Zhang, Y. F., 2017b, Experimental study of thermal effects on a long-span suspension bridge. *Journal of Bridge Engineering* **22(7)**: 04017034.

Xia, Q., Zhou, L. M. and Zhang, J., 2018, Thermal performance analysis of a long-span suspension bridge with long-term monitoring data. *Journal of Civil Structural Health Monitoring* **8(4)**: 543–553.

Xia, Y., Chen, B., Zhou, X. Q. and Xu, Y. L., 2013, Field monitoring and numerical analysis of Tsing Ma Suspension Bridge temperature behavior. *Structural Control and Health Monitoring* **20(4)**: 560–575.

Xu, X., Forde, M. C., Ren, Y. and Huang, Q., 2021, A Bayesian approach for site-specific extreme load prediction of large scale bridges. *Structure and Infrastructure Engineering*: 1–14.

Xu, X., Huang, Q., Ren, Y., Zhao, D. Y., Yang, J. and Zhang, D. Y., 2019, Modeling and separation of thermal effects from cable-stayed bridge response. *Journal of Bridge Engineering* **24(5)**: 04019028.

Xu, X., Ren, Y., Huang, Q., Fan, Z. Y., Tong, Z. J., Chang and W. J., Liu, B., 2020a, Anomaly detection for large span bridges during operational phase using structural health monitoring data. *Smart Materials and Structures* **29(4)**: 045029.

Xu, X., Ren, Y., Huang, Q., Zhao, D. Y., Tong, Z. J. and Chang, W. J., 2020b, Thermal response separation for bridge long-term monitoring systems using multi-resolution wavelet-based methodologies. *Journal of Civil Structural Health Monitoring* **10(3)**: 527–541.

Xu, Y. L., Chen, B., Ng, C. L., Wong, K. Y. and Chan, W. Y., 2010, Monitoring temperature effect on a long suspension bridge. *Structural Control and Health Monitoring* **17(6)**: 632–653.

Xu, Y. L. and Xia, Y., 2012, *Structural Health Monitoring of Long-Span Suspension Bridges*. CRC Press, London.

Yang, D. H., Yi, T. H., Li, H. N. and Zhang, Y. F., 2018a, Correlation-based estimation method for cable-stayed bridge girder deflection variability under thermal action. *Journal of Performance of Constructed Facilities* **32(5)**.

Yang, D. H., Yi, T. H., Li, H. N. and Zhang, Y. F., 2018b, Monitoring and analysis of thermal effect on tower displacement in cable-stayed bridge. *Measurement* **115**: 249–257.

Yang, K., Ding, Y. L., Sun, P., Zhao, H. W. and Geng, F. F., 2019, Modeling of temperature time-lag effect for cncrete box-girder bridges. *Applied Sciences* **9(16)**: 3255.

Yarnold, M. T. and Moon, F. L., 2015, Temperature-based structural health monitoring baseline for long-span bridges. *Engineering Structures* **86**: 157–167.

Yarnold, M. T., Moon, F. L., Aktan, A. E., 2015, Temperature-based structural identification of long-span bridges. *Journal of Structural Engineering* **141(11)**.

Yarnold, M. T., Moon, F. L., Dubbs, N. C. and Aktan, A. E., 2012, Evaluation of a long-span steel tied arch bridge using temperature-based structural identification. In *International Association for Bridge Management and Safety*, Stresa, Italy.

Yu, S. and Ou, J. P., 2017, Structural health monitoring and model updating of Aizhai suspension bridge. *Journal of Aerospace Engineering* **30(2)**: B4016009.

Yuksel, S. B., 2009, Behavior of symmetrically haunched non-prismatic members subjected to temperature changes. *Structural Engineering and Mechanics* **31(3)**: 297–314.

Zhang, F., Shen, J. and Liu, J., 2021a, Effect of encased concrete on section temperature gradient of corrugated steel web box girder. *Advances in Structural Engineering* **24(11)**: 2321–2335.

Zhang, H. H., Wang, P. Z., He, S. H., Li, Y., Chen, K. F. and Sun, N. N., 2021b, Research of thermal effect of cable-stayed bridge with a separated side-box steel-concrete composite girder under solar radiation. *Advances in Civil Engineering* **2021**: 8812687 .

Zhang, H. P., Liu, Y. and Deng, Y., 2021c, Temperature gradient modeling of a steel box-girder suspension bridge using Copulas probabilistic method and field monitoring. *Advances in Structural Engineering* **24(5)**: 947–961.

Zhao, H. W., Ding, Y. L., Nagarajaiah, S., Li, A. Q., 2019, Behavior analysis and early warning of girder deflections of a steel-truss arch railway bridge under the effects of temperature and trains: case study. *Journal of Bridge Engineering* **24(1)**: 05018013.

Zhou, G. D., Yi, T. H. and Chen, B., 2017a, Innovative design of a health monitoring system and its implementation in a complicated long-span arch bridge. *Journal of Aerospace Engineering* **30(2)**: B4016006.

Zhou, G. D., Yi, T. H., Chen, B. and Chen, X., 2018, Modeling deformation induced by thermal loading using long-term bridge monitoring data. *Journal of Performance of Constructed Facilities* **32(3)**: 04018011.

Zhou, G. D., Yi, T. H., Chen, B. and Zhang, H., 2017b, A generalized Pareto distribution–based extreme value model of thermal gradients in a long-span bridge combining parameter updating. *Advances in Structural Engineering* **20(2)**: 202–213.

Zhou, L., Xia, Y., Brownjohn, J. and Koo, K. Y., 2015, Temperature analysis of a long-span suspension bridge based on field monitoring and numerical simulation. *Journal of Bridge Engineering* **21**: 04015027.

Zhou, L. R., Chen, L., Xia, Y. and Koo, K. Y., 2020a, Temperature-induced structural static responses of a long-span steel box girder suspension bridge. *Journal of Zhejiang University-Science A* **21(7)**: 580–592.

Zhou, Y., Sun, L. M., 2019a, A comprehensive study of the thermal response of a long-span cable-stayed bridge: From monitoring phenomena to underlying mechanisms. *Mechanical Systems and Signal Processing* **124**: 330–348.

Zhou, Y. and Sun, L. M., 2019b, Insights into temperature effects on structural deformation of a cable-stayed bridge based on structural health monitoring. *Structural Health Monitoring* **18**(3): 778–791.

Zhou, Y., Sun, L. M., Fu, Z. H., Jiang, Z. and Ren, P. J., 2020b, General formulas for estimating temperature-induced mid-span vertical displacement of cable-stayed bridges. *Engineering Structures* **221**: 111012.

Zhou, Y., Xia, Y., Chen, B. and Fujino, Y., 2020c, Analytical solution to temperature-induced deformation of suspension bridges. *Mechanical Systems and Signal Processing* **139**: 106568.

Zhou, Y., Xia, Y. and Fujino, Y., 2021a, Analytical formulas of beam deflection due to vertical temperature difference. *Engineering Structures* **240**: 112366.

Zhou, Y., Xia, Y., Fujino, Y. and Yamaguchi, K., 2021b, Analytical formulas of thermal deformation of suspension bridges. *Engineering Structures* **238**: 112228.

Zhou, Y., Xia, Y., Sun, Z. and Fujino, Y., 2022, Analytical formulation of the temperature-induced deformation of multispan suspension bridges. *Structural Control and Health Monitoring* **29(6)**: e2937.

Zhu, J. S. and Meng, Q. L., 2017, Effective and fine analysis for temperature effect of bridges in natural environments. *Journal of Bridge Engineering* **22(6)**: 04017107.

Zhu, Q. X., Wang, H., Xu, Z. D., Spencer, B. F., Mao, J. X. and Gong, Z. H., 2021, Mapping temperature contours for a long-span steel truss arch bridge based on field monitoring data. *Journal of Civil Structural Health Monitoring* **11(3)**: 725–743.

Zuk, W., 1965, Thermal behaviour of composite bridges-insulated and uninsulated. *Highway Research Record* **76(1)**: 231–253.

3 Numerical Analysis of Bridge Temperature Behaviors

NOTATIONS

A_{tu}, B_{tu}	Parameters used to determine the turbidity coefficient
C_e	Stefan-Boltzman constant
I	Total solar radiation
I_d, I_i, I_r	Direct, diffuse and reflected solar radiations
I_{d0}, I_{i0}, I_{r0}	Energy reaching the Earth's surface by direction radiation, diffuse radiation, and reflected radiation
I_{ex}	Extraterrestrial radiation
I_{sc}	Solar constant
I_{dh}, I_{ih}, I_{th}	Direct, diffuse and total radiation received by a horizontal surface
K_0	Coefficient to convert degree Celsius to degree Kelvin
K_T	Transmittance factor
L_k	Length of overhanging cantilever slab
L_s	Length of shadow
N	Number of the day of the year
T	Temperature of solid
T^*	Equivalent air temperature
T_a, T_s	Air temperature and bridge surface temperature
T_{ave}, T_{dif}	Average temperature and temperature difference in a day
c	Specific heat of the solid (J/kg/°C)
h_c, h_r	Convection and irradiation heat-transfer coefficients
k	Thermal conductivity (W/m/°C)
k_a	Ratio of atmospheric pressure to the pressure at sea level
k_T	Clearance index
m	Air mass factor
n_x, n_y	Direction cosines of the unit outward vector normal to the boundary surfaces
q	Boundary heat input or loss per unit area (W/m^2)
q_c, q_r, q_s	Convection, irradiation, and solar radiation components of boundary heat input/loss per unit area (W/m^2)
r_e	Reflected coefficient
t	Time

DOI: 10.1201/9781003480136-3

t_u	Turbidity factor
v, v_0	Effective wind speed and original wind speed
x, y, z	Cartesian coordinates
α	Absorptivity coefficient of the surface material
β	Tilt angle
ε	Relative emissivity coefficient
φ	Wind incidence angle with respect to the bridge surface
γ	Azimuth angle with respect to the normal direction of the surface
γ_s	Azimuth angle with respect to the south direction
ρ	Density
θ	Incidence angle of sunrays
ψ	Solar altitude

3.1 PREFACE

In addition to field monitoring studies, numerical simulations are also frequently adopted to investigate the temperature behaviors of bridges. Although field monitoring can provide valuable information about temperature behaviors, including both the temperature distribution and temperature-induced structural responses, it is still incapable of providing complete information on the temperature distribution and structural responses due to sensor number limitations. Typically, only a limited number and types of sensors are installed at the key components of the bridge. In contrast, numerical simulations can provide complete information regarding temperature behaviors. In this chapter, the methods of numerical simulation of bridge temperature behaviors will be presented.

3.2 BASIS OF HEAT-TRANSFER ANALYSIS

3.2.1 General Formulation

With the rapid development of computational methods and computer technology, a number of 1D to 3D FE models have been developed since the 1970s (Elbadry and Ghali, 1983; Wang, 1994). Finite difference heat flow models have also been employed to determine the temperature distribution of bridge members (Riding et al., 2007). These models, however, are not introduced in this book.

The flow of heat in an isotropic solid is governed by the Fourier partial differential equation (Whitaker, 2013):

$$\rho c \frac{\partial T}{\partial t} = k\left(\frac{\partial^2 T}{\partial x^2} + \frac{\partial^2 T}{\partial y^2} + \frac{\partial^2 T}{\partial z^2}\right) \tag{3.1}$$

where x, y, z are the Cartesian coordinates; t is time; T is the temperature of the solid; ρ is density; c is the specific heat of the solid (J/kg/°C); and k is thermal conductivity (W/m/°C). For a bridge subjected to solar radiation, it can be assumed that thermal variation in the direction of the longitudinal axis is normally not significant (Elbadry and Ghali, 1983; Fu et al., 1990). Consequently, the temperature field T of a bridge cross section at any time t can be expressed by a two-dimensional (2D) heat flow equation as

$$\rho c \frac{\partial T}{\partial t} = k\left(\frac{\partial^2 T}{\partial x^2} + \frac{\partial^2 T}{\partial y^2}\right) \tag{3.2}$$

3.2.2 Thermal Boundary Conditions

When the bridge temperature at the boundary is known, the condition is referred to as the First Kind thermal boundary condition (Lienhard IV and Lienhard V, 2020). When heat flow at the boundary is known, the Third Kind thermal boundary condition can be employed (Elbadry and Ghali, 1983):

$$k\left(\frac{\partial T}{\partial x} n_x + \frac{\partial T}{\partial y} n_y\right) + q = 0 \tag{3.3}$$

where n_x and n_y are direction cosines of the unit outward vector normal to the boundary surfaces, and q is boundary heat input or loss per unit area (W/m^2).

For a bridge surface subjected to solar radiation, as illustrated in Figure 3.1 (Branco and Mendes, 1993), q consists of convection q_c, thermal irradiation q_r, and solar radiation q_s, as

$$q = q_c + q_r + q_s \tag{3.4}$$

3.2.2.1 Convection Heat Transfer

The rate of heat transfer by convection q_c is associated with the movement of the air particles. It can be approximated by Newton's convection law with the convection heat transfer coefficient h_c and the difference between the air temperature T_a and the bridge surface temperature T_s as

$$q_c = h_c(T_a - T_s) \tag{3.5}$$

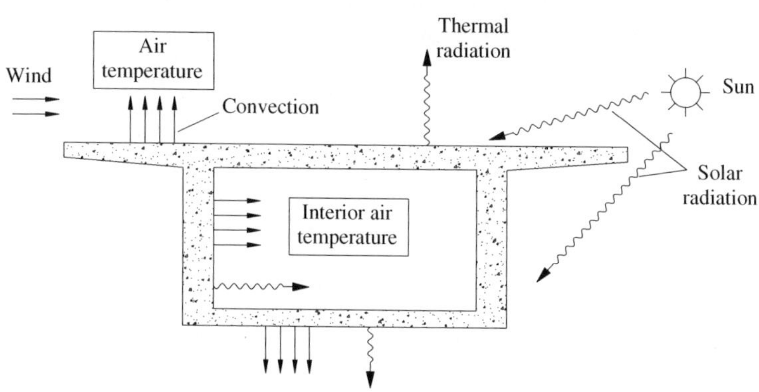

FIGURE 3.1 Energy exchange of a bridge section.

The convection heat transfer coefficient h_c (W/m^2/°C) is related to many parameters such as wind speed, surface roughness, and geometric configuration of the exposed surface and is usually determined experimentally or calculated based on empirical formulae (Branco and Mendes, 1993; Froli et al., 1996):

$$h_c = \begin{cases} 4v + 5.6 & (v < 5\,\text{m/s}) \\ 7.15v^{0.78} & (v \geq 5\,\text{m/s}) \end{cases} \tag{3.6}$$

where v is the wind speed. Note that different bridge surfaces might be exposed to different wind incidence angles and thus different wind speeds. Consequently, the wind speed may need revision before being applied to the convection heat transfer coefficient calculations. For example, Zhou et al. (2015) modified the effective wind speeds on different structural surfaces with respect to the wind incidence angle φ as

$$v = \begin{cases} 0.8v_0 & \varphi \leq 45° \\ 0.7v_0 & 45° < \varphi \leq 90° \\ 0.6v_0 & \varphi > 90° \end{cases} \tag{3.7}$$

with v_0 being the original wind speed. It should also be noted that two values of T_a should be considered for the ambient air outside and inside a bridge section for the box girder sections.

3.2.2.2 Irradiation Heat Transfer

The heat transfer between the bridge surface and the surrounding environment due to thermal irradiation can be modeled by Stefan-Boltzmann radiation law as

$$q_r = C_e \varepsilon \left[(T_a + K_0)^4 - (T - K_0)^4 \right] \tag{3.8}$$

where C_e=5.667 × 10^{-0} W/(m^{2}K^4) is the Stefan-Boltzman constant, ε is the emissivity coefficient relating the radiation of the bridge surface (a grey body) to that of an ideal black body, and K_0=273.15 converts the temperature in degree Celsius to degree Kelvin. Equation (3.8) can be rewritten in a quasi-linear form as

$$q_r = h_r (T_a - T_s) \tag{3.9}$$

where h_r denotes the irradiation heat transfer coefficient and is defined as

$$h_r = C_e \varepsilon \left[(T + K_0)^2 + (T_a + K_0)^2 \right] (T + T_a + 2K_0) \tag{3.10}$$

This coefficient cannot be calculated unless T is known. However, in a time incremental solution, an approximate value of h_r at any instant can be calculated by using earlier values of T. For the usual difference between air temperatures and surface temperatures, h_r can be approximately expressed as (Branco, 1986)

$$h_r = \varepsilon\left[4.8 + 0.075(T_a - 5)\right] \tag{3.11}$$

Thus, the effects of heat flow by convection and irradiation can be combined in terms of an overall heat transfer coefficient h (W/m²/°C) as

$$q = h(T_a - T_s) \tag{3.12}$$

$$h = h_c + h_r \tag{3.13}$$

3.2.2.3 Solar Radiation

Solar radiation is affected by several factors such as the hour of the day, the day of the year, the latitude and the altitude of the bridge, the orientation of the bridge, and the cloudiness of the sky. The spatial position of the Sun and a bridge is illustrated in Figure 3.2, where ψ is the solar altitude, θ is the incidence angle of sunrays, γ_s is the azimuth angle, the vector $\vec{n}$ is perpendicular to the surface, surface azimuth angle γ is measured from the south and is positive toward the west, β is the tilt angle from the horizontal and is positive for south-facing surface, solar incidence angle θ is defined as the angle between the surface normal $\vec{n}$ and a line collinear with the sunrays. The rate of heat absorbed by the bridge surface due to solar radiation q_s is (Rohsenow et al., 1985; Branco et al., 1992)

$$q_s = \alpha I \tag{3.14}$$

where α ($0 < \alpha < 1$) is the absorptivity coefficient of the surface material; I is the total solar radiation (W/m²), including direct solar radiation I_d, diffuse solar radiation I_i, and reflected solar radiation I_r on a surface. These quantities can be obtained through field measurement or empirical formulae.

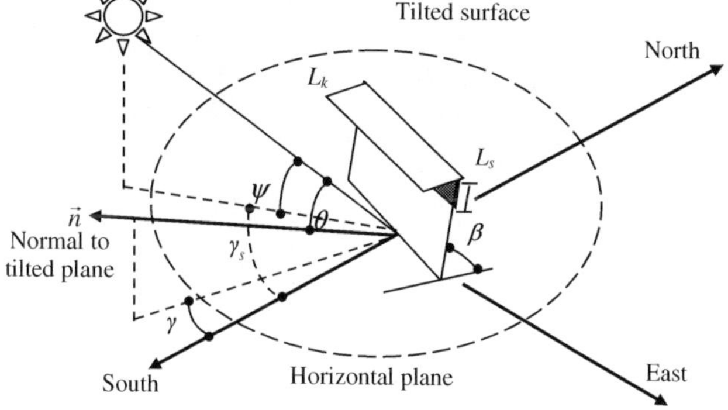

FIGURE 3.2 Spatial position of the Sun and structure.

3.2.2.4 Direct Solar Radiation

The direct radiation energy to heat the bridge depends on the solar constant (I_{sc}) and the absorption of the solar energy into the atmosphere. The solar constant is the amount of solar electromagnetic radiation received by a unit area of a plane perpendicular to the sunrays, at the mean distance from the Sun to the Earth. As the Earth's orbit around the Sun is elliptical, solar constant I_{sc} is suggested to be (Rohsenow et al., 1985):

$$I_{sc} = 1367\left[1 + 0.033\cos\left(360° \times N/365\right)\right] \tag{3.15}$$

where N is the number of the day of the year, counting from January 1. The solar constant varies by about 7.0% during a year (from 1.412 kW/m^2 in early January to 1.321 kW/m^2 in early July) due to the Earth's varying distance from the Sun.

The Earth's atmosphere acts as a filter for the solar radiation so that only a fraction of the total solar radiation reaches the surface of the Earth. The energy reaching the Earth's surface by direct radiation I_{d0} can be expressed as

$$I_{d0} = I_{sc}K_T \tag{3.16}$$

where K_T is the dimensionless transmittance factor, which depends on the atmospheric conditions such as the scattering of the light in a pure atmosphere and the absorption of certain wavelengths by the atmosphere. K_T can be calculated as

$$K_T = 0.9^{k_a t_u m} \tag{3.17}$$

where k_a is the ratio of the atmospheric pressure to the pressure at sea level (refer to Table 3.1), t_u is the turbidity factor indicating the attenuation of radiation in different atmospheric conditions, and m is the air mass factor.

Air pollution increases the absorption of the solar radiation, and the turbidity factor t_u can be expressed as

$$t_u = A_{tu} - B_{tu}\cos\left(360° \times N/365\right) \tag{3.18}$$

where parameters A_{tu} and B_{tu} are suggested to take the values in Table 3.2. The air mass factor, m, gives the relative path length of the radiation through the atmosphere, which is expressed in terms of the solar altitude ψ as

$$m = 1\Big/\left[\sin\psi + 0.15\left(\psi + 3.885\right)^{-1.253}\right] \tag{3.19}$$

TABLE 3.1
Ratio of the Atmospheric Pressure to the Pressure at Sea Level (k_a)

Elevation to the Sea Level (m)	0	500	1000	1500	2000	2500	3000
k_a	1.0	0.94	0.89	0.84	0.79	0.74	0.69

TABLE 3.2
Parameters of the Turbidity Factor

	Mountain Areas	Village	City	Industrial Areas
A_{tu}	2.2	2.8	3.7	3.8
B_{tu}	0.5	0.6	0.5	0.6

The above equation can be simplified as

$$m = 1/\sin\psi \tag{3.20}$$

When the Sun has an incidence angle θ to the surface, the direct solar radiation is

$$I_d = I_{d0}\cos\theta = I_{sc}K_T\cos\theta \tag{3.21}$$

In the zone where an overhanging cantilever slab of length L_k provides shade (Figure 3.2), no direct radiation is present. The length of the shadow, L_s, can be calculated as

$$L_s = L_k \tan\psi \Big/ \left[\sin\left(90^\circ + \gamma - \gamma_s\right)\sin\beta - \cos\beta\tan\psi\right] \tag{3.22}$$

The direct solar radiation given above is derived from the macro weather condition models and thus may be slightly different from the actual local regional scenarios. To address this issue, the actual radiation can be directly measured by pyranometers. For example, Figure 3.3 illustrates a set of pyranometers for measuring the diffuse, reflected, and total solar radiation.

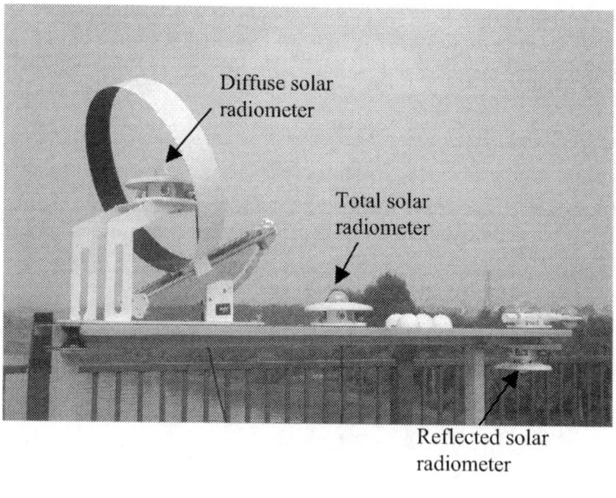

FIGURE 3.3 Configuration of a set of pyranometers.

3.2.2.5 Diffuse Solar Radiation

The direct solar radiation is accompanied by radiation that has been scattered or remitted, called diffuse solar radiation. The energy reaching the Earth's surface by the diffuse solar radiation can be estimated as (Berlage, 1928)

$$I_{i0} = 0.5 \times \left(1 - P^m\right) / \left(1 - 1.4 \ln P\right) \times I_{sc} \sin \psi \tag{3.23}$$

where

$$P = 0.9^{k a t_u} \tag{3.24}$$

The diffuse solar radiation on a surface with a tilt angle β is

$$I_i = 0.5 \times \left(1 + \cos\beta\right) \times I_{i0} \tag{3.25}$$

Another issue that needs to be considered is that only the summation of direct and diffuse solar radiations received by a horizontal surface (I_{th}) is available, as the Hong Kong Observatory does. In this case, the direct and diffuse radiations need to be separated according to the diffuse ratio as

$$I_{\text{ih}} / I_{\text{th}} = \begin{cases} 1 - 0.435 k_T & 0 \le k_T < 0.325 \\ 1.41 - 1.685 k_T & 0.325 \le k_T \le 0.679 \\ 0.259 & 0.679 < k_T \le 1 \end{cases} \tag{3.26}$$

$$k_T = I_{\text{th}} / I_{\text{ex}} \tag{3.27}$$

$$I_{\text{ex}} = I_{\text{sc}} \sin \psi \tag{3.28}$$

where I_{ih} is the diffuse radiation received by a horizontal surface, k_T is the clearance index, I_{ex} is the extraterrestrial radiation, then the direct solar radiation received by a horizontal surface (I_{dh}) can be separated as

$$I_{\text{dh}} = I_{\text{th}} - I_{\text{ih}} \tag{3.29}$$

Then the direct and diffuse solar radiations on a tilted surface can be estimated by Equations (3.21) and (3.25).

3.2.2.6 Reflected Solar Radiation

The direct and diffuse solar radiations may also be reflected onto a nearby surface through the reflected solar radiation (Berlage, 1928)

$$I_{r0} = r_e \left(I_{d0} + I_{i0}\right) \tag{3.30}$$

where r_e is the reflected coefficient of the ground, usually set as 0.2 for ordinary Earth surfaces and 0.7 for snow-covered surfaces. The reflected solar radiation on a surface with a tilt angle β is

$$I_r = 0.5 \times (1 - \cos\beta) \times I_{r0} \tag{3.31}$$

3.2.2.7 Diurnal Temperature Variation

For the computer analysis of the structural temperature field, the daily variation of the ambient temperature can be measured continuously and expressed as a function of the time of day

$$T_a = f(t) \tag{3.32}$$

If only the maximum and minimum temperature in a day are recorded instead of the entire history of ambient temperature, it is customary to assume a sinusoidal variation between the minimum temperature minT_a, assumed to occur at 3:00 and the maximum value maxT_a, assumed to occur at 15:00. With these assumptions, the temperature at any hour, t, of the day can be expressed by the following equation

$$T_a(t) = T_{\text{ave}} + T_{\text{dif}} \cdot \sin\left[\pi/12 \cdot (t - 9)\right] \tag{3.33}$$

where

$$T_{\text{ave}} = \left[\max T_a(t) + \min T_a(t)\right]/2 \tag{3.34}$$

$$T_{\text{dif}} = \left[\max T_a(t) - \min T_a(t)\right]/2 \tag{3.35}$$

3.2.2.8 Equivalent Temperature

The total solar radiation on a surface is then the summation of the direct, diffuse, and reflected solar radiations, that is

$$I = I_d + I_i + I_r \tag{3.36}$$

Based on Equations (3.12)–(3.14), the heat flow of a surface considering the solar radiation is

$$q = h(T_a - T_s) + \alpha I \tag{3.37}$$

or a more convenient form

$$q = h\left(T^* - T_s\right) \tag{3.38}$$

where

$$T^* = T_a + \alpha I / h \tag{3.39}$$

T^* is usually termed "equivalent air temperature" as it includes both the air temperature and solar radiation.

The equivalent air temperature and the overall heat transfer coefficient form the complete thermal boundary conditions for temperature analysis of bridges.

3.2.3 Initial Temperature Conditions

In addition to the thermal boundary conditions, the initial temperature conditions are also required in FE analyses. However, the initial temperatures of all degrees of freedom are unavailable in practice as the thermometers are insufficient. In this case, a pre-analysis for one or more consecutive days can be performed with assumed uniform initial temperatures and real thermal boundary conditions. After the pre-analysis, the temperatures of different components become non-uniform and approximate to the real temperature distribution. This distribution will be used as the initial conditions of the subsequent heat-transfer analysis (Zhou et al., 2015). Typically, pre-analysis for 24 hours is sufficient for an acceptable initial temperature condition estimation.

3.2.4 Summary

In this section, the basis of heat-transfer analysis is presented. In particular, the methods of determining the thermal boundary and initial temperature conditions are introduced in detail. A heat-transfer analysis can then be performed to obtain the temperature field of a bridge. After obtaining the temperature distribution of the bridge, the temperature results can be converted to thermal loads to calculate the temperature-induced structural responses, which will be introduced in the following sections with examples.

3.3 SEPARATED HEAT-TRANSFER AND STRUCTURAL ANALYSES

Generally, there are two approaches to calculating temperature-induced structural behaviors: separated and integrated heat transfer and structural analyses.

The separated approach is widely adopted in literature (Xia et al., 2013; Li et al., 2023). A bridge generally has a complicated structural configuration; thus, calculating the temperature distribution of the entire structure needs a massive FE model, which takes a large amount of computational resources. Therefore, in practice, many researchers build refined 2D or 3D FE models for the main components (such as deck and tower) and conduct heat-transfer analysis to calculate temperature distributions of the components separately. The temperatures are then input into another global bridge model to calculate temperature responses via the structural analysis. This divide-and-conquer approach is inefficient because it requires considerable manual intervention. The results are also inaccurate because the simplified 2D/3D models fail to simulate the real temperature distribution of the entire bridge, particularly for the tower-deck connection region and the varying cross sections.

Until recently, the integrated approach emerged due to the rapid development of personal computers, making modeling and analyzing an entire bridge possible. The approach only establishes one consistent FE model with different element types for the two analyses. The calculated temperature results of the global model can be directly transformed into thermal loads and applied to the same FE model to calculate the temperature-induced bridge responses. No manual interventions are required in this scheme.

A separate approach will be introduced in this section, with the Humber Bridge as an example. The integrated one will be described in the next section.

3.3.1 Humber Bridge and its Monitoring System

The Humber Bridge, completed in June 1981, has a total length of 2220 m with an asymmetric layout comprising the 280-m Hessle side span, 1410-m main span, and 530-m Barton (south) side span, as shown in Figure 3.4. The main span had been the longest in the world for 17 years from its inauguration.

The bridge girder is not continuous through the towers, having a complex arrangement of bearings and expansion joints to accommodate movement due to wind, traffic, and thermal loadings. It was assembled from 124 box segments, each typically 18.1 m long, 22 m wide, and 4.5 m high, with four stiffened bulkheads and has cantilevered footpaths and cycle tracks, bringing the total width to 28 m. The upper

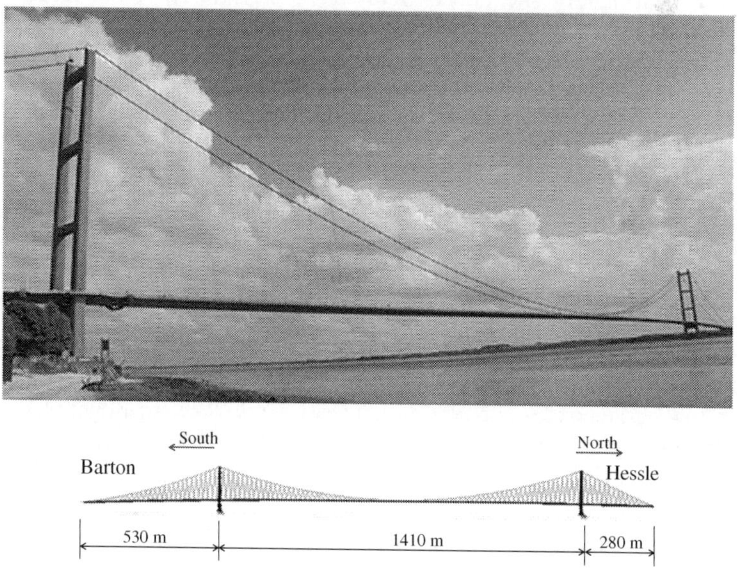

FIGURE 3.4 Configuration of the Humber Bridge.

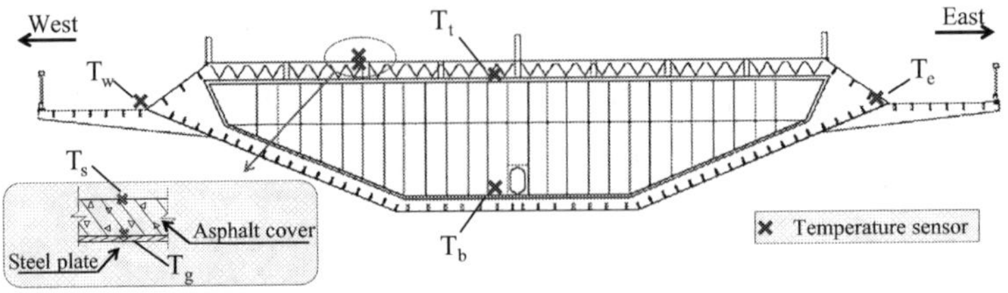

FIGURE 3.5 Thermometers in one section of the Humber Bridge.

roadway surface of the box is an 18.2 m wide orthotropic steel deck originally covered with 41 mm thick rubberized bitumen asphalt, which has been replaced at least once, while the cantilevers retain a thin asphalt surfacing (Brownjohn et al., 1987).

The temperature-related monitoring system of the Humber Bridge was installed in 2011. Specifically, a weather station was installed at the mid-span of the bridge, which supplements data from anemometers and air temperature sensors installed previously by the bridge operator. Meteorological data from nearby Humberside Airport are also available and are integrated into the database of the SHM system. Four thermocouples were installed to record the temperature of the box girder at the mid-span located on the top, bottom, east, and west surfaces of the box girder. Moreover, real-time kinematic position data are available for GPS antennae fixed to each main cable at mid-span, with correction data provided by a reference station at the Hessle anchorage. This system provides the primary position data for the bridge. Four laser extensometers, one on each column of the Hessle and Barton towers, provide bearing movement data. Figure 3.5 depicts the two temperature sensors installed during construction in the middle of the left lanes of the northbound carriageway to collect the surface temperature of the asphalt and the ground temperature in the interface between the steel surface and paved asphalt.

3.3.2 FE Model for Heat-Transfer Analysis

The temperature along the longitudinal direction of the Humber Bridge is assumed to be constant. Therefore, the FE model of a typical box girder section is constructed using ANSYS (ANSYS Inc., 2009) to investigate the temperature distribution of the box girder. Figure 3.6 shows the details of the box girder section and the FE model, which consists of 38,620 elements. The model uses PLANE55 elements for several materials: steel for the stiffened plate, asphalt for the roadway, and air filling the inside hollow of the box section. PLANE55 is a type of 2D element with four nodes, each having a single degree of freedom of temperature and endowed with thermal conduction capability, making it suitable for the 2D, steady-state or transient thermal analysis.

The thermal boundary conditions of the exterior and interior structural surfaces are separately applied in the conventional thermal analysis of box girders. Also, the surrounding air of the exterior and interior of the box girder are considered independent. Consequently, the thermal equilibrium of the entire system can hardly be

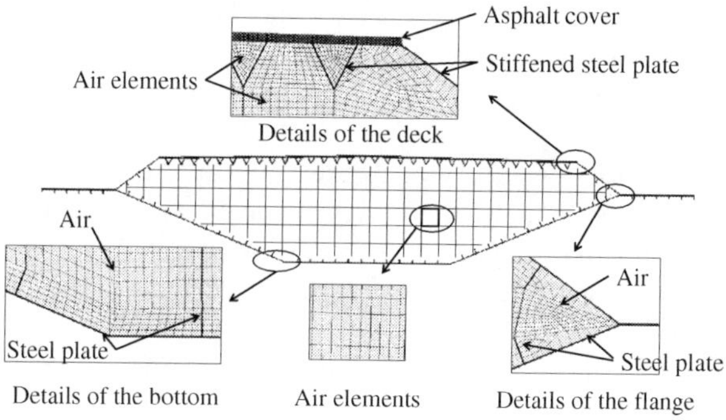

FIGURE 3.6 FE model of a girder section of the Humber Bridge.

maintained. In the thermal analysis of the Humber Bridge, the air filling inside the box is modeled by PLANE55 elements; thus, the thermal boundary conditions of convection on the interior surfaces of the box are not necessary. The interaction of the thermal radiation of the interior surfaces is calculated using the AUX12 radiation matrix, and the results are applied to the inside surface using the super-element MATRIX50 in ANSYS (ANSYS Inc., 2009).

3.3.3 Temperature Variation of Box Girder Section

Extreme environmental conditions, including strong solar radiation, high air temperature, and low wind speed, normally generate high-temperature gradients throughout the box girder. Thus, a typical sunny day on July 24, 2012 with relatively high solar radiation and air temperature was selected. Figure 3.7 shows the environmental conditions of the bridge, including the measured wind speed, wind direction, air temperature, and cloud cover. The cloud cover condition was obtained from the meteorological measurements at the Humberside Airport. The maximum cloud cover in the daytime is only 25%; therefore, the day can be considered a clear day.

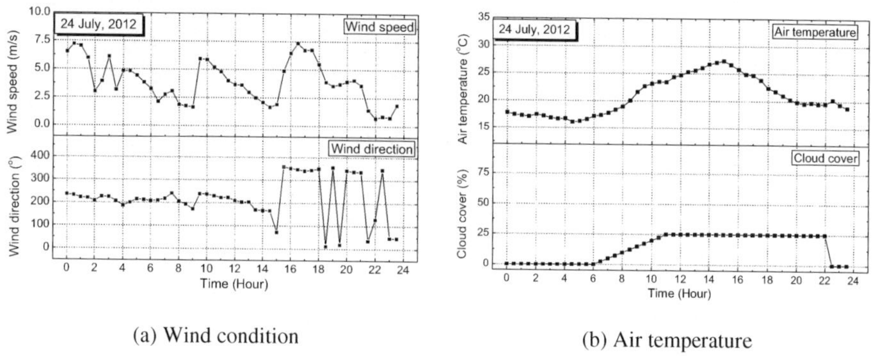

FIGURE 3.7 Wind and temperature conditions of the Humber Bridge: (a) wind condition and (b) air temperature.

Based on the measured meteorological data, the thermal boundary conditions are calculated and applied to the FE model for transient heat transfer analysis. The initial temperature of the box girder is obtained from the final results of a pre-analysis of the previous day.

The temperature variations of the box girder on July 24, 2012 are calculated and compared with the corresponding measurements in Figure 3.8. The structural temperature slightly decreased and reached the minimum in the early morning. The temperature then increased to the maximum in the early afternoon and decreased in the evening and midnight.

The temperature of the entire bridge reached a minimum of approximately 10°C at around 5:00. This finding indicates that the entire bridge had an approximately uniform temperature distribution at this moment. However, different components reached their maximum temperature at different time instants. The asphalt cover had a maximum temperature at approximately 15:00, the top of the inner surface of the box girder reached the maximum temperature at around 16:00, and the bottom

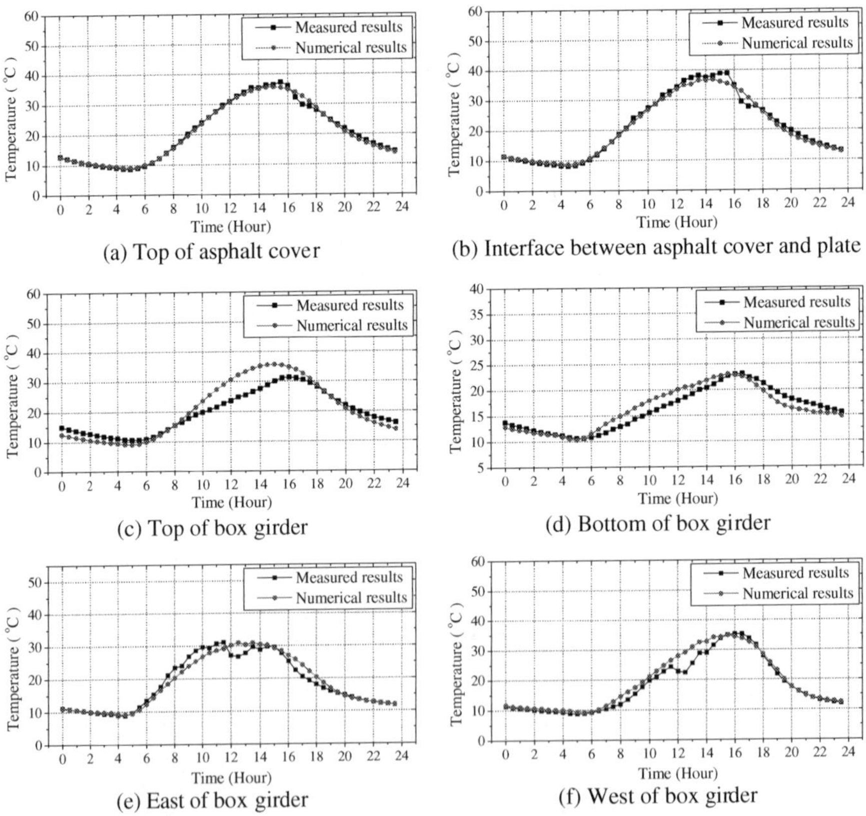

FIGURE 3.8 Measured and simulated temperatures of the bridge deck on July 24, 2012: (a) top of asphalt cover, (b) interface between asphalt cover and plate, (c) top of box girder, (d) bottom of box girder, (e) east of box girder, and (f) west of box girder.

surface a little later. The structural temperature of the east side reached its maximum at around 11:00, 5 hours earlier than the west side, at approximately 16:00. For this type of box girder bridge, the exterior reaches the maximum temperature earlier than the interior, the top earlier than the bottom, and the east earlier than the west. The maximum temperature values were approximately 35°C, 34.5°C, 34°C, and 26°C for the asphalt, interface, girder top, and girder bottom, respectively.

The simulated temperatures at the observed points agree well with the field measurements. Therefore, the effectiveness of the heat transfer analysis is verified. Some measured temperatures exhibit abrupt changes, which have not been well predicted in the numerical simulation. This condition is most likely attributed to the transient local cloud cover affecting solar radiation.

3.3.4 Vertical Temperature Gradient of Box Girder

To study the typical seasonal temperature behavior of the bridge, four sunny days in different seasons are selected for thermal analysis. The selected days are February 11, May 16, July 24, and October 6, 2012. According to the measured air temperature, February 11 was the coldest day in winter, whereas July 24 was the hottest in summer. May 16 and October 6 can represent spring and autumn weather conditions, respectively.

The vertical temperature differences of the box girder in different seasons are obtained from the numerical analysis and plotted in Figure 3.9. Figure 3.9a depicts the variation of the temperature difference of the cross section over time. The temperature difference in this simulation refers to the difference between the maximum and minimum temperatures of the section along the web on one side, which may occur at different points at different time instances. The temperature difference of the section was considerable, and the difference in daytime was significantly larger than the difference at night. The vertical temperature difference was small before sunrise, increasing to a maximum at noon and decreasing until after sunset. The east side had the largest temperature difference at about 9:00, and the west side had the maximum at approximately 15:00. Moreover, the maximum vertical temperature

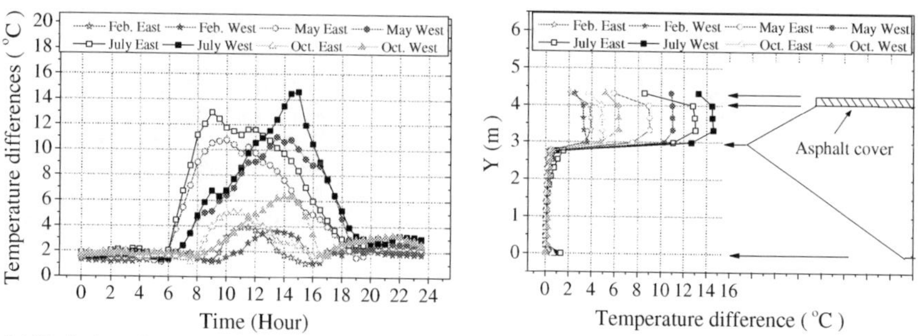

(a) Variation of temperature difference over time (b) Largest temperature gradients in one day

FIGURE 3.9 Temperature difference of a box girder cross section along the vertical direction.

difference on the west side was larger than that on the east side. The vertical temperature difference was largest in summer and smallest in winter.

The vertical temperature gradient profile is shown in Figure 3.9b. The upper web had higher temperatures than other components because it received more direct solar radiation. The lower web and bottom deck had the lowest temperatures because they were blocked from direct solar radiation. The east and west sides of the box girder reached peak temperatures at different times, as shown in Figure 3.9a. This finding indicates that transversal temperature difference (TTD) existed in the east and west upper webs.

Owing to the analysis efficiency of the 2D model, factors influencing TTD are further explored.

3.3.5 TTD of Box Girder

Eurocode 1 for thermal actions (European Committee for Standardization CEN, 2003) states that only the vertical temperature difference must be considered in usual cases. However, TTD must also be considered in particular cases such as that in which one side is more exposed to sunlight than the other. The box girder of the Humber Bridge is streamlined with inclined webs on the east and west sides that may cause significant TTD due to receiving different solar radiation.

In this section, TTD between the sensors installed on the east and west sides of the bridge deck (Figure 3.5) is investigated first. Figure 3.10a shows the absolute value of TTD in two clear days, representing the situations in the summer and spring. In the morning, the east side received much more solar radiation than the west, and the temperature difference reached the maximum of 12°C at 8:00 in summer and 5°C at 11:00 in winter. On the contrary, in the afternoon, the west side received much more solar radiation than the east, and the TTD reached the maximum of 13°C at 17:00 in summer and 5°C at 15:00 in winter. Therefore, summer has a more significant TTD. Figure 3.10b shows the daily maximum TTD in 2012, which could be as high as 18°C. In general, significant TTD occurred from March to September.

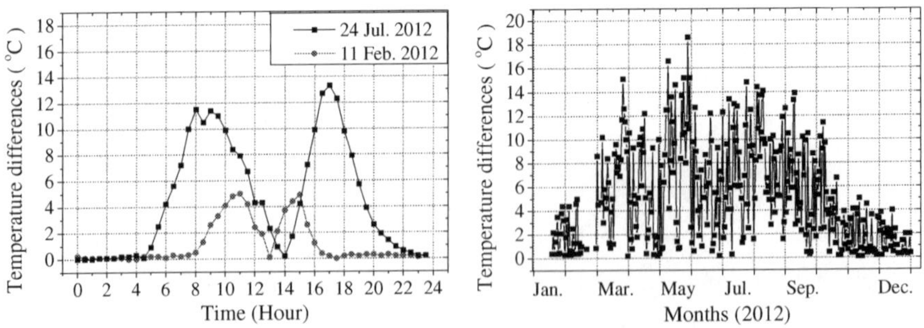

(a) Variation in one day (b) Daily maximum value

FIGURE 3.10 Measured absolute TTD of the box girder: (a) variation in one day and (b) daily maximum value.

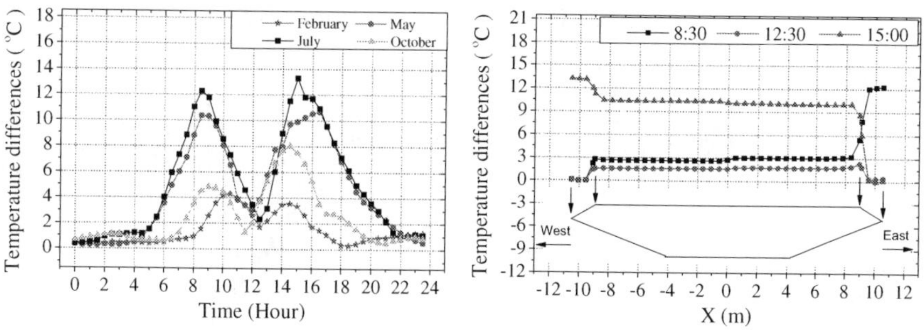

(a) Variation in one day

(b) Distribution on 24 July 2012

FIGURE 3.11 Simulated absolute TTD of the box girder: (a) variation in one day and (b) distribution on July 24, 2012.

Eurocode 1 (European Committee for Standardization CEN, 2003) recommends 5°C as the linear TTD between the outer edges of a bridge if no other information is available and no indications of higher values exist. However, the present results show that the TTD of the Humber Bridge reaches as high as 18°C, and that value over a longer period of monitoring is likely higher. As described previously, the maximal TTD observed at the Cleddau Bridge was 15°C (Kromanis et al., 2015). Cleddau is also a steel box girder bridge despite having a very different section from Humber. However, the need to consider TTD for such bridges seems clear.

The numerical results are then examined to investigate the TTD of the entire box girder. Figure 3.11a shows the maximal TTD during the four chosen days. The peaks occurred around 9:00 and 15:00. Similar to the measurement data, summer and spring have larger TTD than winter and autumn.

The distribution of the TTD along the deck on July 24, 2012 is illustrated in Figure 3.11b. The east upper web had a much higher temperature in the morning, the top deck had a slightly higher temperature than both upper webs at noon, and the west web took the maximum temperature in the afternoon. An abrupt temperature change occurred at the connection between the deck and webs. Therefore, the inclined upper webs play a critical role in the TTD.

3.3.6 Effect of Box Girder Profile on TTD

Three different box girder shapes are investigated to study the effect of inclined upper webs on TTD. Case 1 is the current cross section of the Humber Bridge; in Case 2, the footpath is at the mid-height of the section; and in Case 3, the upper webs are flush with the top deck. The TTD variations on July 24, 2012 for three cases are shown in Figure 3.12. The three cases have similar distributions of TTD. However, Case 2, with the deepest upper webs, shows the largest TTD. In Case 3, the upper webs are located at almost the same level as the upper deck but have higher temperatures than the latter. This observation is because the upper webs are made of steel and

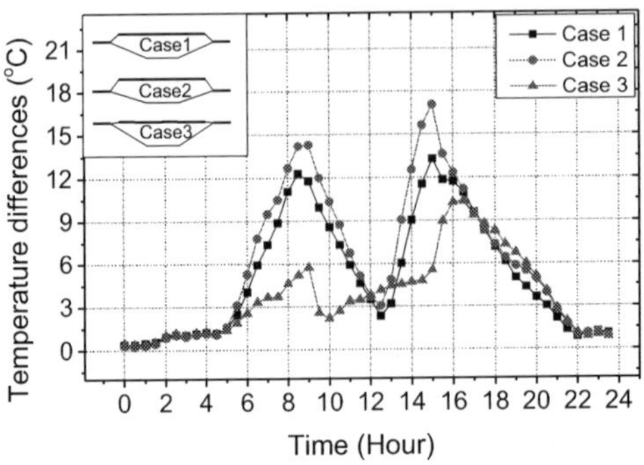

FIGURE 3.12 Simulated absolute TTD of the box girder with different shapes on July 24, 2012.

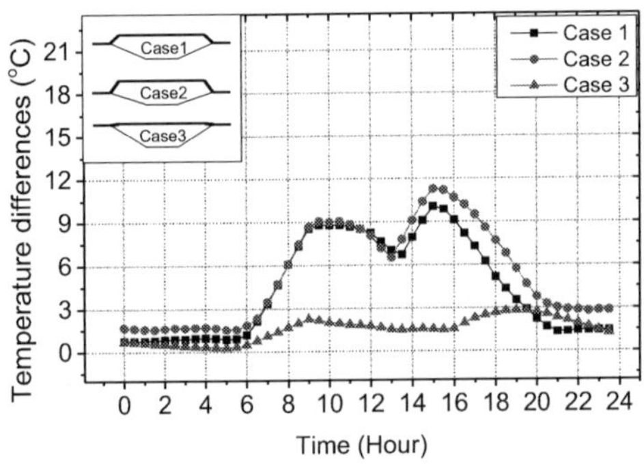

FIGURE 3.13 Simulated absolute TTD of the box girder with asphalt cover in upper webs.

are directly exposed to an open environment. In contrast, the upper deck is covered by a 40mm thick asphalt pavement with a low thermal conductivity. The effect of asphalt cover on TTD will be investigated next.

The upper webs are assumed to be covered with 40mm of asphalt, the same as the upper deck. The corresponding TTDs for the above three shapes are illustrated in Figure 3.13. The TTDs decrease significantly as compared with those without asphalt cover. In particular, the maximum TTD decreases from 17°C to 11°C in Case 2 and from 10°C to 3°C in Case 3.

The effects of bridge orientation with respect to the Sun (the bridge azimuth) on TTD are also investigated based on the current box girder section of the Humber Bridge. Four typical orientations, 0°, 45°, 90°, and −45° as defined in Figure 3.14 are considered. The actual Humber Bridge is almost in the south-north direction, corresponding to the 0° case. Here, the wind speed variation on different surfaces is not considered for simplicity. Thus, the convection coefficient is uniform for all surfaces. The TTD on the day is illustrated in Figure 3.14. Although the time variation of TTD

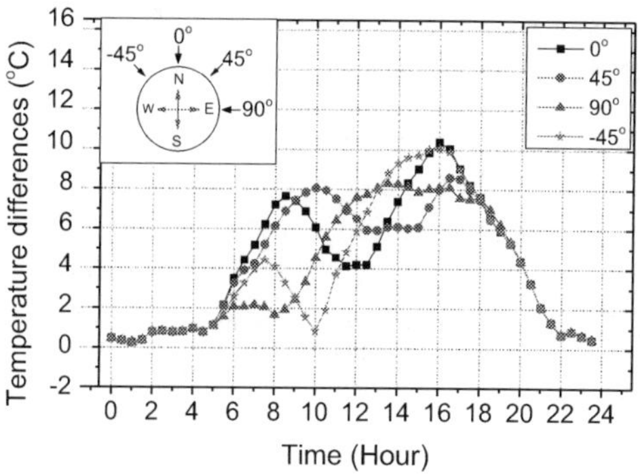

FIGURE 3.14 TTD of the box girder with different orientations on July 24, 2012.

is different for different orientations, the maximal TTDs are almost similar. Detailed investigations show that when one upper web receives direct solar radiation in the morning, the opposite upper web is blocked from solar radiation by the upper deck, regardless of the bridge orientation. Therefore, the bridge orientation slightly affects the maximum value of TTD in one day.

3.3.7 Summary

In this part, the temperature distribution pattern across the girder section in the Humber Bridge was investigated with an FE model consisting of 2D elements by ignoring the temperature variation in the longitudinal directions. The simulated results are in good agreement with the measurement counterparts, verifying the effectiveness and accuracy of simulation using 2D elements.

As an extension, it is further found that the streamlined steel box girder with inclined upper webs may exhibit significant vertical and transversal temperature differences because the upper webs on each side receive very different solar radiation in the daytime. In particular, the TTD may exceed the Eurocode specifications. Such a large TTD may cause in-plane bending of the box girder (or deck) and thus generate rotations at the bearings. On this basis, parametric studies with the established numerical model show that deep upper webs play a critical role in TTD, and asphalt cover can reduce TTD considerably. The bridge orientation has a limited effect on the maximum TTD.

In other studies, 3D models of various components are also established. For example, Xia et al. (2013) built FE models of the Tsing Ma Bridge's girder, tower, and cross-frame. Heat-transfer analyses were conducted to calculate their temperatures, which were close to the measurements. These temperature results were input into a 3D FE model of the entire bridge to analyze the temperature-induced responses. The displacement results agreed well with the measurement counterparts. The temperature-induced stresses were separated from the total strain and found to be generally small.

3.4 INTEGRATED 3D GLOBAL HEAT-TRANSFER AND STRUCTURAL ANALYSIS

Unlike the approaches using separated FE models for heat-transfer analysis and structural analyses introduced in the previous section, the integrated heat-transfer and structural analysis adopts a unified FE model. The heat-transfer analysis and structural analysis use the same FE model with different element types, which results in different degrees of freedom and output responses. The procedures are as follows and illustrated in Figure 3.15 (Shan et al., 2023).

1. Thermal elements are assigned to the unified global FE model first.
2. According to previously presented procedures, the thermal boundary conditions can be determined from the measured environmental conditions or the theoretical atmosphere model.
3. By applying the thermal boundary conditions, the temperature field in the model can be obtained from the heat-transfer analysis.
4. Then, the structural elements are assigned to the same FE model to replace the thermal elements. The thermal degrees of freedom are then changed to displacement degrees of freedom accordingly.
5. The calculated temperatures of all nodes are converted to thermal loads and applied to the model for the structural analysis.

This integrated numerical simulation with the unified FE model calculates the 3D global temperature distribution of the entire bridge, reducing the error in the varying cross sections along the longitudinal direction. Also, the calculated temperatures at different components are automatically transferred into the FE model for the structural analysis, avoiding manually inputting the temperature into the global

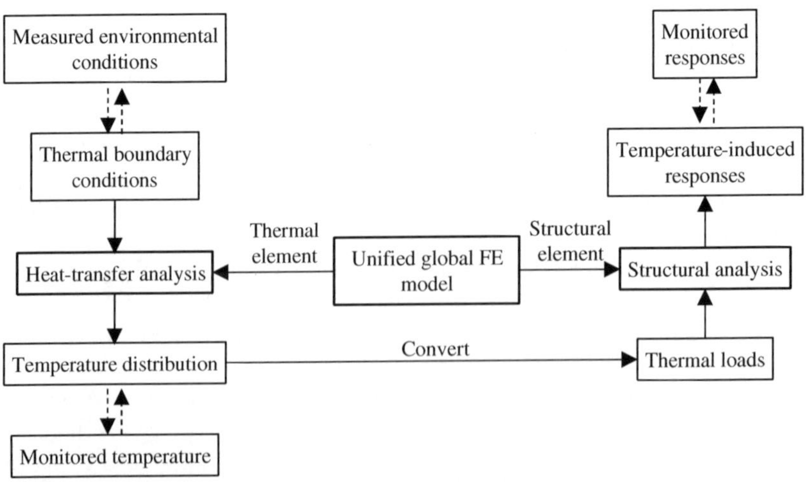

FIGURE 3.15 Framework of the integrated analysis.

FE model. Therefore, the integrated framework enables an efficient and accurate analysis of bridges' global 3D thermal behavior, as compared with the conventional divide-and-conquer approach.

Examples of the integrated framework will be demonstrated in later chapters, such as Sections 5.3 and 6.3.

3.5 CHAPTER SUMMARY

In this chapter, the basis of the heat-transfer analysis of bridges is given, particularly the method of determining the proper thermal boundary and initial temperature conditions. Then, the separated and integrated heat-transfer and structural analysis frameworks are presented. In the following chapters, the thermal behaviors of different types of bridges will be presented from the perspectives of field monitoring, numerical investigation, and theoretical analysis.

REFERENCES

ANSYS Inc., 2009, *ANSYS 12.0*. Canonsburg, PA, ANSYS Inc.

Berlage, H., 1928, Zur Theorie der Beleuchtung einer horizontalen Flache durch Tageslicht. *Meteorologische Zeitshrift* **45**: 174.

Branco, F., 1986, Thermal effects on composite box girder bridges during construction. In *Proceedings of 2nd International Conference on Short and Medium Span Bridges*. Canadian Society for Civil Engineering, Ottawa, Ontario.

Branco, F. A., Mendes, P. and Mirambell, E., 1992, Heat of hydration effects in concrete structures. *Materials Journal* **89(2)**: 139–145.

Branco, F. A. and Mendes, P. A., 1993, Thermal actions for concrete bridge design. *Journal of Structural Engineering* **119(8)**: 2313–2331.

Brownjohn, J., Dumanoglu, A., Severn, R. and Taylor, C., 1987, Ambient vibration measurements of the Humber suspension bridge and comparison with calculated characteristics. Proceedings of the Institution of Civil Engineers **83**: 561–600.

Elbadry, M. M. and Ghali, A., 1983, Temperature variations in concrete bridges. *Journal of Structural Engineering* **109(10)**: 2355–2374.

European Committee for Standardization CEN, 2003, *Eurocode 1: Actions on Structures, Part 1-5: General Actions — Thermal Actions*. EN1991-1-5: 2003E. Brussels, European Committee for Standardization CEN.

Froli, M., Hariga, N., Nati, G. and Orlandini, M., 1996, Longitudinal thermal behaviour of a concrete box girder bridge. *Structural Engineering International* **6(4)**: 237–242.

Fu, H. C., Ng, S. F. and Cheung, M. S., 1990, Thermal behavior of composite bridges. *Journal of Structural Engineering* **116(12)**: 3302–3323.

Kromanis, R., Kripakaran, P. and Harvey, B., 2015, Long-term structural health monitoring of the Cleddau bridge: evaluation of quasi-static temperature effects on bearing movements. *Structure and Infrastructure Engineering* **12(10)**: 1342–1355.

Li, L., Chen, B., Zhou, L., Xia, Q., Zhou, Y., Zhou, X. and Xia, Y., 2023, Thermal behaviors of bridges: A literature review. *Advances in Structural Engineering* **26(6)**: 985–1010.

Lienhard IV, J. H. and Lienhard V, J. H., 2020, *A Heat Transfer Textbook*, Fifth Edition. Cambridge, MA, Phlogiston Press.

Riding, K. A., Poole, J. L., Schindler, A. K., Juenger, M. C. G. and Folliard, K. J., 2007, Temperature boundary condition models for concrete bridge members. *ACI Materials Journal* **104(4)**: 379–387.

Rohsenow, W. M., Hartnett, J. P. and Ganic, E. N., 1985, *Handbook of Heat Transfer Applications*. New York, McGraw-Hill.

Shan, Y., Jing, Q., Li, L., Gao, W., Xia, Z. and Xia, Y., 2023, Temperature behavior of cable-stayed bridges. Part II— temperature actions by using unified analysis. *Advances in Structural Engineering* **26(9)**: 1600–1620.

Wang, C., 1994, *Temperature Effects in Hardening Concrete*, Calgary, University of Calgary.

Whitaker, S., 2013, *Fundamental Principles of Heat Transfer*, Amsterdam, Elsevier.

Xia, Y., Chen, B., Zhou, X. Q. and Xu, Y. L., 2013, Field monitoring and numerical analysis of Tsing Ma Suspension Bridge temperature behavior. *Structural Control and Health Monitoring* **20(4)**: 560–575.

Zhou, L., Xia, Y., Brownjohn, J. and Koo, K. Y., 2015, Temperature analysis of a long-span suspension bridge based on field monitoring and numerical simulation. *Journal of Bridge Engineering* **21**: 04015027.

4 Temperature Behaviors of Suspension Bridges

NOTATIONS

A_i	Cross-sectional area of the *i*th span cable
D_i	Mid-span elevation of the main-span cable
E	Young's modulus
E_C	Young's modulus of main cables
F, **H**, **L**, **Q**	Vectors of variables f_i, h_i, l_i and q_i
G	Coefficient matrix of the system of linear equations used for solving **δF**, **δL**, and **δH** (see Equation 4.59)
H_i	Horizontal component of the main cable's tension force
H_0	Initial horizontal component of cable tension force in each span
L	Total span of the suspension bridge (between two end anchorages, see Figure 4.28)
L_E	Temperature equivalent length
L_G	Total girder length of self-anchored suspension bridge (see Figure 4.35)
M_i, Q_i	Intermediate variables defined in Equations (4.12) and (4.13)
S_i	Cable length of *i*th span
δS_i^{Free}	Stress-free thermal deformation of the *i*th span cable
δS_i^{Elas}	Elastic deformation of the *i*th span cable
T_C, T_P	Temperatures of cables and towers
W_i	Total vertical load
c	Specific heat capacity
c_{ni}, c_{li}, $c_{\alpha i}$	Coefficients related to differentiations of S_i with respect to n_i, l_i, and α_i used in Equations (4.6)–(4.8)
f_i	Cable sag of the *i*th span
h_{Pi}	Tower height
h_c	Heat-transfer coefficient
h_i	Height differences between two ends of the main cable
k	Thermal conductivity
k_{Pi}	Flexural stiffness of tower *i*
l_i	Girder length of the *i*th span
l_{mx}	Longest span length of a multi-span suspension bridge
n_i	Sag-to-span ratio
q_i	Uniformly distributed vertical load
u	Number of spans of suspension bridges
z_i	Parameter defined by Equation (4.18)
α	Absorptivity coefficient
α_T	Thermal expansion coefficient

DOI: 10.1201/9781003480136-4

α_i	Cable-chord inclination
δX	Change of variable X
ε	Emissivity coefficient
ε_i	Intermediate variable used in Equation (4.78)
κ_i	Parameter related to the ith tower's flexural stiffness and the initial horizontal tension of the main cables, $\kappa_i = k_{Pi}/H_0$
ρ	Density
ζ_i	Span ratio between side span to mid-span, $i = 1, 3$
θ_C, θ_T	Thermal expansion coefficients of cables and towers

4.1 PREFACE

Suspension bridges are an efficient bridge type. A suspension bridge consists of main cables, towers, stiffening girders/trusses, hangers (or suspenders), and anchorages. The main cables carry most loads of the bridge deck through the hangers/suspenders and transfer their tensions to the anchorages at both ends. Suspension bridges can be classified into earth-anchored and self-anchored types according to the anchorage conditions of the main cables. They are the most cost-effective bridge type as the main span exceeds 1000 m. Most suspension bridges are arranged in the twin-tower three-span format. Still, more spans can also be arranged, such as the Maanshan Yangtze River Bridge, and Yingwuzhou Yangtze River Bridge, which are triple-tower four-span suspension bridges. The 1915 Çanakkale Bridge, which opened on March 18, 2022 with a main span of 2023 m, is the longest of all types at present.

The temperature behaviors of cable-suspension bridges are complicated. The main cables play an essential role in the structural system. The main towers have different temperature distributions and actions from those of the main cables and bridge deck. In practice, massive thermometers are usually incorporated into the SHM system to monitor both the structural temperature distribution and the ambient air temperature. For example, 115 thermometers were installed in the SHM system of the Tsing Ma Bridge (Xu and Xia, 2011), including 6 for ambient temperature, 86 for deck temperature, and 23 for cable temperatures. However, capturing the full temperature distribution of the suspension bridge is difficult as the thermometer number is much lesser than the structural components. In this regard, numerical investigations are usually conducted to reconstruct the temperature field in suspension bridges, especially at those representative and important components. Very few theoretical analyses (Zhou et al., 2020, 2021, 2022) have also been conducted to reveal the underlying mechanisms of suspension bridge deformations due to temperature change, including both three- and multi-span suspension bridges.

This chapter studies the temperature behaviors of suspension bridges from field monitoring, numerical simulations, and theoretical analysis aspects. The Tsing Ma Bridge will be used as an example for field monitoring and numerical investigations. The Akashi Kaikyo Bridge and the Maanshan Yangtze River Bridge will be studied to verify the analytical formulas of the temperature-induced structural deformations for three- and multi-span suspension bridges, respectively.

4.2 FIELD MONITORING STUDIES OF THE TSING MA BRIDGE

Located at latitude N22.2° and longitude E114.1°, the Tsing Ma Suspension Bridge has a total span of 2132 m. It carries a dual three-lane highway on the upper deck and two railway tracks and two protected carriageways on the lower deck. The main span, across the Tsing Yi Island and Ma Wan Island, is 1377 m long, as shown in Figure 4.1. The angle between the bridge's longitudinal axis and the south is 73°.

The bridge deck is a hybrid steel structure consisting of Vierendeel cross frames supported on two longitudinal trusses acting compositely with stiffened steel plates. The bridge deck at the main span and the Ma Wan side span is a suspended deck, and the structural configuration is typical for every 18-m segment. Two outer longitudinal trusses, spacing at 26 m, link up the cross frames along the bridge's longitudinal axis and act as the main girders of the bridge. The structural translational movements at the Ma Wan abutment are restrained in three translational directions. At the Tsing Yi abutment, the vertical (z-axis) and lateral (y-axis) movements of the bridge deck are restrained, while the deck can move freely along the longitudinal direction (x-axis).

The two reinforced concrete towers are almost identical, made of two rectangular hollow concrete tower legs and four hollow portal beams, except that the Ma Wan tower (west tower) is 150 mm higher than the Tsing Yi tower (east tower). Both towers are 206.4 m high with respect to the base level. The two side spans on the Ma Wan and Tsing Yi side are supported by two and three piers, respectively, as shown in Figure 4.1. All supporting piers in the side spans are reinforced concrete structures. Piers M2 and T1 provide lateral restraint to the bridge deck.

The cable system of the Tsing Ma Bridge consists of two main cables, 95 pairs of suspender units, and 95 pairs of cable bands. The two main cables are accommodated by the four saddles located at the top of the tower legs. The two parallel main cables are fixed at the tower saddles, move together with the towers, and transfer the loadings into the anchorages. The Tsing Yi side span cables are seen as inclined sagged cables, with the top ends fixed at the tower saddles and the lower ends fixed at the main anchorage. The Ma Wan side span cables are also inclined sagged cables with the top ends fixed at the tower saddles, with the lower ends fixed at the main anchorage through the saddles on Pier M2. On each main cable, there are 19 suspender

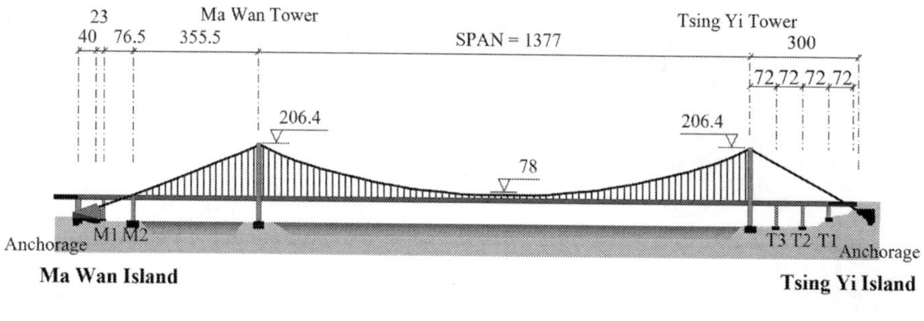

FIGURE 4.1 Configuration of the Tsing Ma Bridge (unit: m).

units within the Ma Wan approach span and 76 suspender units in the main span. In the Tsing Yi approach span, there are no suspender units. The distance between two suspender units is 18 m along the longitudinal axis of the bridge deck. The two anchorages are gravity structures, which are integral to the deck abutments.

4.2.1 SHM System on the Bridge

The SHM system for the Tsing Ma Bridge is shown in Figure 4.2. The system is composed of six integrated modules, totally 276 sensors in 8 types, including 6 anemometers, 115 thermometers, 7 weigh-in-motion stations, 2 displacement transduces, 14 GPS stations, 9 level sensing stations, 13 accelerometers, and 110 strain gauges.

4.2.1.1 Temperature Monitoring

The locations of the temperature sensors are shown in Figure 4.3. In the figure, the first two letters in the tag numbers, TC and P1–P6, denote the sensor type of thermometers; the third letter "T" stands for Tsing Ma Bridge; the fourth letter indicates the deck section location (from Sections A–R); the last letter indicates the sensor locations within its girder section. Specifically, as the last letter, "A" stands for abutment, "C" for the main cable, "N" for the north side of the deck, "S" for the south side of the deck, and "T" for the tower. The numbers in the brackets indicate the quantities of the sensors. The collected temperature data from the SHM system can be grouped into three categories: (1) ambient temperature (P3 type, 6 in quantity);

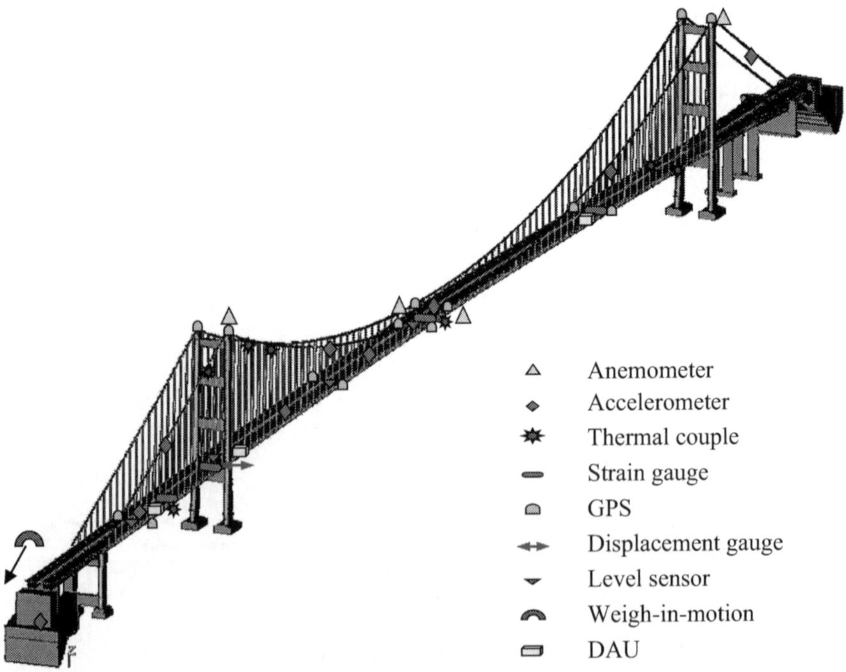

FIGURE 4.2 Layout of sensors on the Tsing Ma Bridge.

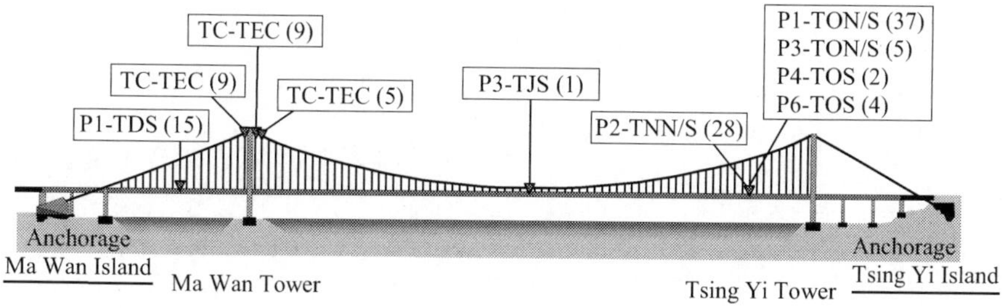

FIGURE 4.3 Distribution of temperature sensors in the Tsing Ma Bridge (Xu and Xia, 2011).

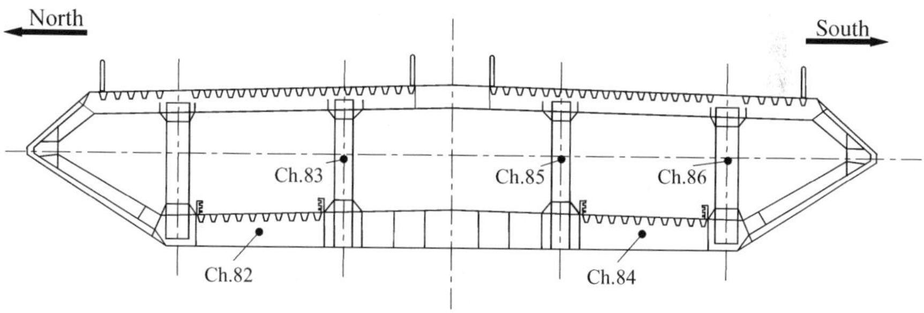

FIGURE 4.4 Ambient temperature sensors at Section O (Xu and Xia, 2011).

(2) section temperature (P1, P2, P4, and P6 types, 86 in quantity); and (3) cable temperature (TC type, 23 in quantity). The TC type is the thermocouple sensor, and P1–P6 are PT100 Platinum resistance temperature sensors. For example, as shown in Figure 4.3, there are five thermometers at section P3-TON/S, tagged as Ch. 82 to Ch. 86, and their respective locations are shown in Figure 4.4. One air temperature sensor (Ch. 81, P3-TJS) is approximately at the middle section of the main span and attached to a sign gantry. In particular, 23 thermocouples (TC-TEC) were embedded inside the main cables at three different locations to measure the cable temperature and those of one section are shown in Figure 4.5. As the cross-section of the main cable is relatively small, the densely deployed sensors can capture the temperature of the cable sufficiently.

The deck section in the Ma Wan side span is equipped with 15 sensors (P1-TDS), and another deck section close to the Tsing Yi tower is equipped with 71 sensors. The latter can be found in Figures 4.6 and 4.7. In particular, Figure 4.6 shows the position of 37 sensors installed on the cross frame (P1-TON/S), four of which are located on the vertical members (Nos. 10, 17, 20, and 27) and others on the chord members. Figure 4.7 illustrates seven groups of temperature sensors (details 10–16, P2-TNN/S, 28 sensors in total) mounted on the orthotropic deck plates 2.25 m away from the section. At each detail, one sensor measures the temperature of the steel plate, and the other three measure the temperature distribution over the deck trough. In addition,

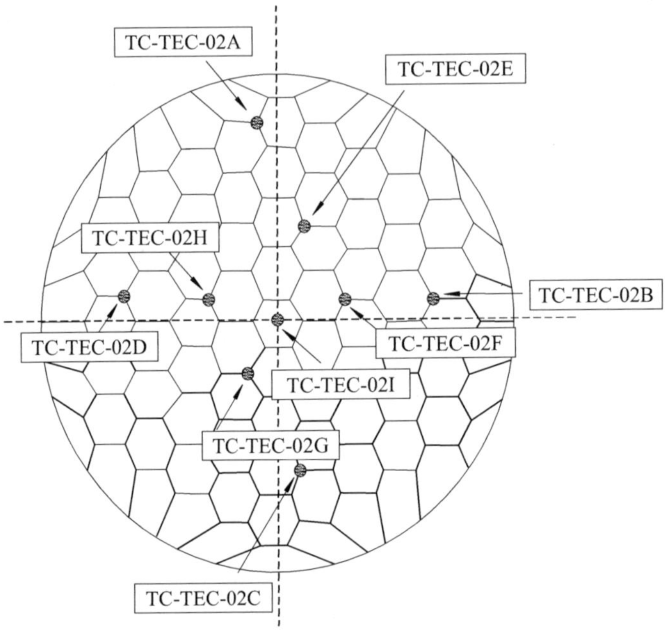

FIGURE 4.5 Temperature sensors at the main cable (TC-TEC).

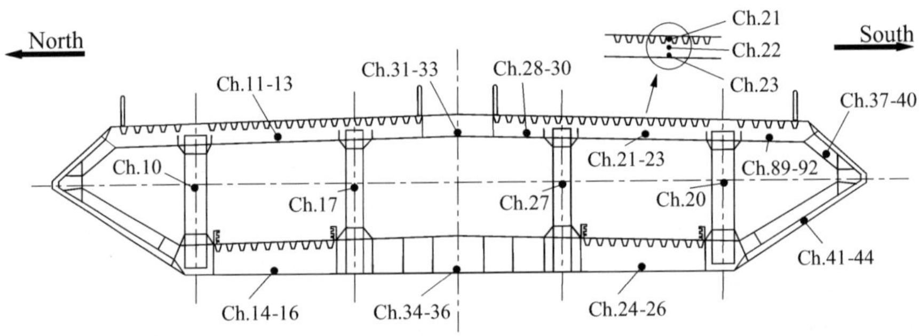

FIGURE 4.6 Temperature sensors on the cross frame (P1-TON/S) (Xu and Xia, 2011).

four temperature sensors (P6-TOS) were embedded in the south carriageway, and two temperature sensors (P4-TOS) were installed on the corrugated sheets on the south side of the section, which are not shown here for brevity. The sampling frequency of all the temperature sensors is 0.07 Hz.

The bridge also has six anemometers: two at the middle of the main span, two at the middle of the Ma Wan side span, and one each at the Tsing Yi tower and the Ma Wan tower.

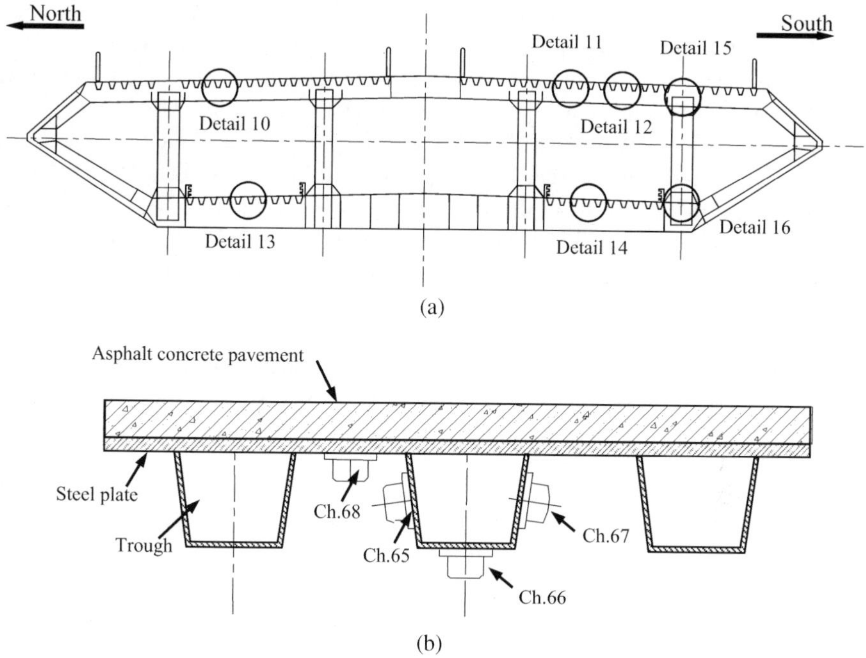

FIGURE 4.7 Temperature sensors on the deck plate (P2-TNN/S).

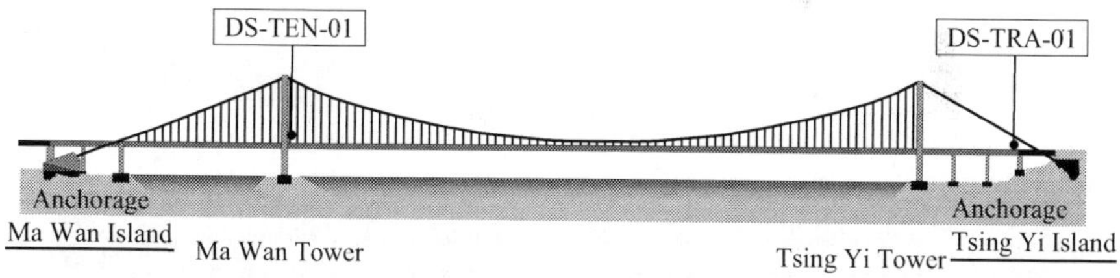

FIGURE 4.8 Distribution of displacement transducers in the Tsing Ma Bridge.

4.2.1.2 Structural Response Monitoring

Multiple types of sensors are equipped in the SHM system to monitor the bridge responses thoroughly. The displacement of the bridge in three orthogonal directions—longitudinal (x-direction), (2) lateral (y-direction), and (3) vertical (z-direction)—is measured by displacement transducers, level sensing stations, and GPS stations. The lateral and longitudinal movements of the bridge deck are measured by two displacement transducers at two locations, as shown in Figure 4.8. "DS" in the tags stands for displacement transducers, "T" stands for the Tsing Ma Bridge, the fourth letter "E" (or "R") indicates the section, and the fifth letter "N" stands for the north side of the deck, while "A" indicates the location of the abutment. For example, transducer DS-TEN-01 measures the lateral motion, which is positive if the deck sways to the north side. The transducer is connected between the bearing frame and the tower leg of the Ma Wan tower, which measures the relative lateral movement of the deck to the Ma Wan tower. The longitudinal movement of the bridge deck at the Tsing Yi abutment is recorded by another displacement transducer DS-TRA-01.

The vertical motion of the bridge deck is monitored by the level sensing stations, which give positive values when the deck moves downward. The level sensing stations are mounted at six locations along the bridge deck, as displayed in Figure 4.9. Three level sensing stations were installed on one side of the bridge deck at the location of the Ma Wan abutment (LV-TAA-01), Ma Wan tower (LV-TEN-01), and Tsing Yi tower (LV-TPS-01). The vertical displacement at the Ma Wan approach span is monitored by one pair of level sensing stations (LV-TBN/S-01), and that at the main span is monitored by two pairs (LV-THN/S-01 and LV-TJN/S-01). The torsion of the deck sections is calculated by the difference of vertical displacements between the two level sensing stations divided by their distance. The level sensing stations measure the vertical displacements at a sampling rate of 2.56 Hz with a measurement accuracy of approximately 2 mm.

GPS receivers are placed on bridge towers, main cables, and bridge deck, as shown in Figure 4.10. The Ma Wan tower and the Tsing Yi tower are respectively equipped with a pair of GPS receivers (TM-MWN/S and TM-TYN/S), and they are mounted at the top of saddles on each of the tower legs. The displacement of the main cables is monitored through a pair of GPS receivers at mid-span (TM-02N/S). The mid-span of the Ma Wan side span is equipped with a pair of receivers (TM-05N/S). Three

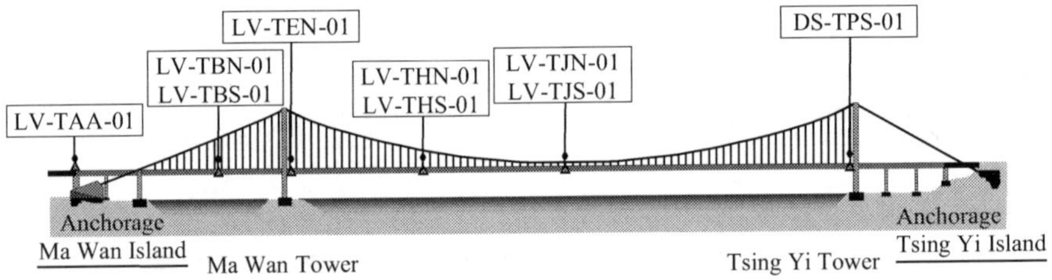

FIGURE 4.9 Distribution of level sensing stations in the Tsing Ma Bridge.

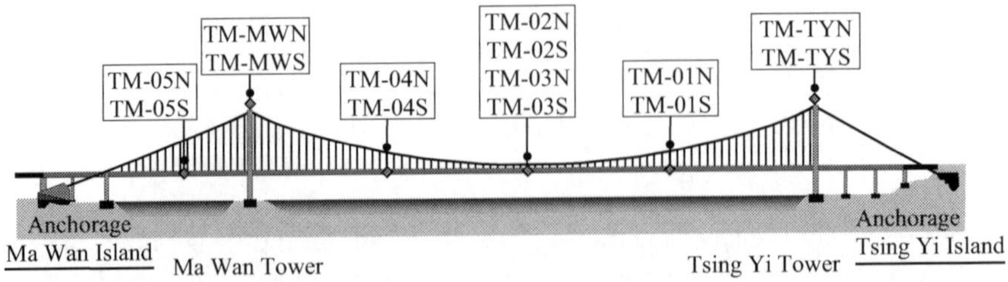

FIGURE 4.10 Distribution of GPS receivers in the Tsing Ma Bridge.

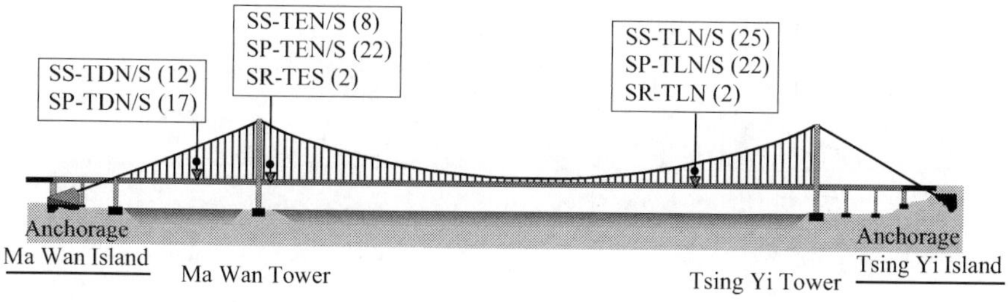

FIGURE 4.11 Distribution of strain gauges in the Tsing Ma Bridge: (a) cross frame at Section E and (b) view A.

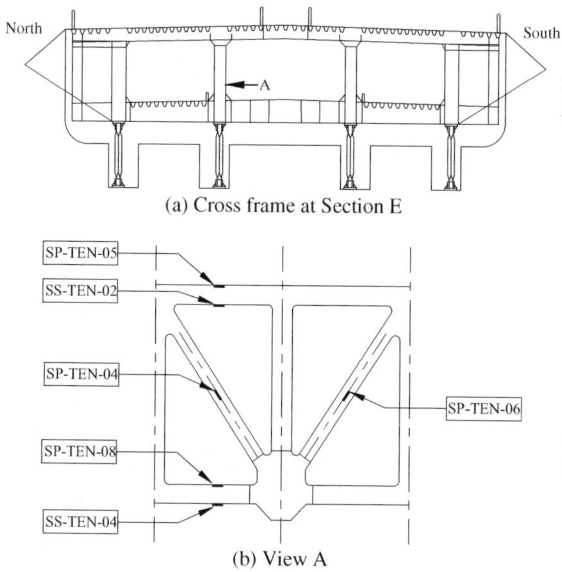

FIGURE 4.12 Distribution of strain gauges at Section E.

pairs of GPS receivers are located at one-quarter (TM-01N/S), one-half (TM-03N/S), and three-quarters (TM-04N/S) of the main span of the bridge deck. The displacements in three directions are positive if the bridge component moves toward the Tsing Yi Island, sways to the north side, and goes upward, respectively.

110 strain gauges are distributed at three sections of the Tsing Ma Bridge, as shown in Figure 4.11. The strain gauges have three configurations: 44 single gauges, 62 pairs, and four rosettes. The first two letters in the sensor tags, "SS/P/R" indicate the strain gauge types, the third letter "T" stands for the Tsing Ma Bridge, the fourth letter indicates the section location, the fifth letter indicates the detailed locations of the sensors within their respective sections, and the number in the parentheses indicates the quantities of strain gauges. For example, 32 strain gauges were installed at the cross frame of Section E, near the Ma Wan tower. The configuration of the section is shown in Figure 4.12a. The north outer truss, north inner truss, south outer

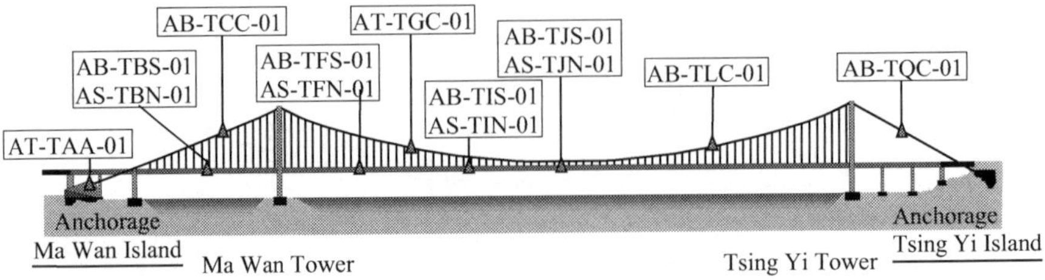

FIGURE 4.13 Distribution of accelerometers in the Tsing Ma Bridge.

truss, and south inner truss were equipped with four pairs of gauges (SP type) and two single gauges (SS type). Figure 4.12b illustrates the position of six strain gauges (four SP type and two SS type) at the north inner truss.

In addition, the SHM system of the bridge includes 13 accelerometers: four uni-axial and four bi-axial accelerometers at four sections of the bridge deck, three bi-axial and one tri-axial on the main cables, and one tri-axial at the Ma Wan abutment, as shown in Figure 4.13. The first two letters in the sensor tags "AS/B/T" respectively indicate the uni-, bi-, and tri-axial accelerometers, the third letter "T" stands for the Tsing Ma Bridge, the fourth letter indicates the section, and the fifth letter indicates the locations within the section. For example, in each of the four deck sections, one bi-axial accelerometer was installed on the south side to measure the accelerations in the vertical and lateral directions, while one uni-axial accelerometer was installed on the north side to measure the vertical acceleration. Three bi-axial accelerometers were installed on the main cables to measure the accelerations in the vertical and horizontal directions. Tri-axial accelerometers AT-TGC-01 and AT-TAA-01 measure the acceleration of the main cable and Ma Wan abutment, respectively, in the longitudinal, vertical, and lateral directions.

The various monitoring data facilitate the comprehensive evaluation of the bridge condition in practice. The measured temperature distribution and temperature-induced structural responses will be presented in the following sections.

4.2.2 Measured Temperature Distribution

The monitoring results of four representative days, each in one season, namely January 17, April 18, July 17, and October 27, 2005, are presented in this section. The summer day, July 17, will be presented in detail.

4.2.2.1 Temperature of Deck Plate

The measured temperature of the deck plate at Detail 12 (Ch. 66 to Ch. 68 in Figure 4.7) on July 17, 2005 is shown in Figure 4.14. The three temperatures were similar from midnight till 8:00 in the morning. Then the temperatures started to rise at different rates, resulting in the highest temperatures of the three channels in the afternoon having the relation that Ch. 68 (≈50°C at 16:00) > Ch.67 (≈46°C at 16:00) > Ch. 66 (≈42°C at 17:00). The three components started to cool down and the temperatures gradually became similar at 20:00. Therefore, the temperatures of

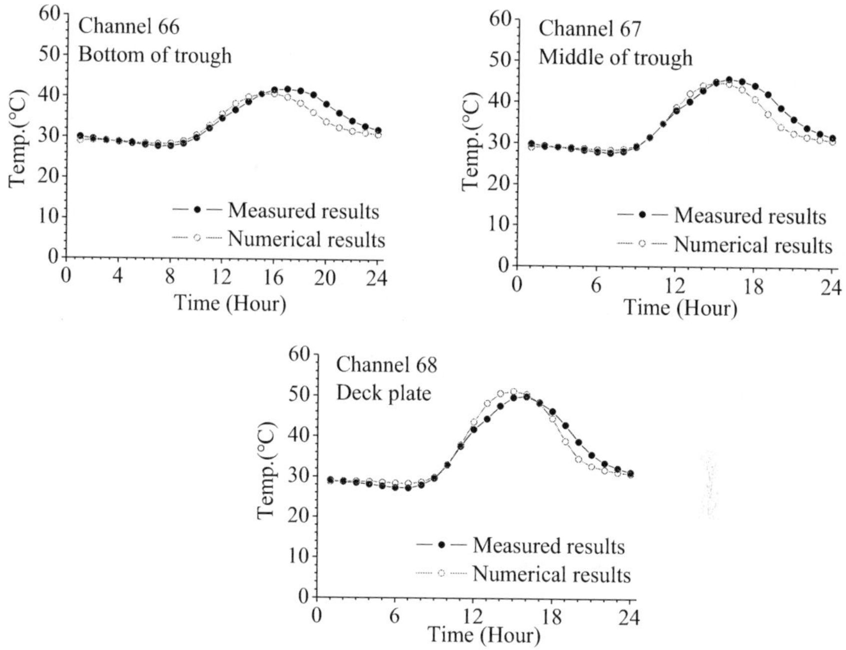

FIGURE 4.14 Measured and simulated temperature at Detail 12, on July 17, 2005.

the three locations were similar during the night, and vertical temperature gradients between Ch. 68 and Ch. 66 could reach 8°C in the daytime.

The temperatures of Ch. 68 in the other three days are shown in Figure 4.15. In spring and autumn, the highest temperatures were similar and around 44°C–46°C, while in the winter, the highest temperature was around 28°C only.

4.2.2.2 Temperature of Cross Frame

The measured temperatures of different components in the cross frame (Figure 4.6) on July 17, 2005 are shown in Figure 4.16. The daily variations of the upper chord temperatures were more significant than those of the lower chords and vertical trusses. Similar to the deck plate cases, the temperatures of the three components from midnight to 8:00 and from 20:00 to midnight were similar, while the daytime temperatures followed the pattern: Ch. 12/22 > Ch. 17/20 > Ch. 15/25. This indicates that the structural temperature decreased from the top to the bottom. As for the transverse direction, only minimal temperature differences among each pair can be observed, with the left part having a slightly higher temperature than the right counterpart.

As no temperature sensor has been installed on the tower, no measurement results are presented in this section. The measured temperatures at different components on July 17, 2005 are listed in Table 4.1. The minimum temperatures of different components of the entire bridge were similar (less than a 3°C difference), while the maximum temperatures varied significantly (more than 10°C). The minimum

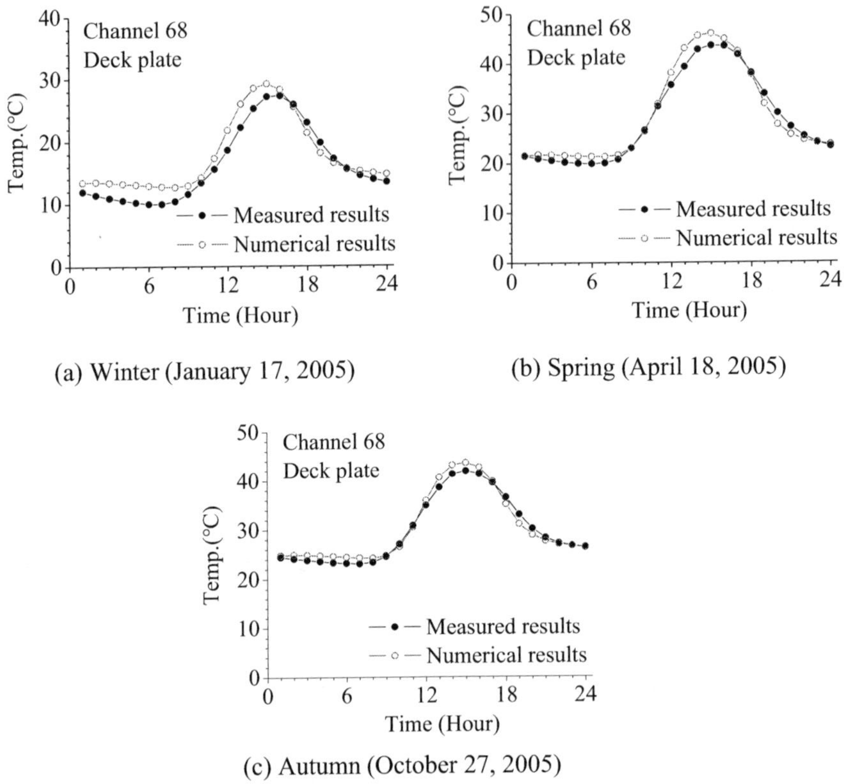

FIGURE 4.15 Comparison of the measured and simulated temperature at Detail 12 in different seasons: (a) winter (January 17, 2005), (b) spring (April 18, 2005), and (c) autumn (October 27, 2005).

TABLE 4.1
Peak Temperature at Different Components on July 17, 2005 (Unit: °C)

	Minimum		Maximum	
Component	**Measurement**	**Simulation**	**Measurement**	**Simulation**
Deck plate (Ch. 68)	27.12	28.25	49.95	51.28
Cross frame (Ch. 12)	27.98	28.04	38.84	39.05
Tower section (NE)	N.A.	28.37	N.A.	43.86
Cable (Ch. 115)	29.36	N.A.	36.75	N.A.

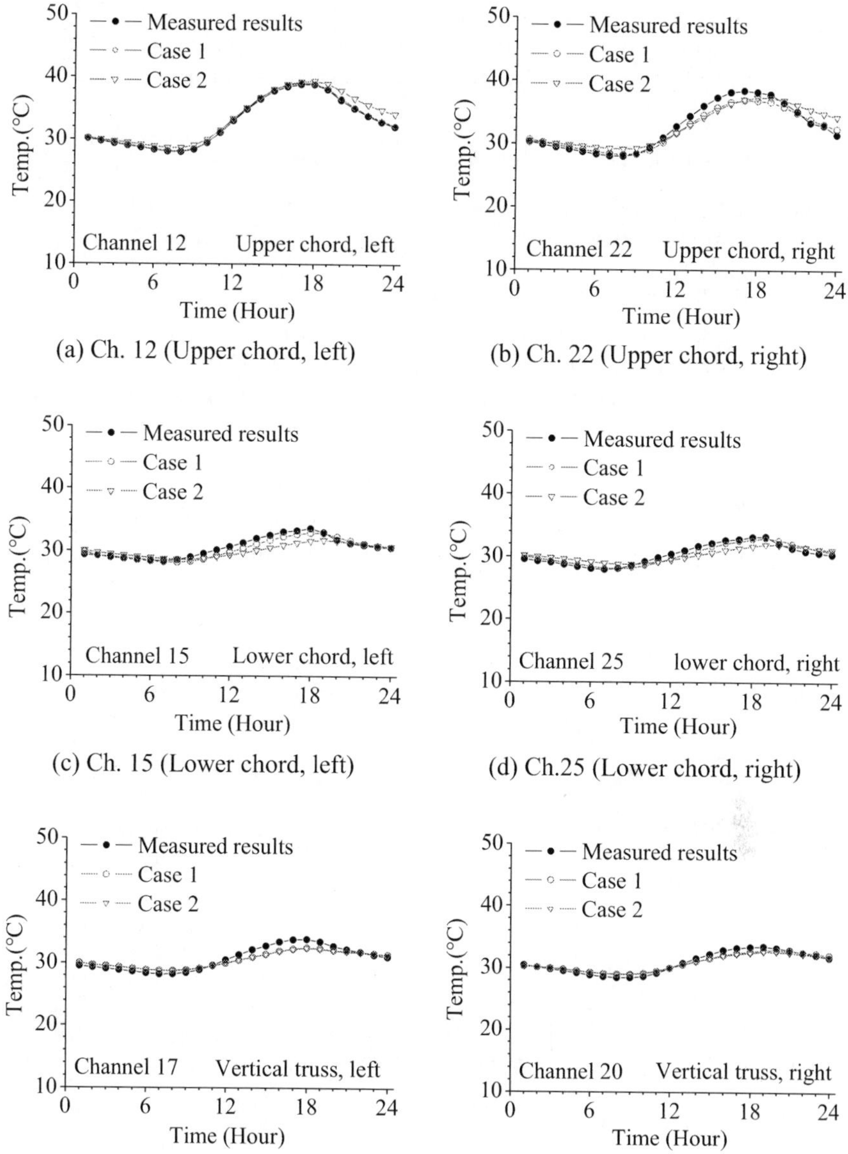

(a) Ch. 12 (Upper chord, left)
(b) Ch. 22 (Upper chord, right)
(c) Ch. 15 (Lower chord, left)
(d) Ch.25 (Lower chord, right)
(e) Ch. 17 (Vertical truss, left)
(f) Ch.20 (Vertical truss, right)

FIGURE 4.16 Measured and simulated temperatures at different parts of the cross frame (July 17, 2005): (a) Ch. 12 (upper chord, left), (b) Ch. 22 (upper chord, right), (c) Ch. 15 (lower chord, left), (d) Ch. 25 (lower chord, right), (e) Ch. 17 (vertical truss, left), and (f) Ch. 20 (vertical truss, right).

temperatures typically were measured in the early morning before sunrise, while the maximum temperatures were measured in mid-afternoon, indicating that the temperature differences were typically induced by the solar radiation differences.

4.2.3 Temperature-Induced Structural Responses

The measured structural responses, including the displacements of different components and stresses of the deck components, are investigated in this section.

4.2.3.1 Displacement of Bridge Deck

Figure 4.17 presents the GPS measured displacement of the bridge deck in the middle of the main span on July 17, 2005. The displacements at 0:00 are set as baselines, and the displacements shown here are the relative ones. The displacements indicate positive movements in the corresponding three directions if the bridge component moves toward the Tsing Yi Island, sways to the north side, and goes upward. As the longitudinal movement of the deck is constrained at the Ma Wan abutment, the positive displacement in the x-direction indicates the expansion of the deck. As shown in the figure, the deck moved slightly toward the Ma Wan Island (negative values) in

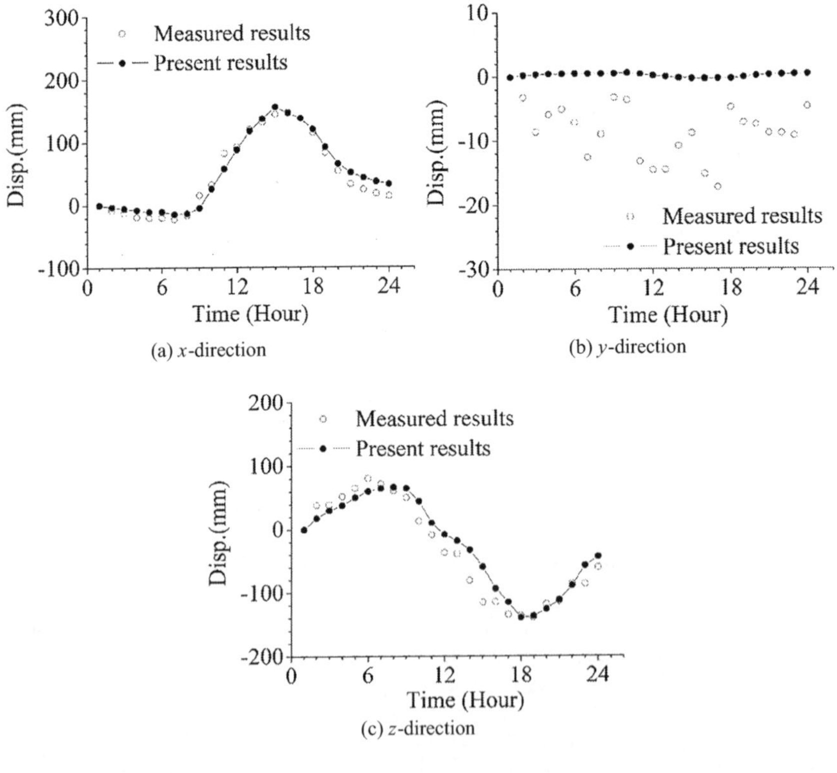

FIGURE 4.17 Displacement of the bridge deck in the middle of the main span on July 17, 2005: (a) x-direction, (b) y-direction, and (c) z-direction.

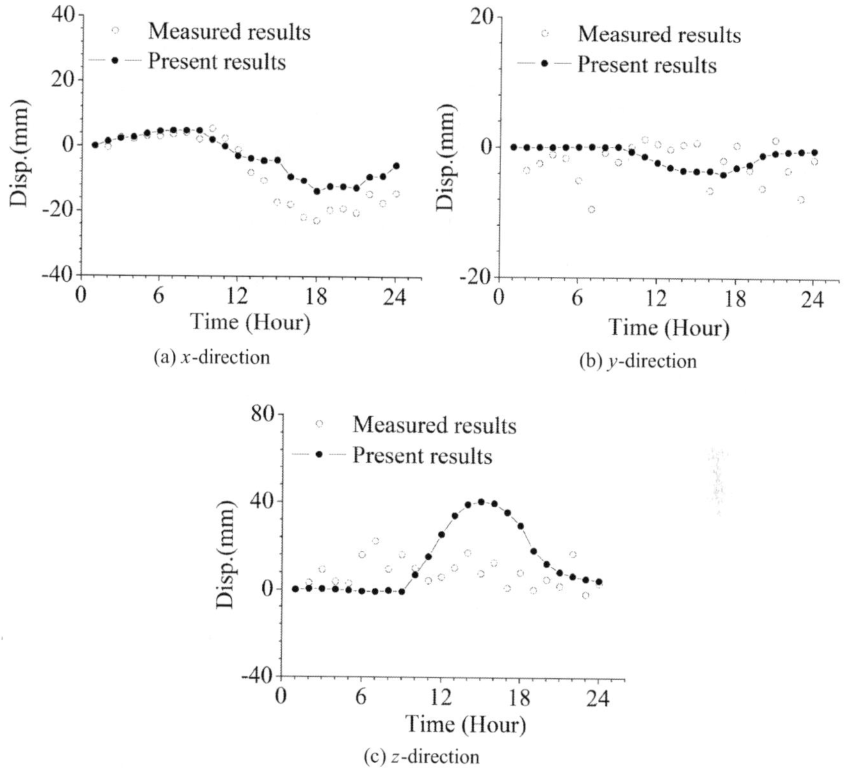

FIGURE 4.18 Displacement of the Ma Wan Tower on July 17, 2005: (a) x-direction, (b) y-direction, and (c) z-direction.

the early morning because the temperature might decrease slightly from midnight to early morning. It then moved toward the Tsing Yi Island (positive values) when the temperature increased after sunrise. In the vertical direction, the deck first moved upwards from 0:00 to 7:00 and then downwards as the temperature increased. In the evening to midnight, the deck went up again as the temperature decreased. At the same time, the deck generally swayed slightly to the south side in a day, but it does not show any pattern with the temperature change.

4.2.3.2 Displacement of Bridge Towers

The displacement at the top of the Ma Wan tower on July 17, 2005 is shown in Figure 4.18. The Ma Wan tower swayed toward the Ma Wan side with the temperature increase. The measured displacements in the lateral and vertical directions are relatively scattered and irregular compared to the longitudinal ones.

4.2.3.3 Displacement at Girder End

The longitudinal expansion displacement of the bridge girder at the Tsing Yi abutment on July 17, 2005 is drawn in Figure 4.19. The daily temperature-induced

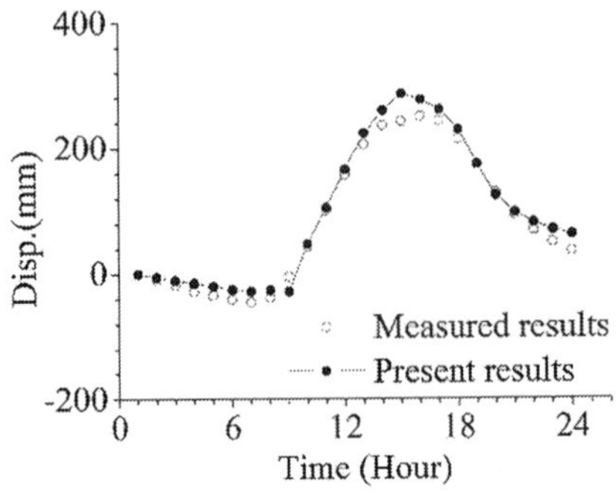

FIGURE 4.19 Longitudinal displacement of the girder at the Tsing Yi abutment on July 17, 2005.

variation in the longitudinal displacement at the girder end was about 300 mm on a summer day.

4.2.3.4 Displacement of Main Cables

The displacement of the main cables is monitored by a pair of GPS stations mounted in the middle of the main cables (see Figure 4.10). The hourly mean displacement of the main cables in three directions is shown in Figure 4.20. The main cables

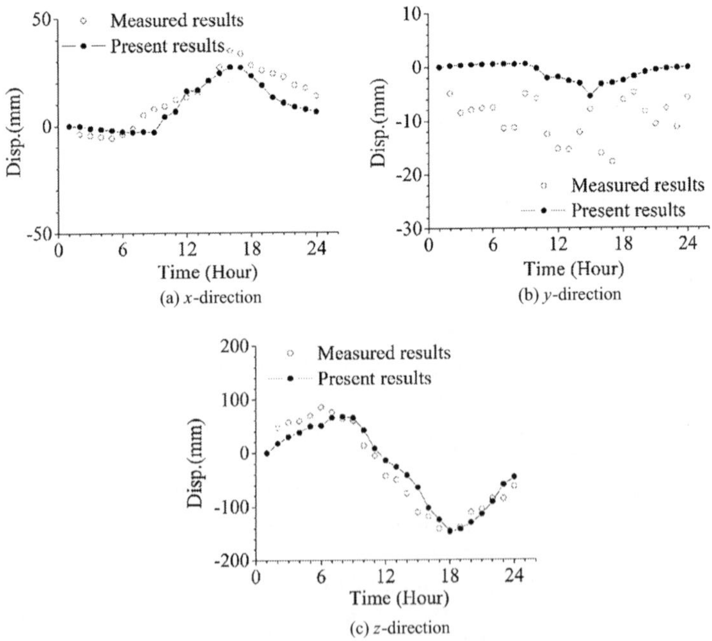

FIGURE 4.20 Displacement of the middle point at the main cable on July 17, 2005: (a) *x*-direction, (b) *y*-direction, and (c) *z*-direction.

moved toward the Tsing Yi Island when the temperature rose. In addition, the main cables first moved upwards as the temperature decreased from 0:00 to 7:00. Then they moved downwards as the temperature increased. From the evening to midnight, the main cables went up again as the temperature decreased. The daily peak-to-peak vertical displacement of the main cables was about 220 mm.

4.2.3.5 Temperature-Induced Stress

The variations in stresses of bridge members are investigated. Considering that temperature typically causes static rather than dynamic deformation of bridges, the attention is primarily placed on the mean stresses rather than the maximum and minimum stresses. This is because the transient responses are caused by dynamic loadings such as highway and railway traffic. For the axially loaded members, the stress is calculated from the strain multiplied by Young's modulus of steel, 2.01×10^{11} Pa, after removing the stress-free component caused by temperature variation.

The Tsing Ma Bridge has 110 strain gauges installed across three sections (see Figure 4.11). The hourly averaged strains at the upper chord, diagonal braces, and bottom chord of Section E (see Figure 4.12) on July 17, 2005 are shown in Figure 4.21. Note that the strain induced by moving train loadings has been removed from the original time histories. Therefore, the stresses are induced purely by the temperatures. Within a single day, the stress variation was more than 10 MPa.

In this section, the measured temperature distribution and temperature-induced structural responses are presented. The measured structural responses were observed

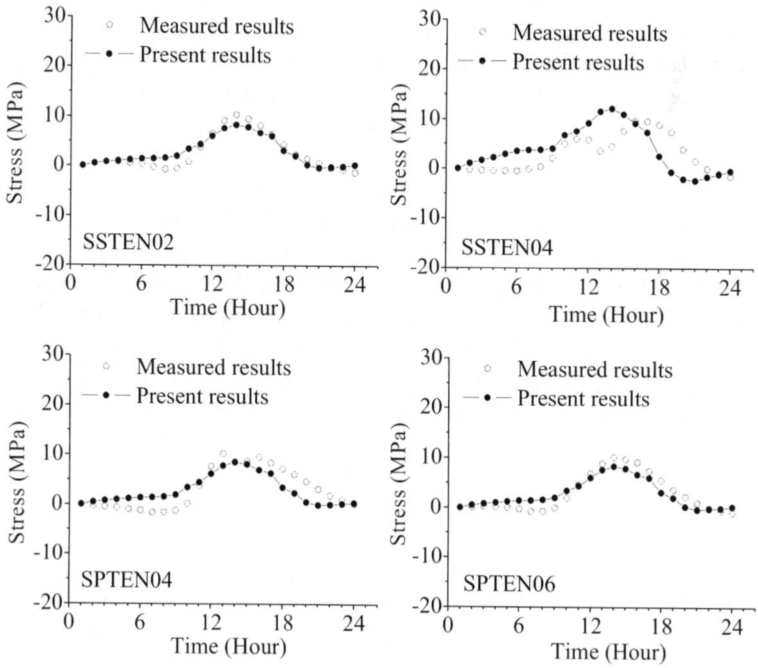

FIGURE 4.21 Axial stresses at Section E on July 17, 2005.

to be closely related to temperature changes. In the next section, numerical simulations will be conducted to examine temperature behaviors that are unrevealed by the monitoring results due to the limited sensors.

4.3 NUMERICAL SIMULATION OF THE TSING MA BRIDGE

4.3.1 Heat-Transfer Analysis

Because of the complicated geometrical configuration of the bridge, it is very difficult to establish one single FE model for the entire bridge and conduct transient thermal analysis. Thermal variation of the bridge deck in the longitudinal direction of the bridge is assumed to be insignificant, and thus, only a typical segment is studied. This assumption is based on the fact that air temperature, solar radiation, and segment configurations along the longitudinal direction of the bridge are similar. Similarly, the thermal variation of the bridge towers in the vertical direction is assumed to be small, and one section of the tower is thus analyzed. By this means, different components are modeled separately to simplify the modeling and analysis. Consequently, FE models of a typical deck plate, a cross frame, and a segment of the bridge tower are established using 3D solid elements in the commercial FE package ANSYS (ANSYS Inc., 2009).

4.3.1.1 FE Models

The orthotropic bridge deck plate consists of steel deck plates, asphalt concrete pavement, and deck troughs. The upper and the lower deck plates have the same thickness of 13 mm. The upper deck trough and lower deck trough are 8 mm thick and 10 mm thick, respectively. All the steel deck plates are covered by a 40 mm thick asphalt concrete cover. The top surface of the deck plates is subjected to direct solar radiation while the other parts of the deck plates are not. The deck plates, troughs, and deck covers are modeled using thermal-type solid elements (SOLID90 in ANSYS). Figure 4.22a shows the FE model around Detail 12 (see Figure 4.7) for heat-transfer analysis, consisting of 8932 nodes and 6720 SOLID90 3D elements. The FE models of deck plates at other positions are also constructed similarly.

Appropriate thermal boundary conditions are critical for transient thermal analysis. Considering the complicated configuration of the large-scale bridge, different surfaces are subjected to different heat flow and air temperatures. In particular, the upper surface of the deck is directly subjected to solar radiation, and thus the equivalent air temperature is applied. For other boundary surfaces without solar radiation, only convection and irradiation are employed as the surface heat flow. The temperature measured from P3-TJS (see Figure 4.3) is used as the air temperature. The temperatures measured at the steel plate and deck troughs are utilized to verify the numerical results.

A detailed 3D FE model of one cross frame is constructed, consisting of 24,242 nodes and 3259 SOLID90 elements, as shown in Figure 4.22b. The cross frame is modeled based on the design with minor simplification by neglecting accessory components. An enlarged detail of one connection of the cross frame is shown in Figure 4.22c.

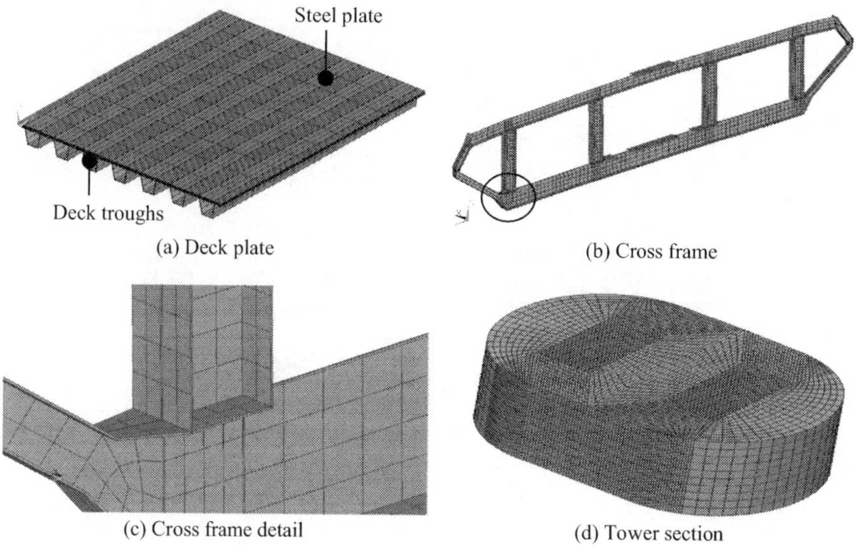

FIGURE 4.22 FE models of the bridge components (Xu and Xia, 2011): (a) deck plate, (b) cross frame, (c) cross frame detail, and (d) tower section.

Because the cross frame is enclosed by the deck plates and corrugated sheets, heat exchange occurs between the plates and the cross frame. Consequently, the temperature measured at the interface between the frame and enclosed plates (sheets) is used as the thermal boundary condition (the First Kind thermal boundary condition). Other surfaces of the frame that are exposed to air do not receive solar radiation directly. Consequently, the air temperatures measured from the sensors close to the frame (P3-TON and P3-TOS in Figure 4.3) are used as the ambient temperature (the Third Kind boundary condition). Take the upper chord of the frame that sensors Ch. 21 to Ch. 23 (P1-TOS) are mounted on as an example. The sensor Ch. 21 was installed at the upper side of the member and is adjacent to the deck plate, and Ch. 23 was installed at the lower side of the member and is exposed to air.

Two cases of boundary conditions are employed here. In Case 1, both monitoring data from Ch. 21 and Ch. 23 are used as the boundary temperature. Consequently, the temperature at Ch. 22 can be calculated with the heat conductivity analysis and compared with the measurement data. In this situation, the heat convection condition is not necessary. In Case 2, the field monitoring data from Ch. 21 are used as the boundary temperature of the upper surface. In the bottom surface, the heat flow condition with air temperature measured from Ch. 86 is used as the boundary condition. In this case, temperature from Ch. 22 and Ch. 23 will be employed to verify the numerical results.

A detailed FE model of one tower section is constructed using 23,969 nodes and 20,000 solid elements, as shown in Figure 4.22d. The cross-section of the bridge tower and its orientation are illustrated in Figure 4.23. The tower section is enclosed by two semicircles and one rectangular with two rectangular holes in the middle.

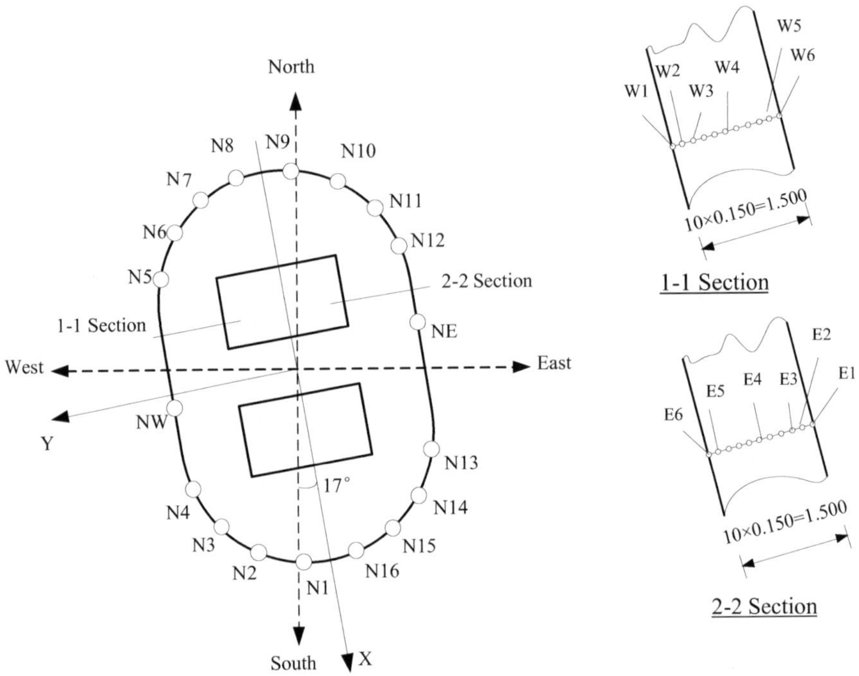

FIGURE 4.23 Configuration of the tower section (Xu and Xia, 2011).

The length of the tower section is 5.0 m. As the outer surface of the tower is a curve, different points receive different solar radiations according to their orientations. In this regard, the outer surface of the tower is divided into 18 segments indicated by the 18 points (N1–N16, NN, and NS). Each segment receives the same level of solar radiation calculated according to its orientation. The determined heat flow and air temperature from P3-TJS are applied to the outer surfaces of the tower section as the boundary conditions.

With the FE models and corresponding boundary conditions described above, the transient heat-transfer analysis is carried out to obtain the temperature distribution of each component. The material parameters of the FE models for the thermal analysis are listed in Table 4.2. The convection heat-transfer coefficient h_c depends on the wind speed and geometric configuration, and an empirical formula (Froli et al., 1996) is employed. The wind speed just above the deck is used in this section.

4.3.1.2 Temperature of Deck Plate

The calculated temperature of the deck plate at Detail 12 on July 17, 2005 is compared with the measurement counterparts (Ch. 66 to Ch. 68) in Figure 4.14. The predicted temperatures of the three points agree with the measurements very well at all times, which validates the effectiveness of the numerical model and the heat-transfer

TABLE 4.2
Thermal Properties of Materials

Parameters		Asphalt	Concrete	Steel
Thermal conductivity	k (W/m/°C)	2.50	1.54	55
Density	ρ (kg/m^3)	2100	2400	7800
Specific heat capacity	c (J/kg/°C)	960	950	460
Absorptivity coefficient	α	0.90	0.65	0.685
Emissivity coefficient	ε	0.92	0.88	0.8
Elasticity modulus	E (Pa)	1.3×10^{9}	3.25×10^{10}	2.01×10^{11}
Thermal expansion coefficient	α_T (1/°C)	1.0×10^{-5}	1.0×10^{-5}	1.3×10^{-5}

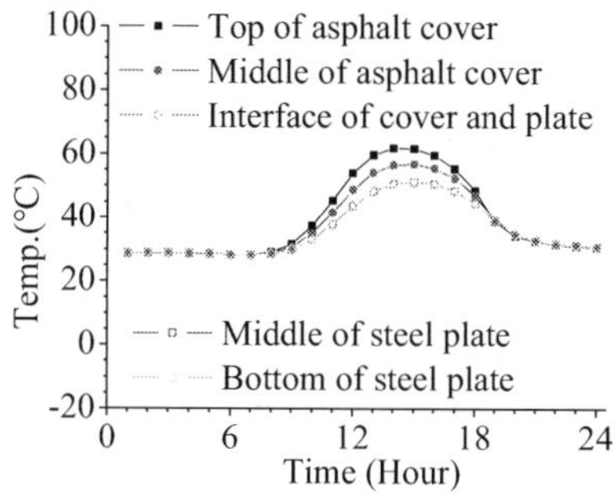

FIGURE 4.24 Measured and simulated temperature of the steel plate and pavement on July 17, 2005.

analysis. With this model, the calculated temperatures of the steel plate and pavement at Detail 12 are plotted in Figure 4.24. The temperature at the asphalt concrete pavement could reach as high as about 61°C in the summer. In addition, the temperature of the asphalt concrete pavement was much higher than that of the steel plate in the afternoon and the maximum difference was 11°C. The temperature difference between the top and bottom of the steel plate was little during the whole day. This is because the thermal conductivity of steel is much larger than that of asphalt.

The temperature of the deck plate in the other three seasons is also calculated and compared with the measurement counterparts in Figure 4.15. The parameters of the model are the same as those adopted in the summer, except the radiation intensity varied in different seasons. Similar to the summer day, the temperatures in other seasons agree well with the measurements. Table 4.3 lists the daily peak temperature at the top of the pavement (Detail 12) in different seasons. The temperature distribution of the bridge plate at other deck details has also been obtained but is not shown here for brevity.

TABLE 4.3
Calculated Daily Peak Temperature at the Top Surface of the Pavement (Detail 12) in Different Seasons (Unit: °C)

Season	Winter	Spring	Summer	Autumn
Date	January 17, 2005	April 18, 2005	July 17, 2005	October 17, 2005
Maximum	35.41	56.52	61.85	51.38
Minimum	12.67	21.31	28.22	24.23

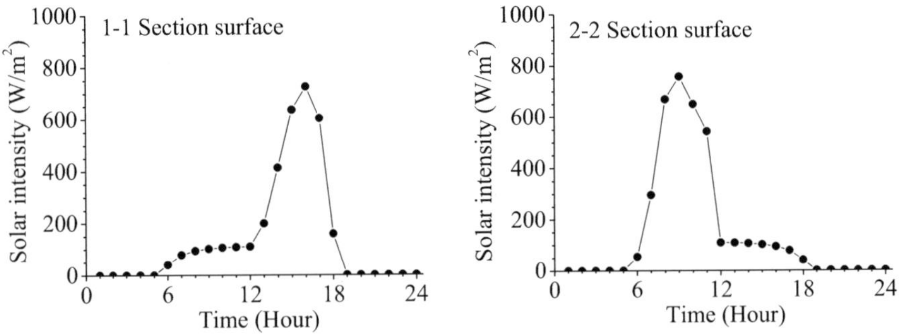

FIGURE 4.25 Solar intensity of the surfaces of the bridge tower on July 17, 2005.

4.3.1.3 Temperature of Cross Frame

The temperature distribution across the cross frame is investigated under two types of boundary conditions (Cases 1 and 2) as described in Section 4.3.1.1. The calculated temperature variation at Ch. 22 (the upper right chord member) on July 17, 2005 for the two cases is compared with the measured results in Figure 4.16. The numerical temperature results in both cases agree with the measurement counterparts very well at different times, and using the first boundary condition (Case 1) can predict the temperature better than Case 2. This is because the temperature at the bottom of the member (Ch. 23) is already known in Case 1 and is more accurate than the numerical calculation in Case 2. Similar observations can be made for the upper left chord, lower chord, and vertical truss members, which are not presented here.

4.3.1.4 Temperature of Bridge Tower

The solar intensity on the tower surfaces on July 17, 2005 is calculated based on the solar model, and the corresponding results on the west surface (1–1 Section in Figure 4.23) and the east surface (2–2 Section in Figure 4.23) are shown in Figure 4.25. It indicates that the peak solar intensity on the east surface occurred at about 9:00, and on the west surface at about 16:00. As for the east surface, there was no direct solar radiation in the afternoon due to the block of the bridge tower, and only a small amount of diffuse solar radiation was received. The situation of the west surface was the opposite.

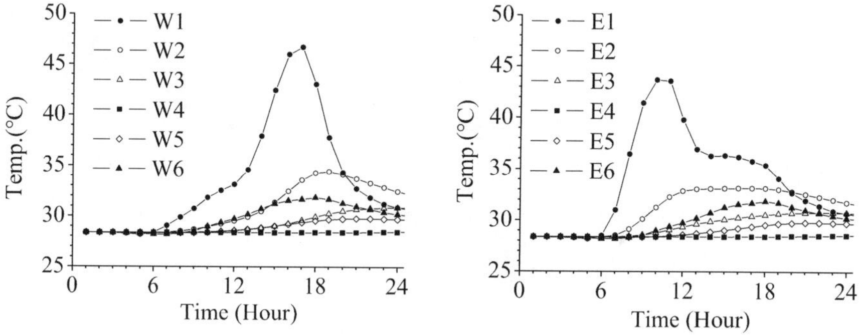

(a) 1-1 Section (b) 2-2 Section

FIGURE 4.26 Variation of temperature at the tower section on July 17, 2005: (a) 1–1 Section and (b) 2–2 Section.

The temperature variation of the tower section on July 17, 2005 is calculated and shown in Figure 4.26. As no temperature sensor was installed on the tower, only numerical results are shown here. The 2-2 Section facing the east received more solar radiation in the morning than in the afternoon. Therefore, the structural temperature at the surface of the 2-2 Section (E1) reached its peak value (about 44°C) at about 11:00 and then dropped gradually in the afternoon. On the other hand, the temperature at the surface of 1-1 Section (W1) reached the peak value (about 47°C) at about 17:00. The temperature at points E2 and W2, which is 0.15 m away from the tower surface, was much smaller than that of surface point (E1 and W1). The temperatures at inner points dramatically decreased with the increasing distance to the tower surface. No obvious temperature variation can be observed at the middle points (E4 and W4) during the whole day because the heat conductivity of concrete is small.

4.3.1.5 Summary

The measured and simulated temperatures at different components on the summer day are compared in Table 4.1. The simulation results agree well with those from the measurement, and the maximum difference is about 1.3°C only. Therefore, the numerical models and heat-transfer analysis are able to simulate the temperature distribution throughout the bridge. These calculated temperatures will be extracted and applied to the structural analysis in the next section.

4.3.2 Structural Analysis

4.3.2.1 Structural Model of the Tsing Ma Bridge

An FE model of the entire Tsing Ma Bridge is constructed via ABAQUS (Abaqus, 2011) to compute the temperature-induced displacement and strain. Components of the cross frames and longitudinal trusses are modeled as beam elements that are appropriately meshed for connections with highway decks and railway tracks. Since the highway decks and railway tracks are at different heights, multi-point constraints

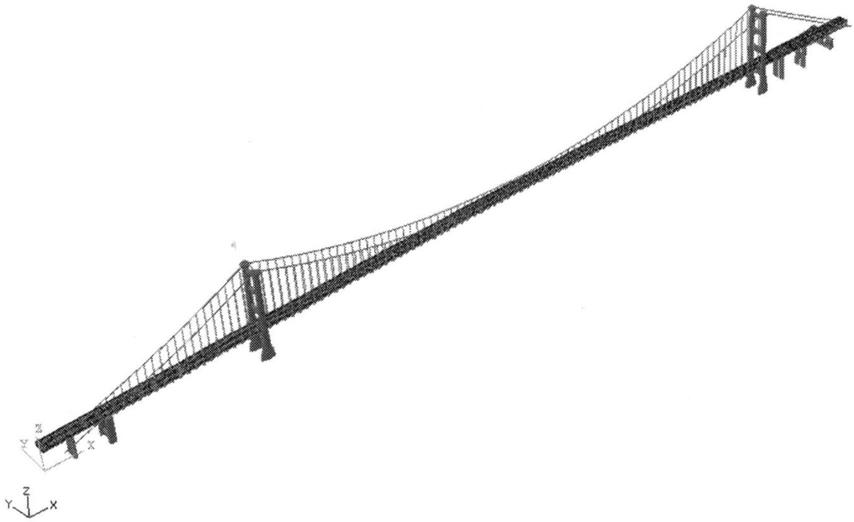

FIGURE 4.27 The FE model of the Tsing Ma Bridge.

are used to connect them to simulate the master-slave relation. The bridge towers and piers are simulated using 3D beam elements, the deck plates are modeled using shell elements, and the suspenders and main cables are modeled using bar elements. The entire model has 12,898 nodes and 21,946 elements. A total of 4788 multi-point constraints connect the orthotropic deck plates and the chords of cross frames. The FE model of the entire Tsing Ma Bridge is shown in Figure 4.27. The model has been updated. Frequencies calculated from the updated FE model agree well with the measurement results, with an average difference less than 4% for the first 18 frequencies.

The calculated temperature distribution at a given time is applied to the entire model for static analysis. Displacement and strain at various components can then be calculated for each hour. The temperature, strain, and displacement at midnight (00:00) are set as the baselines, and the results at later instants are relative changes to the baselines. The numerical results will be compared with the measurement counterparts. To reduce the random error of the temperature and the responses, the average temperature over each hour is employed as the input in the numerical analysis, and the averaged response measurements over the hour are used for comparison. Without further clarification, the displacements measured by the pairs of GPS located in the north and south of the bridge are averaged and regarded as the displacements at the corresponding locations.

4.3.2.2 Displacement of Different Bridge Components

The calculated displacements of the bridge deck, tower, abutment, and main cable are respectively shown in Figures 4.17–4.20. All these displacement results are in good agreement with their measured counterparts. However, the calculated lateral

(y) displacements do not agree well with the measurements, possibly due to a relatively large signal-to-noise ratio in the measurements.

4.3.2.3 Stress of the Bridge Deck

The calculated strains at the upper chord, diagonal braces, and bottom chord of Section E (Figure 4.12) on July 17, 2005 are compared with their measurement counterparts in Figure 4.21. The calculated hourly stress variation agrees well with the field measurement, which verifies that the stress variation is mainly due to temperature changes.

4.3.3 Summary

In this section, two sets of FE models of the Tsing Ma Bridge were established to reconstruct the temperature field in important bridge components and calculate the temperature-induced structural responses. The calculated results are in good agreement with the field measured counterparts. This approach balanced the calculation burden and simulation accuracy in practice.

4.4 ANALYTICAL FORMULATIONS OF THREE-SPAN SUSPENSION BRIDGE

This section will theoretically analyze the temperature-induced displacements of the three-span suspension bridge (one main span and two side spans). A general methodology for calculating the global thermal deformation of the girders, towers, and main cables of three-span suspension bridges is derived. The Akashi Kaikyo Bridge will be used as an example to discuss the mechanisms of temperature effects on the bridge displacements. The strain/stress responses of the bridges are highly localized structural features and will not be analyzed in this section.

4.4.1 Complete Formulation

Consider a common two-tower, three-span ground-anchored suspension bridge (Figure 4.28a). The main cable is divided into three spans of l_i (i = 1, 2, and 3 represent the left side-span, main-span, and right side-span cables, respectively) by the two upright towers with a height of h_{Pi} (i = 1, 2). The two-tower tops, which act as the intermediate supports to the main cable, are not necessarily located at the same level. The sag and cable-chord inclination of each span cable are denoted as f_i and α_i (i = 1, 2, 3), respectively, where f_i is a positive value measured downward from the midpoint of the cable chord, and α_i is positive for the counterclockwise rotation from the horizontal. Note that f_i is the sag at the midpoint rather than the lowest point of the cable, and a three-span suspension bridge always has $\alpha_1 > 0$ and $\alpha_3 < 0$ for calculation convenience. For each span cable, the elevation difference between the two end supports is $h_i = l_i \cdot \tan\alpha_i$ (i = 1, 2, 3). The total horizontal distance between the two anchorages is $L = \sum_{i=1}^{3} l_i$.

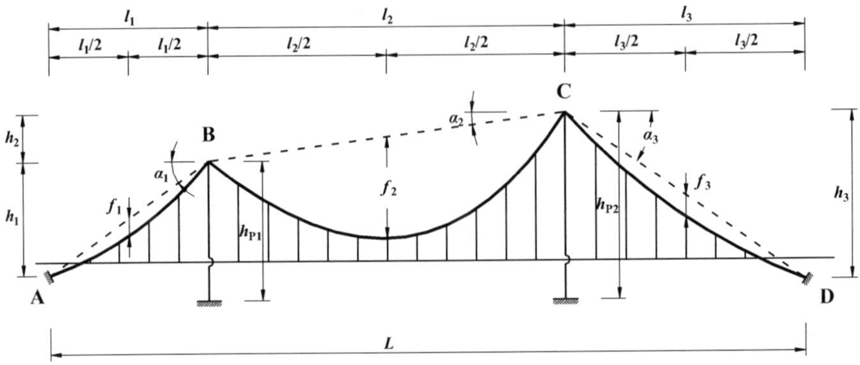

(a) Structural configuration

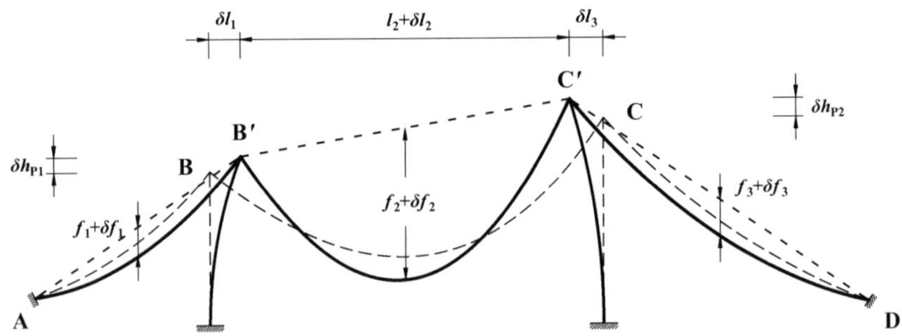

(b) Structural deformation

FIGURE 4.28 Analytical model of a general two-tower ground-anchored suspension bridge: (a) structural configuration and (b) structural deformation.

The deformed configuration of the bridge is illustrated in Figure 4.28b when main cables and bridge towers experience a uniform temperature variation. The following assumptions, justified in Appendix A, are adopted in the bridge deformation analysis:

1. The cable curve between two consecutive supports is a parabola, and its length varies because of temperature changes only (this implies that the length change of the cable due to the tension force change is ignored).
2. The horizontal tensions of main cables on two sides of a tower top are always equal.
3. The stiffening girder and suspenders do not prevent movement of the main cable.
4. The main cables are firmly fixed to the anchorages and tower tops.

The sag of the ith span cable can be expressed as (Gimsing and Georgakis, 2012; Zhou et al., 2020)

$$f_i = q_i l_i^2/(8H_i) = W_i l_i/(8H_i) \tag{4.1}$$

where q_i is the uniformly distributed vertical load, H_i is the horizontal component of the tension force, and $W_i = q_i l_i$ is the total vertical load acting on the cable. Because $\delta W_i = 0$, the differentiation of Equation (4.1) produces

$$\delta f_i / f_i = \delta l_i / l_i - \delta H_i / H_i \tag{4.2}$$

A horizontal force equilibrium at the tower top is always maintained (that is, $H_1 = H_2 = H_3$ and $\delta H_1 = \delta H_2 = \delta H_3$); thus, the following equations for $i = 1$ and 2 can be obtained:

$$\delta f_i / f_i - \delta f_{i+1} / f_{i+1} = \delta l_i / l_i - \delta l_{i+1} / l_i \tag{4.3}$$

The cable length S_i ($i = 1, 2, 3$) is (Zhou et al., 2020)

$$S_i = l_i \left(\sec \alpha_i + 8/3 \cdot n_i^2 \cos^3 \alpha_i \right) \tag{4.4}$$

where $n_i = f_i / l_i$ is the sag-to-span ratio.

Differentiating Equation (4.4) yields

$$\delta S_i = c_{ni} \cdot \delta n_i + c_{li} \cdot \delta l_i + c_{\alpha i} \cdot \delta \alpha_i \tag{4.5}$$

where the coefficients c_{ni}, c_{li}, and $c_{\alpha i}$ are

$$c_{ni} = \partial S_i / \partial n_i = 16/3 \cdot n_i l_i \cos^3 \alpha_i \tag{4.6}$$

$$c_{li} = \partial S_i / \partial l_i = \sec \alpha_i + 8/3 \cdot n_i^2 \cos^3 \alpha_i \tag{4.7}$$

$$c_{\alpha i} = \partial S_i / \partial \alpha_i = l_i \left(\sin \alpha_i / \cos^2 \alpha_i - 8 n_i^2 \sin \alpha_i \cos^2 \alpha_i \right) \tag{4.8}$$

Differentiating $n_i = f_i / l_i$ with respect to f_i and l_i provides

$$\delta n_i = \delta \left(f_i / l_i \right) = \left(l_i \cdot \delta f_i - f_i \cdot \delta l_i \right) / l_i^2 \tag{4.9}$$

As $h_i = l_i \cdot \tan \alpha_i$ ($i = 1, 2, 3$), differentiating h_i with respect to l_i and α_i leads to $\delta h_i = \delta l_i \cdot \tan \alpha_i + l_i \cdot \sec^2 \alpha_i \cdot \delta \alpha_i$. Because δh_i is equal to the difference between the change in tower height, that is, $\delta h_i = \delta h_{Pi} - \delta h_{P(i-1)}$, $\delta \alpha_i$ is obtained as

$$\delta \alpha_i = -\sin 2\alpha_i / (2 l_i) \cdot \delta l_i + \cos^2 \alpha_i / l_i \cdot \left(\delta h_{Pi} - \delta h_{P(i-1)} \right) \tag{4.10}$$

where $\delta h_0 = \delta h_3 = 0$ because they correspond to unmovable anchorage blocks. Substituting Equations (4.9) and (4.10) into Equation (4.5) yields the following equations ($i = 1, 2, 3$):

$$c_{ni} / l_i \cdot \delta f_i + M_i \cdot \delta l_i = Q_i \tag{4.11}$$

where

$$M_i = -c_{ni} n_i / l_i + c_{li} - c_{\alpha i} \sin 2\alpha_i / (2 l_i) \tag{4.12}$$

$$Q_i = \delta S_i - c_{\alpha i} \cos^2 \alpha_i / l_i \cdot \left(\delta h_{Pi} - \delta h_{P(i-1)} \right) \tag{4.13}$$

Because the distance between the two anchorages is constant, the following is true:

$$\delta l_1 + \delta l_2 + \delta l_3 = 0 \tag{4.14}$$

Consequently, Equations (4.3), (4.11), and (4.14) constitute a set of linear equations of δf_i and δl_i ($i=1, 2, 3$):

$$\left[\begin{array}{ccc|ccc} c_{n1} l_1^{-1} & 0 & 0 & M_1 & 0 & 0 \\ 0 & c_{n2} l_2^{-1} & 0 & 0 & M_2 & 0 \\ 0 & 0 & c_{n3} l_3^{-1} & 0 & 0 & M_3 \\ \hline -f_1^{-1} & f_2^{-1} & 0 & l_1^{-1} & -l_2^{-1} & 0 \\ 0 & -f_2^{-1} & f_3^{-1} & 0 & l_2^{-1} & -l_3^{-1} \\ 0 & 0 & 0 & 1 & 1 & 1 \end{array} \right] \cdot \left[\begin{array}{c} \delta f_1 \\ \delta f_2 \\ \delta f_3 \\ \hline \delta l_1 \\ \delta l_2 \\ \delta l_3 \end{array} \right] = \left[\begin{array}{c} Q_1 \\ Q_2 \\ Q_3 \\ \hline 0 \\ 0 \\ 0 \end{array} \right] \tag{4.15}$$

If the second-order terms of n in c_{li} and $c_{\alpha i}$ ($i=1, 2, 3$) are omitted, then Equation (4.15) can be solved as (refer to Appendix B)

$$\begin{aligned} \delta f_i &= n_i \cdot \delta S_i / \cos \alpha_i - n_i \tan \alpha_i \left(\delta h_{Pi} - \delta h_{P(i-1)} \right) \\ &\quad + n_i \left(l_i - z_i \right) \Big/ \left(\sum_{j=1}^{3} z_j \right) \cdot \sum_{j=1}^{3} \left[\delta S_j / \cos \alpha_j - \tan \alpha_j \left(\delta h_{Pj} - \delta h_{P(j-1)} \right) \right] \end{aligned} \tag{4.16}$$

$$\begin{aligned} \delta l_i &= \delta S_i / \cos \alpha_i - \tan \alpha_i \left(\delta h_{Pi} - \delta h_{P(i-1)} \right) \\ &\quad - z_i \Big/ \sum_{j=1}^{3} z_j \cdot \sum_{j=1}^{3} \left[\delta S_j / \cos \alpha_j - \tan \alpha_j \left(\delta h_{Pj} - \delta h_{P(j-1)} \right) \right] \end{aligned} \tag{4.17}$$

where

$$z_i = 16/3 \cdot l_i n_i^2 \cos^2 \alpha_i \tag{4.18}$$

Equations (4.16) and (4.17) are the sag change of the main cables and horizontal displacement of the tower tops, respectively. Because the main cables and towers are subjected to the temperature changes δT_C and δT_P, respectively, and θ_C and θ_P are the corresponding coefficients of linear expansion, one can obtain

$$\delta S_i = S_i\theta_C \cdot \delta T_C = l_i\theta_C \sec\alpha_i \cdot \delta T_C \tag{4.19}$$

$$\delta h_{Pj} = h_{Pj}\theta_P \cdot \delta T_P \tag{4.20}$$

where δS_i ($i=1, 2, 3$) and δh_{Pj} ($j=0, 1, 2, 3$) represent the contribution from the temperature change in cables and towers, respectively; and the second-order term of n in S_i (Equation 4.4) is neglected.

The change in the mid-span elevation of the main-span cable is the sum of the sag and tower height changes. Let the upward movement be positive. It has

$$\delta D_2 = -\delta f_2 + (h_{P1} + h_{P2})/2 \cdot \theta_P \cdot \delta T_P \tag{4.21}$$

Because the suspender at the mid-span is usually short, it undergoes negligible thermal deformation, and the main cable and the girder at the mid-span exhibit nearly the same vertical displacement δD_2.

As both anchorages are unmovable, δl_1 and δl_3 equal the tower-top horizontal displacement, which takes positive values when the tower-top moves toward the central span. δl_2 corresponds to the variation in the horizontal distance of the tower tops.

4.4.2 Simplified Formulation

Equations (4.16) and (4.17) are rather complicated. They are simplified in this section for practical bridges.

For side-span cables hanging with a girder, q_i of the side-span and main-span cables are assumed identical. Equation (4.1) can be rewritten as

$$n_i = f_i/l_i = q_i l_i/(8H_i) \tag{4.22}$$

Then n_i is proportional to the span length l_i. Introducing a span ratio ζ_i into Equation (4.22) yields

$$\zeta_i = l_i/l_2 = n_i/n_2 \tag{4.23}$$

For the main-span cable, $\zeta_2=1$, and for the side-span cable, ζ_i ($i=1, 3$) is usually smaller than 0.5 and $n_i=\zeta_i n_2 \le 1/2$ ($i=1, 3$) because $n_2 \approx 1/10$ (Ministry of Transport of the People's Republic of China, 2015).

From Equation (4.18),

$$z_i/l_i = 16 \cdot l_i n_i^2 \cos^2\alpha_i/(3l_i) = 16/3 \cdot n_2^2\zeta_i^2 \cdot \cos^2\alpha_i \le 16/3 \cdot n_2^2\zeta_i^2 \approx 4/75 \cdot \zeta_i^2 \tag{4.24}$$

Consequently, $z_i/l_i \le 1.3\%$ ($i=1$ and 3) and $z_2/l_2 \le 5.3\%$. Therefore, $l_i - z_i \approx l_i$. This approximation is also feasible for side-span cables without suspenders. Substituting

$l_i - z_i \approx l_i$, $\delta S_i = l_i \theta_C \cdot \sec\alpha_i \cdot \delta T_C$, $\sec\alpha_i = \sqrt{l_i^2 + h_i^2}/l_i$, $\tan\alpha_i = h_i/l_i$ $(i = 1, 2, 3)$, and $\delta h_{Pj} = h_{Pj}\theta_P \cdot \delta T_P$ $(j = 0, 1, 2, 3)$ into Equation (4.16), one obtains the following:

$$\delta f_i = n_i\left[\left(l_i^2 + h_i^2\right)/l_i\,\theta_C \cdot \delta T_C - h_i \cdot \left(h_{Pi} - h_{P(i-1)}\right)/l_i \cdot \theta_P \cdot \delta T_P\right]$$
$$+ l_i n_i \Big/ \sum_{k=1}^{3} z_k \cdot \sum_{j=1}^{3}\left[\left(l_j^2 + h_j^2\right)/l_j \cdot \theta_C \cdot \delta T_C - h_j \cdot \left(h_{Pj} - h_{P(j-1)}\right)/l_j \cdot \theta_P \cdot \delta T_P\right] \tag{4.25}$$

For most suspension bridges, the following four approximations are applicable: (1) $h_{P1} \approx h_1$, (2) $h_{P2} \approx |h_3|$, (3) $\theta_C \cdot \delta T_C \approx \theta_P \cdot \delta T_P$, and (4) $h_{P2} - h_{P1} \approx h_1$ (both tower bottoms at the same level) or $h_2 = 0$ (both tower tops at the same level, or $\alpha_2 = 0$). Equation (4.25) then becomes:

$$\delta f_i = n_i l_i \theta_C \cdot \delta T_C + n_i l_i \Big/ \left(\sum_{k=1}^{3} z_k\right) \cdot \theta_C \cdot \delta T_C \cdot \sum_{j=1}^{3} l_j \tag{4.26}$$

The ratio of the first to second items in Equation (4.26) is

$$\sum_{i=1}^{3} z_i \Big/ \sum_{i=1}^{3} l_i = 16/3 \cdot \sum_{i=1}^{3}\left(l_i n_i^2 \cos^2\alpha_i\right) \Big/ \sum_{i=1}^{3} l_i \tag{4.27}$$

Because $n_2 \approx 1/10$ and $n_i \leq 1/20$ $(i = 1, 3)$, Equation (4.27) is considerably lower than 1. Therefore, the first item of Equation (4.26) is negligible compared with the second item. Consequently, Equation (4.16) can be simplified as follows:

$$\delta f_i = \left(f_i \Big/ \sum_{j=1}^{3} z_j\right) \cdot L\theta_C \cdot \delta T_C \tag{4.28}$$

Similarly, Equation (4.17) for the tower-top horizontal displacement δl_i $(i = 1, 2, 3)$ can be simplified as

$$\delta l_i = \left[l_i - z_i \Big/ \left(\sum_{j=1}^{3} z_j\right) \cdot L\right]\theta_C \cdot \delta T_C \tag{4.29}$$

Equations (4.28) and (4.29) exhibit a unified form similar to 1D thermal deformation $L_E\theta_C \cdot \delta T_C$, where L_E is the temperature equivalent length and can be expressed as

$$L_E = \begin{cases} \left(f_i \Big/ \sum_{j=1}^{3} z_j\right) \cdot L & \text{for } \delta f_i, i = 1, 2, 3 \\ l_i - \left(z_i \Big/ \sum_{j=1}^{3} z_j\right) \cdot L & \text{for } \delta l_i, i = 1, 2, 3 \end{cases} \tag{4.30}$$

When the sags of both side-span cables are neglected, that is, $f_1 = f_3 = 0$, the temperature equivalent length can be further simplified as

$$L_E = \begin{cases} 0 & \text{for } \delta f_1, \delta f_3 \\ 3L/\left(16 n_2 \cos^2 \alpha_2\right) & \text{for } \delta f_2 \\ l_1 & \text{for } \delta l_1 \\ -\left(l_1 + l_3\right) & \text{for } \delta l_2 \\ l_3 & \text{for } \delta l_3 \end{cases} \tag{4.31}$$

The sag effect of the side-span cables can be quantified by comparing the two temperature equivalent lengths in Equations (4.30) and (4.31), which will be studied in Section 4.4.4.

4.4.3 Verification of Formulas

The Akashi Kaikyo Bridge is used to verify the derived formula. The bridge is a three-span suspension bridge across the Akashi Strait in Japan with a steel truss-stiffened girder and diagonally braced steel towers (Kitagawa, 2004). It was opened to the public in April 1998. Since then, it has had the longest span of any bridge in the world till 2022, when the 1915 Çanakkale Bridge in Turkey was constructed with a main span of 2023 m. An SHM system was installed on the bridge to measure wind, earthquake, traffic, and temperature loads and their effects. Figure 4.29 shows the layout of the temperature sensors and GPS stations used in this study. The cable temperature was measured at three locations: near the anchorage 1A (T1), at the top of tower 2P (T2), and near the mid-span of the central span (T3). Moreover, a thermometer was installed near the midpoint of the central span girder (T4) to measure air temperature. Structural displacement was measured using GPS with a reference station at the 1A anchorage and two rovers, namely one on the 2P tower top (GPS1) and the other on the mid-span of the central span girder (GPS2).

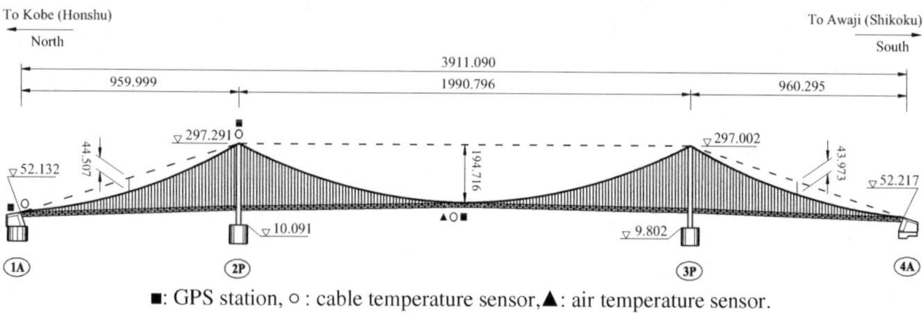

FIGURE 4.29 Configuration of the Akashi Kaikyo Bridge and GPS and temperature measurement points (unit: m). *Note*: The elevation is measured from the Tokyo Bay mean tide level. ■: GPS station, ○: cable temperature sensor, ▲: air temperature sensor.

The 10-min means of temperature and structural displacement obtained from September 1998 to August 2001 are studied. The averaging can filter out the high-frequency components of the displacement due to dynamic loadings such as wind and traffic. The 10-min means represent the temperature-related structural response.

Figure 4.30 presents the cable and air temperatures, the tower-top horizontal displacement of tower 2P, and the mid-span elevation of the main-span girder during the study period. The discontinuity of the curves in the figure is caused by the malfunction of the monitoring equipment. With the temperature increase, the tower top moved toward the central span (approximately 20 cm in a year), and the midpoint of the main-span girder moved downward (approximately 200 cm), demonstrating that temperature change governed the global deformation of the bridge.

Additionally, Figure 4.30 presents the temperature and displacement measured between July 7 and July 9, 1999. Cable temperatures T1 and T2 varied at the same magnitude and pace. Their daily variations were smaller than that of the cable temperature T3. Moreover, T1 and T2 lagged behind T3 and air temperature T4 by approximately 7 hours. This is because T1 and T2 sensors were installed inside the main cables (underneath the polyethylene tubes), whereas T3 thermometer was attached to the cable band. The latter is affected by air temperature and solar radiation. Because the resolution of GPS data was relatively low 20 years ago, the small swaying movement of the tower exhibited a stepped shape.

The mid-span vertical displacement of the main cable versus the cable temperature is plotted in Figure 4.31, which shows a linear relation. The slope of the linear regression equation (dashed line) is calculated as −0.0701 m/°C, indicating that the

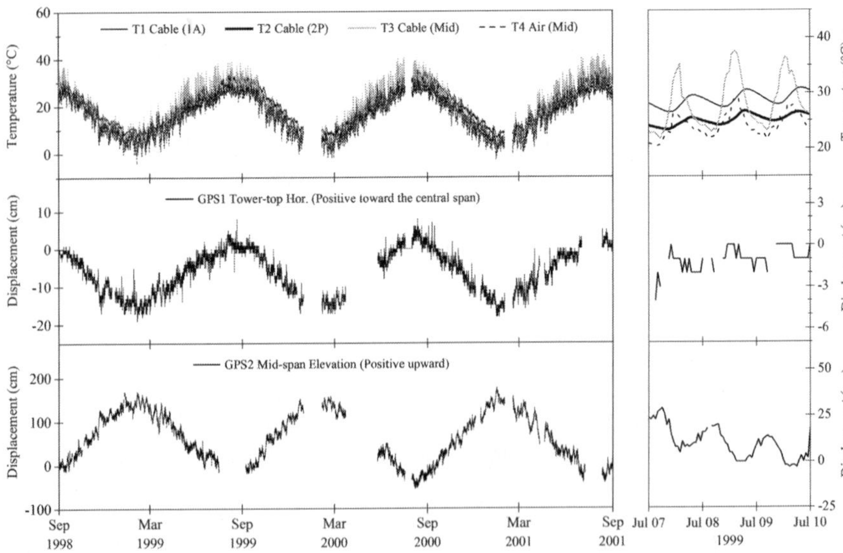

FIGURE 4.30 Temperature and structural displacement measured during 1998–2001.

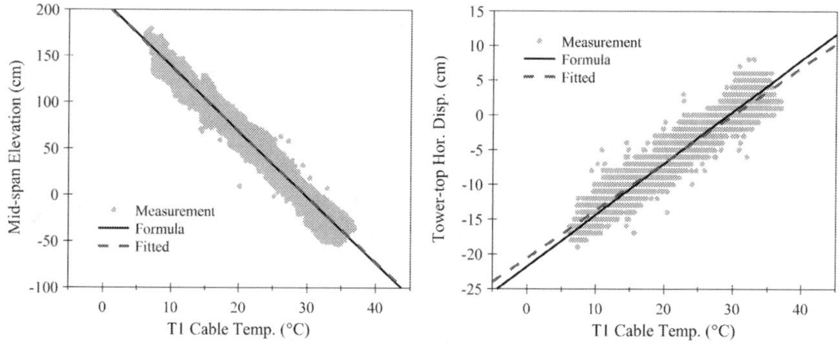

(a) T1 versus mid-span elevation (b) T1 versus tower-top horizontal displacement

FIGURE 4.31 Structural displacement versus temperature change (1998–2001): (a) T1 versus mid-span elevation and (b) T1 versus tower-top horizontal displacement.

mid-span elevation decreased by 0.0701 m, with a unit increase in cable temperature. A similar linear regression analysis shows that the tower top moved 0.0069 m toward the main span, with a unit increase in cable temperature.

The temperature-induced displacement of the bridge is calculated using the formulas provided herein. From Equations (4.16) and (4.17), when the cable temperature increases by 1°C, the elevation of the main cable at the mid-span decreases by 0.0708 m, and the tower top moves 0.0074 m toward the main span. Therefore, the results obtained using the derived formulas agree well with the field measurement data. The calculation is based on the following parameters: (1) $l_1=959.999$ m, $l_2=1990.796$ m, and $l_3=960.295$ m; (2) $n_1=0.046$, $n_2=0.098$, and $n_3=0.046$; (3) $\alpha_1=14.326°$, $\alpha_2=-0.008°$, and $\alpha_3=-14.300°$; (4) $h_{P1}=h_{P2}=287.200$ m; (5) $\theta_C=\theta_P=1.2\times10^{-5}$°C^{-1}; and (6) $\delta T_C=\delta T_P=1$°C. The calculated results are also shown in Figure 4.31 (solid lines passing the center of the data points).

With the simplified Equations (4.28) and (4.29) and $\delta T_C=\delta T_P=1$°C, the elevation of the main cable at the mid-span decreases by 0.0715 m, the tower top moves 0.0075 m toward the main span, and both provide good estimates of bridge deformation. Furthermore, when the sag of side-span cables is neglected, the aforementioned two quantities are 0.0865 and 0.0115 m, according to Equation (4.31). In this case, neglecting the side-span cable sags leads to considerable errors.

The formulas and simplified formulas are also applied to seven other long-span suspension bridges. Figure 4.32 presents δf_2 and δl_i $(i=1, 2, 3)$ of the eight bridges calculated using different formulas, according to the geometric information listed in Table 4.4. Most of the relative errors of Equations (4.28) and (4.29) with respect to the exact formulas (Equations 4.16 and 4.17) are less than 2%, indicating the effectiveness of the simplified formulas. The relative errors of Equation (4.31) are small for all bridges except the Akashi Kaikyo Bridge, which contains a long side span.

In conclusion, Equations (4.28) and (4.29) present an excellent approximation of Equations (4.16) and (4.17). Equation (4.31) provides a satisfactory estimate of most

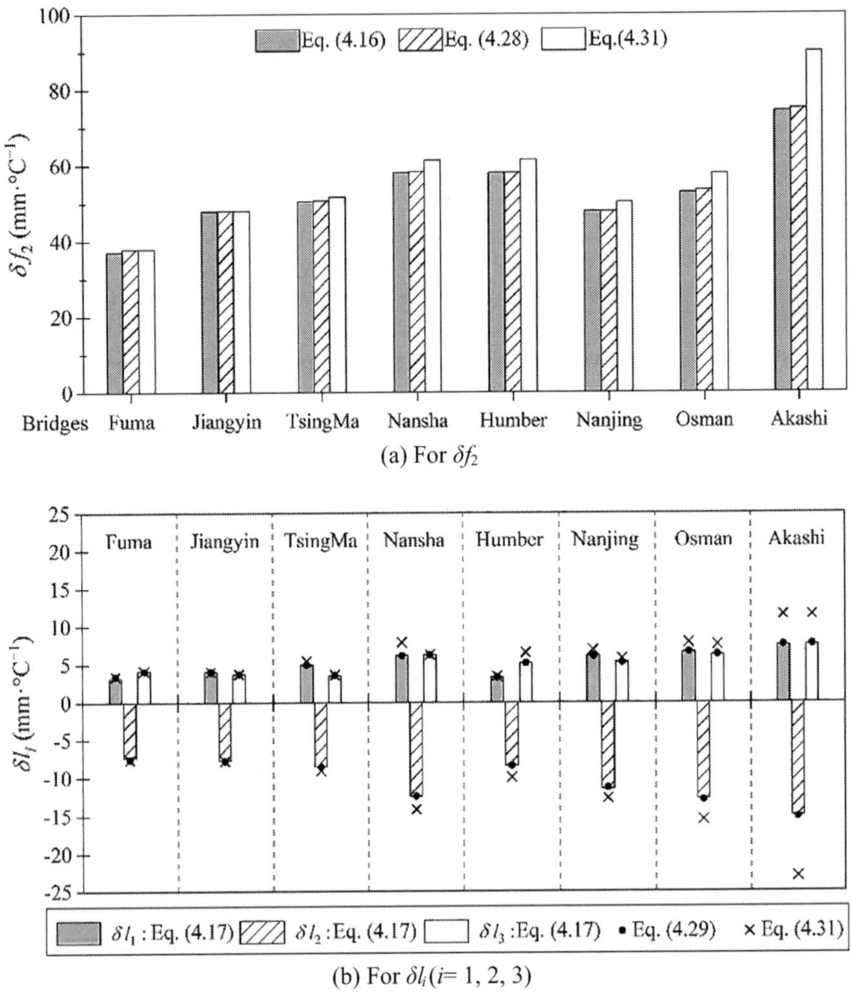

FIGURE 4.32 Thermal sensitivity of structural deformations ($\delta T_C = \delta T_P = 1$°C): (a) For δf_2 and (b) for δl_i (i= 1, 2, 3).

suspension bridges and fair errors for bridges with long side spans. The effect of the sag in side-span cables is studied in the following section.

4.4.4 Sag Effects of Side-Span Cables

4.4.4.1 Side-Span Cables without Suspenders

For side-span cables without hanging a girder, the vertical load q_i ($i = 1, 2, 3$) is the self-weight of the cables and is thus much smaller than that of main-span cables. From Equation (4.1), the sag of the side-span cables is negligible, and Equation (4.31) is valid.

TABLE 4.4
Geometry of Sample Bridges

Bridges	Fuma	Jiangyin	Tsing Ma	Nansha (Nizhou)	Humber	Nanjing No.4	Osman Gazi	Akashi Kaikyo
Location	CHN	CHN	HKG	CHN	GBR	CHN	TUR	JPN
l_1 (m)	285.000	336.506	455.000	658.000	285.000	576.200	658.050	959.999
l_2 (m)	1050.000	1385.042	1377.000	1688.000	1410.000	1418.000	1550.000	1990.796
l_3 (m)	345.000	309.346	300.000	522.000	543.900	481.800	633.250	960.295
Σl_i (m)	1680.000	2030.894	2132.000	2868.000	2238.900	2476.000	2841.300	3911.090
n_1	0	0	0.024	0.041	0.017	0.032	0.036	0.046
n_2	0.100	0.095	0.093	0.105	0.082	0.111	0.111	0.098
n_3	0	0	0.007	0	0.032	0.028	0.037	0.046
α_1 (°)	25.481	25.670	20.972	18.443	23.901	19.954	19.110	14.326
α_2 (°)	0	0	0	0	0	0	0	−0.008
α_3 (°)	−21.056	−25.789	−27.834	−22.474	−14.088	−23.471	−19.075	−14.300
h_{P1} (m)	213.500	191.236	204.400	256.939	155.500	227.200	241.850	287.200
h_{P2} (m)	169.570	191.236	204.400	256.939	155.500	227.200	241.850	287.200
θ_C (10^{-5}/°C)	1.20	1.20	1.20	1.20	1.20	1.20	1.20	1.20
θ_P (10^{-5}/°C)	1.00	1.00	1.00	1.00	1.00	1.00	1.20	1.20
z_1 (m)	0	0	1.215	5.317	0.348	2.803	4.063	10.331
z_2 (m)	56.000	66.997	63.279	99.748	50.460	93.300	101.404	101.572
z_3 (m)	0	0	0.053	0	2.725	1.752	4.234	10.084
Σz_i (m)	56.000	66.997	64.548	105.065	53.533	97.855	109.701	121.987

4.4.4.2 Side-Span Cables with Suspenders

For side-span cables hanging a girder, Equation (4.23) shows that $n_i = n_2\zeta_i$ ($i = 1, 3$). To demonstrate the side-span cable sag effect on δf_2, Equation (4.28) can be rewritten as

$$\delta f_2 = z_2 \Big/ \left(\sum_{j=1}^{3} z_j \right) \cdot 3L\theta_C \cdot \delta T_C \Big/ \left(16 n_2 \cos^2 \alpha_2 \right) \tag{4.32}$$

The sag effect of the side-span cables is represented by a coefficient $\beta_{f2} = z_2 \Big/ \left(\sum_{j=1}^{3} z_j \right)$. In most suspension bridges, tower tops are at the same level, namely $\alpha_2 = 0$, and the two anchorages and the mid-span of the main-span cable are located at a similar elevation level, namely, $|h_i| \approx f_2$ ($i = 1, 3$). Therefore, $\cos^2\alpha_i$ ($i = 1, 3$) can be approximated as

$$\cos^2 \alpha_i = l_i^2 \Big/ \left(l_i^2 + h_i^2 \right) = l_i^2 \Big/ \left(l_i^2 + f_2^2 \right) = l_i^2 \Big/ \left(l_i^2 + n_2^2 \cdot l_2^2 \right) = l_i^2 \Big/ \left(\zeta_i^2 + n_2^2 \right) \tag{4.33}$$

Using Equation (4.33) and $\alpha_2=0$, β_{f2} can be quantified as:

$$\beta_{f2}=\frac{z_2}{\sum_{j=1}^{3}z_j}=\frac{l_2n_2^2\cos^2\alpha_2}{\sum_{j=1}^{3}l_in_i^2\cos^2\alpha_i}=\frac{1}{\zeta_1^5/\left(\zeta_1^2+n_2^2\right)+1+\zeta_3^5/\left(\zeta_3^2+n_2^2\right)} \tag{4.34}$$

Figure 4.33 presents the relationship between β_{f2} and ζ_1 in the case of $\zeta_1=\zeta_3$ and $n_2=1/10$. For the Akashi Kaikyo Bridge with $\zeta_1=\zeta_3=0.48$, the calculated β_{f2} value is 0.83, indicating that Equation (4.31) presents only 83% of the value calculated with Equation (4.28) and causes 17% error. For most long-span suspension bridges, $\zeta_i<0.3$ ($i=1, 3$) and $\beta_{f2}>0.95$; and Equation (4.31) gives an error smaller than 5% error in the calculation of mid-span displacement. This is verified in Figure 4.32a for the other seven long-span suspension bridges.

Subsequently, the sag effect of side-span cables on the tower-top horizontal displacement is investigated. Because $\delta l_2=-(\delta l_1+\delta l_3)$, we only discuss δl_i ($i=1, 3$). According to Equations (4.29) and (4.31), a term $\beta_{l_i}=1-\left(z_i\Big/\sum_{j=1}^{3}z_j\right)L/l_i$ ($i=1, 3$) is introduced. By using the span ratio ζ_i, the total distance L between anchorages can be obtained:

$$L=\left(l_1+l_1/\zeta_1+\zeta_3/\zeta_1\cdot l_1\right) \tag{4.35}$$

Assuming $\zeta_1=\zeta_3$ and $z_1=z_3$, β_{l1} can be quantified as

$$\beta_{l_1}=1-z_1\Big/\sum_{j=1}^{3}z_j\cdot L/l_1=1-\zeta_1^4/\left(\zeta_1^2+n_2^2+2\zeta_1^5\right)\cdot\left(2\zeta_1+1\right) \tag{4.36}$$

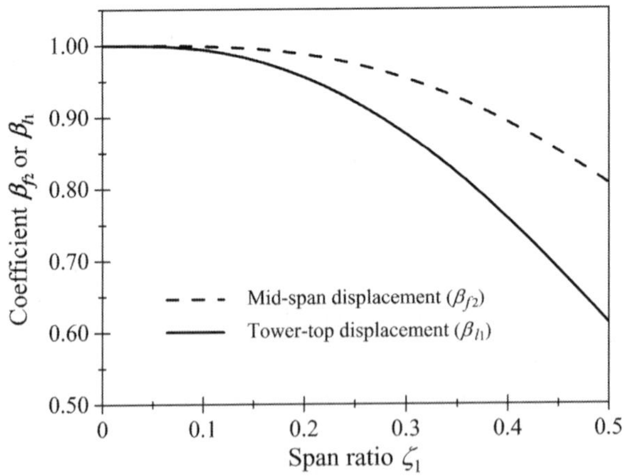

FIGURE 4.33 Influence of side-to-main span ratio on mid-span vertical and tower-top horizontal movements.

Figure 4.33 presents the relation between ζ_1 and β_{l1}. The sag of side-span cables exhibits more influence on the tower-top horizontal displacement than on the mid-span vertical displacement. For the Akashi Kaikyo Bridge with $\zeta_1=\zeta_3=0.48$, β_{l1} is 0.64. Thus, Equation (4.31) causes 36% error. When the two side-span ratios have $\zeta_i=0.3$ ($i=1$, 3), Equation (4.31) leads to 15% error in the calculation of the tower-top displacement.

The sag effect of the side-span cables is quantified by the modification coefficients. Except for the idealized situation of straight anchor cables (equivalent to $\zeta_1=0$ in Figure 4.33), both β_{f2} and β_{l1} are less than 1.0, indicating that the sags of side-span cables reduce the mid-span vertical and tower-top horizontal displacements. Meanwhile, as $\beta_{f2}>\beta_{l1}$, the temperature-induced tower-top movement is more sensitive than the main-span cable sag changes to the sag of side-span cables.

4.4.5 Discussions

4.4.5.1 Comparison of Thermal Deformation

The temperature-induced displacements of different components can be compared using the proposed formulas. For the simplified case ($f_1=f_3=0$, $\alpha_2=0$, $h_{P1}\approx h_1$, $h_{P2}\approx |h_3|$, $\theta_C\cdot\delta T_C\approx\theta_P\cdot\delta T_P$), the magnitude ratio of δf_i and δl_i ($i=1$, 2, 3) is identical to the ratio of L_E, which is presented in Equation (4.31). Because $n_2\approx 1/10$ and the side-to-main span ratio is usually <0.5, δf_2 and δD_2 are much larger than δl_i ($i=1$, 2, 3). For a bridge with a straight side-span cable and $\zeta_1=\zeta_3=0.5$, $\delta f_2:\delta l_1:\delta l_2:\delta l_3=7.5:1:2:1$.

Furthermore, the effects of temperature change in main cables and towers can be separated according to Equations (4.16), (4.17), and (4.21). Their effects on δf_2 and δl_1 for the eight bridges are compared in Figure 4.34, given $\delta T_C=\delta T_P=1$°C. The main-span cable contributes the most (approximately 65%) to δf_2, the side-span cable contributes the second most (each approximately 20%), and towers contribute the least but with an opposite effect (each approximately –5%). For the horizontal movement of the left tower top (δl_1), the temperature change in the side-span cable on the same side contributes the most (approximately 130%), followed by the tower on the same side (approximately –10% to –30%). The subsequent contribution is from the main-span cable (approximately 0 to –30%), followed by the contribution of the other side-span cable (less than –10%), and the opposite tower exhibits little effect.

4.4.5.2 Self-Anchored Suspension Bridges

A self-anchored suspension bridge is a variant of conventional earth-anchored suspension bridges with the main cable anchored on the girder (Figure 4.35) (Ochsendorf and Billington, 1999). Considering the girder is continuous at the girder–tower intersections with the total length of L_G, the distance change between the anchorage points of the main cable can be estimated with $\delta L_G=L_G\theta_G\cdot\delta T_G$, where θ_G and δT_G are the linear expansion coefficient and temperature change of the girder, respectively. For a two-tower self-anchored suspension bridge, the compatibility condition of all

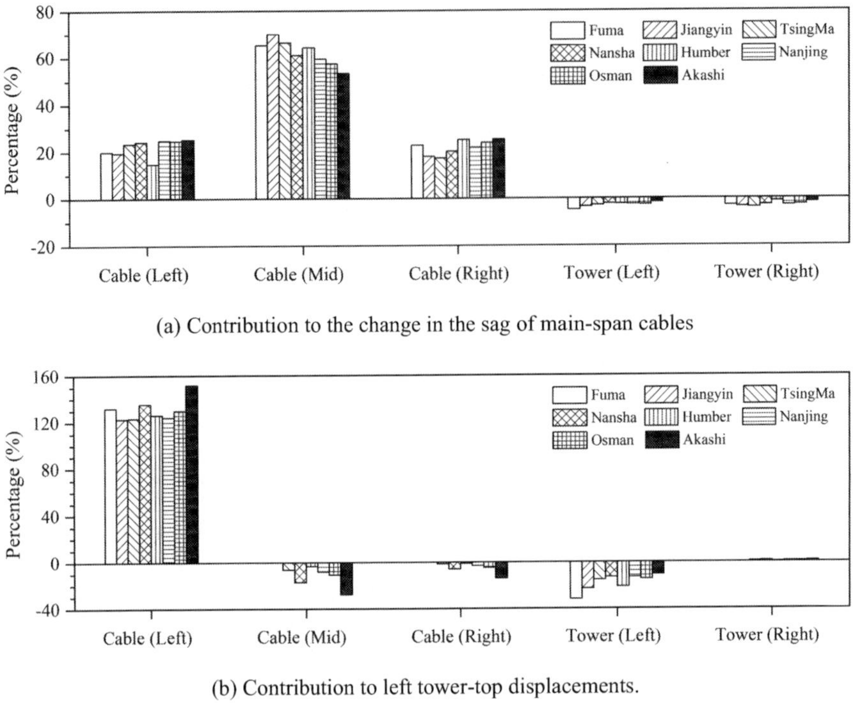

FIGURE 4.34 Contribution of the changes in cable and tower temperatures to structural deformation ($\delta T_C = \delta T_P = 1°C$): (a) Contribution to the change in the sag of main-span cables and contribution to left tower-top displacements.

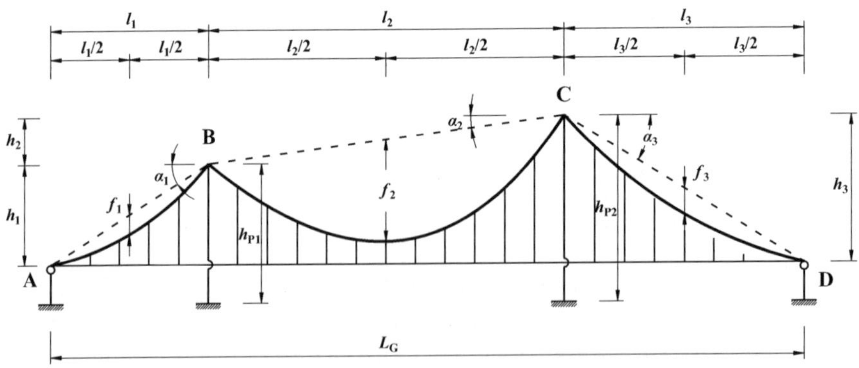

FIGURE 4.35 Self-anchored suspension bridge.

span cables is $\sum_{i=1}^{3} \delta l_i = \delta L_G$. Consequently, the right column vector of Equation (4.15) becomes $[Q_1\ Q_2\ Q_3\ 0\ 0\ \delta L_G]^{\mathrm{T}}$, and the solution to the new system of equations is given as

$$\delta f_i = n_i/\cos\alpha_i \cdot Q_i + n_i\left(l_i - z_i\right)\Big/\left(\sum_{k=1}^{3} z_k\right)\cdot\left[-\delta L_G + \sum_{j=1}^{3}\left(Q_j/\cos\alpha_j\right)\right] \tag{4.37}$$

$$\delta l_i = Q_i/\cos\alpha_i - z_i\Big/\sum_{k=1}^{3} z_k \cdot\left[-\delta L_G + \sum_{j=1}^{3}\left(Q_j/\cos\alpha_j\right)\right] \tag{4.38}$$

From Equation (4.38), the change in the girder length δL_G is distributed to the change in the span length of each span cable in proportion to the parameter z_i.

4.4.6 Summary

In this section, a general analytical formula is derived for quantifying the thermal deformation of three-span suspension bridges. The methodology treats a suspension bridge as an assembly of interconnected members and considers the effects of the side-span cable sags and the cable tension changes with temperature. The resulting formulas have a unified simple form similar to the 1D thermal expansion formula and provide a clear physical interpretation of the thermal deformation of suspension bridges.

It is found that ignorance of the side-span cable sags overestimates the global thermal deformation. The sag effect can be characterized by the modification coefficients β_{f2} and β_{l1} with $\beta_{f2} = \beta_{l1} = 1.0$ corresponding to the results by Zhou et al. (2020). The tower-top movement is more sensitive than the main-span cable sag changes to the sag effect. For side-span cables with suspenders, the sag effect on the bridge displacement increases with an increase in the span ratio. For most suspension bridges with a span ratio less than 0.3, the sag effect is insignificant; however, for bridges with a span ratio of 0.5, the ignorance of side-span cable sags causes 20% and 40% errors for δf_2 and δl_1, respectively. Moreover, the thermal deformation of the main- and side-span cables and towers interacts with each other because cable tension changes with varying temperatures. The closed-form solutions are also available for the self-anchored suspension bridges.

This theoretical approach is more advantageous than the widely used regression and FE analyses because the analytical formulas provide a simple, fast, and accurate quantification of temperature-induced displacement. The approach is scientifically rigorous and concise because the sag effects of the side-span cables and the temperature-induced cable tension changes are taken into consideration. The method can be extended to other cable structures, such as transmission lines or cable car systems. In the next section, the formulas are extended to multi-span suspension bridges.

4.5 ANALYTICAL FORMULATIONS OF MULTI-SPAN SUSPENSION BRIDGE

In this section, the formulas derived in the previous section will be extended to general multi-span (no less than three spans) suspension bridges.

4.5.1 Complete Formulation

Consider a general multi-span ground-anchored suspension bridge, as shown in Figure 4.36. The main cables of the bridge in Figure 4.36a have u ($u \geq 3$) spans, u+1 supports with the two anchorages corresponding to 0 and u, and upright towers from 1 to $u-1$. The ith span cable has a span of l_i, sag of f_i, and inclination angle of α_i ($i=1, 2,\ldots, u$), while the jth tower has a height of h_{Pj} ($j=1, 2,\ldots, u-1$) and the flexural stiffness of k_{Pj} ($j=1, 2,\ldots, u-1$). α_i is positive for the counterclockwise rotation from the horizontal, and f_i corresponds to the midway point rather than the lowest point of the cable. For each span cable, the elevation difference between two consecutive supports is $h_i = l_i \cdot \tan\alpha_i$. The total horizontal distance between the two anchorages is $L = \sum_{i=1}^{u} l_i$.

When the main cable and towers are subjected to a uniform temperature variation δT_C and δT_P, respectively, the bridge deforms, as shown in Figure 4.36b. In particular, the cable span becomes $l_i + \delta l_i$, the cable sag becomes $f_i + \delta f_i$ ($i=1, 2,\ldots, u$), and the

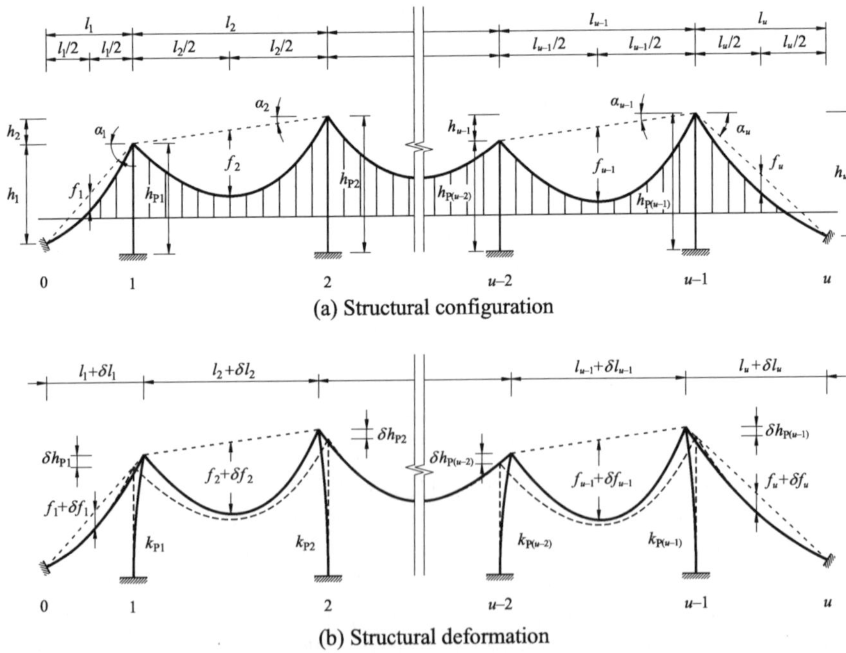

FIGURE 4.36 Analytical model of a general multi-span suspension bridge: (a) structural configuration and (b) structural deformation.

variation in the tower height is δh_{Pj} ($j=1, 2,\ldots, u-1$). For an earth-anchored suspension bridge, the average girder temperature and the temperature differences on the cross-section of the cables, girder, and towers have little influence on the global deformation in the vertical plane (e.g., the mid-span vertical and tower-top horizontal displacements), which has been elaborated by Zhou et al. (2020). These temperature variables are thus neglected herein. The following assumptions are adopted in analyzing the thermal deformation:

1. The cable curve is a parabola between two consecutive supports;
2. The initial horizontal component of the cable tension in each span is equal, denoted as a constant H_0;
3. The stiffening girder and suspenders do not restrain the movement of the main cables and thus not shown in Figure 4.36b; and
4. The main cables are firmly fixed at the anchorages and tower tops.

The derivation for the multi-span cases will be similar to that of the three-span case. Still, for derivation completeness and easy reference, the entire derivation process will be given in this section.

The sag of each span cable can be expressed as (Zhou et al., 2020):

$$f_i = q_i l_i^2/(8H_i) = W_i l_i/(8H_i) \tag{4.39}$$

where q_i ($i=1, 2,\ldots, u$) is the uniformly distributed vertical load, corresponding to the bridge dead load; $H_i=H_0$ is the horizontal component of the tension force; and $W_i=q_i l_i$ is the total vertical load acting on the ith cable. As $\delta W_i=0$, the differentiation of Equation (4.39) leads to:

$$\delta f_i/f_i = \delta l_i/l_i - \delta H_i/H_0 \tag{4.40}$$

The tower-top horizontal displacement of the jth ($j=1, 2,\ldots, u-1$) tower is $\sum_{m=1}^{j} \delta l_m$, which is related to the horizontal tension difference of the adjacent main cables. It can be expressed as:

$$\delta H_{j+1} - \delta H_j = k_{Pj} \sum_{m=1}^{j} \delta l_m \tag{4.41}$$

The cable length S_i ($i=1, 2,\ldots, u$) is expressed as:

$$S_i = l_i\left[\sec\alpha_i + 8/3\cdot n_i^2\cos^3\alpha_i - 32/5\cdot n_i^4\left(5\cos^7\alpha_i - 4\cos^5\alpha_i\right)\right] \tag{4.42}$$

where $n_i=f_i/l_i$ is the sag-to-span ratio.

Differentiating Equation (4.42) yields:

$$\delta S_i = c_{ni}\cdot\delta n_i + c_{li}\cdot\delta l_i + c_{\alpha i}\cdot\delta\alpha_i \tag{4.43}$$

where the coefficients c_{ni}, c_{li}, and $c_{\alpha i}$ are

$$c_{ni} = \partial S_i/\partial n_i = l_i \cdot \left[16/3 \cdot n_i \cos^3 \alpha_i - 128/5 \cdot n_i^3 \left(5\cos^7 \alpha_i - 4\cos^5 \alpha_i\right)\right] \quad (4.44)$$

$$c_{li} = \partial S_i/\partial l_i = \sec \alpha_i + 8/3 \cdot n_i^2 \cos^3 \alpha_i - 32/5 \cdot n_i^4 \left(5\cos^7 \alpha_i - 4\cos^5 \alpha_i\right) \quad (4.45)$$

$$c_{\alpha i} = \frac{\partial S_i}{\partial \alpha_i} = l_i \left[\frac{\sin \alpha_i}{\cos^2 \alpha_i} - 8n_i^2 \sin \alpha_i \cos^2 \alpha_i + 32n_i^4 \cos^4 \alpha_i \sin \alpha_i \left(7\cos^2 \alpha_i - 4\right)\right] \quad (4.46)$$

Differentiating n_i with respect to f_i and l_i yields:

$$\delta n_i = \delta\left(f_i/l_i\right) = \left(l_i \cdot \delta f_i - f_i \cdot \delta l_i\right)/l_i^2 \quad (4.47)$$

For $h_i = l_i \cdot \tan\alpha_i$ $(i = 1, 2, \ldots, u)$, differentiating h_i with respect to l_i and α_i leads to $\delta h_i = \delta l_i \cdot \tan\alpha_i + l_i \cdot \sec^2\alpha_i \cdot \delta\alpha_i$. As δh_i is equal to the difference of the tower height changes, i.e., $\delta h_i = \delta h_{Pi} - \delta h_{P(i-1)}$, $\delta\alpha_i$ is obtained as:

$$\delta \alpha_i = -\sin 2\alpha_i/(2l_i) \cdot \delta l_i + \cos^2 \alpha_i/l_i \cdot \left(\delta h_{Pi} - \delta h_{P(i-1)}\right) \quad (4.48)$$

$$\delta h_{Pi} = h_{Pi}\theta_P \cdot \delta T_P \quad (4.49)$$

where θ_P is the linear expansion coefficient of the tower, and δh_{P0} and δh_{Pi} are 0 because they correspond to the unmovable anchorage blocks.

If the elastic elongation or shrinkage of the cables due to cable tension changes is considered, then the cable length change consists of two items, i.e., one is the stress-free thermal deformation (δS_i^{Free}), and the other is the elastic deformation caused by the cable tension changes (δS_i^{Elas}):

$$\delta S_i = \delta S_i^{\text{Free}} + \delta S_i^{\text{Elas}} \quad (4.50)$$

where

$$\delta S_i^{\text{Free}} = S_i \theta_C \cdot \delta T_C \quad (4.51)$$

$$\delta S_i^{\text{Elas}} = \frac{\delta H_i}{H_i} \cdot \int_0^{l_i} \frac{H_i}{E_C A_i}\left[1 + \left(\frac{dy}{dx}\right)^2\right] \cdot dx = \frac{\delta H_i}{H_0} \cdot \frac{H_0 l_i}{E_C A_i}\left(\sec^2 \alpha_i + \frac{16}{3} n_i^2\right) \quad (4.52)$$

where θ_C is the linear expansion coefficient of the main cable, and E_C and A_i are the elastic modulus and cross-sectional area of the main cable, respectively. Substituting

Equations (4.47), (4.48) and (4.50) into Equation (4.43), we have the following equations ($i = 1, 2, \ldots, u$):

$$P_i \cdot \delta f_i + M_i \cdot \delta l_i + E_i \cdot \delta H_i / H_0 = Q_i \tag{4.53}$$

where

$$P_i = c_{ni} / l_i \tag{4.54}$$

$$M_i = -c_{ni} n_i / l_i + c_{li} - c_{\alpha i} \sin 2\alpha_i / (2 l_i) \tag{4.55}$$

$$E_i = -H_0 l_i / (E_C A_i) \cdot \left(\sec^2 \alpha_i + 16/3 \cdot n_i^2 \right) \tag{4.56}$$

$$Q_i = S_i \theta_C \cdot \delta T_C - c_{\alpha i} \cos^2 \alpha_i / l_i \cdot \left(h_{Pi} - h_{P(i-1)} \right) \theta_P \cdot \delta T_P \tag{4.57}$$

The distance between the two anchorages is constant, which means:

$$\sum_{i=1}^{u} \delta l_i = 0 \tag{4.58}$$

As a result, Equations (4.40), (4.41), (4.53), and (4.58) constitute a set of linear equations of δf_i, δl_i, and $\delta H_i / H_0$ ($i = 1, 2, \ldots, u$):

$$\mathbf{G}_{3u \times 3u} \cdot \begin{bmatrix} \delta \mathbf{F}_{u \times 1} \\ \delta \mathbf{L}_{u \times 1} \\ \delta \mathbf{H}_{u \times 1} \end{bmatrix} = \begin{bmatrix} \mathbf{Q}_{u \times 1} \\ \mathbf{0}_{u \times 1} \\ \mathbf{0}_{u \times 1} \end{bmatrix} \tag{4.59}$$

where

$$\delta \mathbf{F}_{u \times 1} = \begin{bmatrix} \delta f_1 & \delta f_2 & \cdots & \delta f_u \end{bmatrix}^{\mathrm{T}} \tag{4.60}$$

$$\delta \mathbf{L}_{u \times 1} = \begin{bmatrix} \delta l_1 & \delta l_2 & \cdots & \delta l_u \end{bmatrix}^{\mathrm{T}} \tag{4.61}$$

$$\delta \mathbf{H}_{u \times 1} = \begin{bmatrix} \partial H_1 / H_0 & \partial H_2 / H_0 & \cdots & \partial H_u / H_0 \end{bmatrix}^{\mathrm{T}} \tag{4.62}$$

$$\mathbf{Q}_{u \times 1} = \begin{bmatrix} Q_1 & Q_2 & \cdots & Q_u \end{bmatrix}^{\mathrm{T}} \tag{4.63}$$

$$\mathbf{G}_{3u\times 3u}=\left[\begin{array}{c|c|c}
\begin{matrix} P_1 & & \mathbf{0} \\ & P_2 & \\ \mathbf{0} & \ddots & \\ & & P_u \end{matrix} & \begin{matrix} M_1 & & \mathbf{0} \\ & M_2 & \\ \mathbf{0} & \ddots & \\ & & M_u \end{matrix} & \begin{matrix} E_1 & & \mathbf{0} \\ & E_2 & \\ \mathbf{0} & \ddots & \\ & & E_u \end{matrix} \\ \hline
\mathbf{0} & \begin{matrix} \kappa_1 & & & \mathbf{0} & \\ \kappa_2 & \kappa_2 & & & \\ \vdots & \vdots & \ddots & & \\ \kappa_{u-1} & \kappa_{u-1} & \cdots & \kappa_{u-1} & \\ 1 & 1 & \cdots & 1 & 1 \end{matrix} & \begin{matrix} 1 & -1 & & \mathbf{0} & \\ & 1 & -1 & & \\ & & \ddots & \ddots & \\ & \mathbf{0} & & 1 & -1 \\ & & & & 0 \end{matrix} \\ \hline
\begin{matrix} f_1^{-1} & & \mathbf{0} \\ & f_2^{-1} & \\ \mathbf{0} & \ddots & \\ & & f_u^{-1} \end{matrix} & \begin{matrix} -l_1^{-1} & & \mathbf{0} \\ & -l_2^{-1} & \\ \mathbf{0} & \ddots & \\ & & -l_u^{-1} \end{matrix} & \mathbf{I}_u
\end{array}\right] \tag{4.64}$$

$$\kappa_i = k_{\mathrm{P}i}/H_0 \tag{4.65}$$

$\mathbf{0}_{u\times 1}$ is a $u\times 1$ zero vector, and $\mathbf{I}_u$ is a $u\times u$ identity matrix.

The sag change $\boldsymbol{\delta}\mathbf{F}$, span change $\boldsymbol{\delta}\mathbf{L}$, and cable tension (relative) change $\boldsymbol{\delta}\mathbf{H}$ can be solved from Equation (4.59). The detailed procedures are presented in Appendix C, and the solutions are shown in Equations (C37), (C40), and (C41). The equations and solutions will be simplified and presented in Section 4.5.2.

The mid-span elevation change of the ith span cable, δD_i, is the algebraic sum of the sag change and the averaged height change of the adjacent towers. Since the elevation is positive upward, we have

$$\delta D_i = -\delta f_i + \left(h_{P(i-1)} + h_{Pi}\right)/2\cdot\theta_P\cdot\delta T_P \tag{4.66}$$

Meanwhile, the tower-top horizontal movement of the jth tower is $\sum_{m=1}^{j}\delta l_m$.

4.5.2 A Case Study

The Maanshan Yangtze River Bridge in China is used to verify the developed formulas. Opened in December 2013, the bridge is a three-tower suspension bridge with cable spans of 360+1080+1080+360 m (Figure 4.37) and a sag-to-span ratio of 1/9 for both main span cables. The steel box girder is 38.5 m wide and 3.5 m high, accommodating dual three-lane highway traffic. Compared with the conventional two-tower suspension bridges, the additional middle tower reduces the structural rigidity significantly. A stiffened middle tower is required to limit the tower-top movement under the most adverse conditions when only one main span is loaded. However, an over-stiffened mid-tower is unfavorable as the large difference in cable tensions on two sides of the mid-tower may cause cable slip at the tower saddle. As a result, the mid-tower of the Maanshan Yangtze River Bridge adopts a steel-concrete composite tower with an intermediate stiffness. More details about the bridge can be found in the work by Ge and Xiang (2011), Qiang et al. (2012) and Zhang et al. (2013).

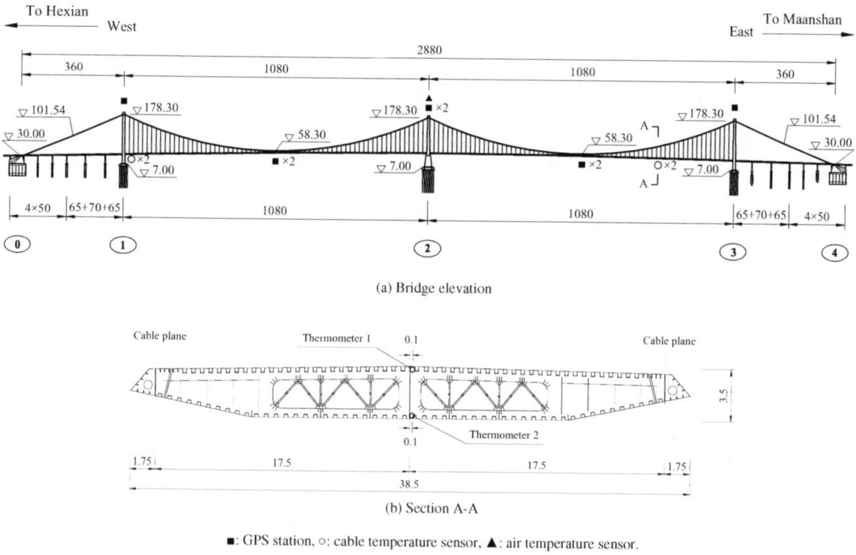

FIGURE 4.37 Geometry of the Maanshan Yangtze River Bridge and the sensor layout (unit: m): (a) bridge elevation and (b) Section A–A. ■: GPS station, ○: cable temperature sensor, ▲: air temperature sensor.

4.5.2.1 Field Measurement of the Bridge

An SHM system was installed on the bridge. The locations of GPS and temperature sensors relevant to this study are shown in Figure 4.37. The available temperature data includes the steel temperature of the top and bottom plates of the east and west main-span girders (sampled at a time interval of 5 minutes, or 1/300 Hz) and the air temperature inside the mid-tower (sampled at 1 Hz). No temperature sensors were installed on the main cables and towers. The elevations of both upstream and downstream sides at the middle of the second and third spans are measured by GPS rovers with a sampling frequency of 1 Hz. Two GPS rovers were installed on the top of the mid-tower (upstream and downstream sides), one on each side tower.

Figure 4.38 presents the hourly mean temperature and displacement of the bridge from January 2018 to December 2019. The averaging can filter out, to some extent, the high-frequency displacements in the raw data caused by dynamic loads, such as wind and traffic loads. Also, past field studies showed that the concerned deformations were insensitive to the static wind and traffic loads compared to the temperature (Xu et al., 2010; Fujino et al., 2019; Zhou et al., 2020). The hourly means, therefore, present the temperature-induced structural responses. The missing data in the figure is because of the malfunction of the sensors. The steel box girder temperature is relatively uniform along the bridge axis because the correlation coefficient between the west and east girder temperatures is greater than 0.997. The first subplot of Figure 4.38 presents the structural temperature of the top and bottom plates of the

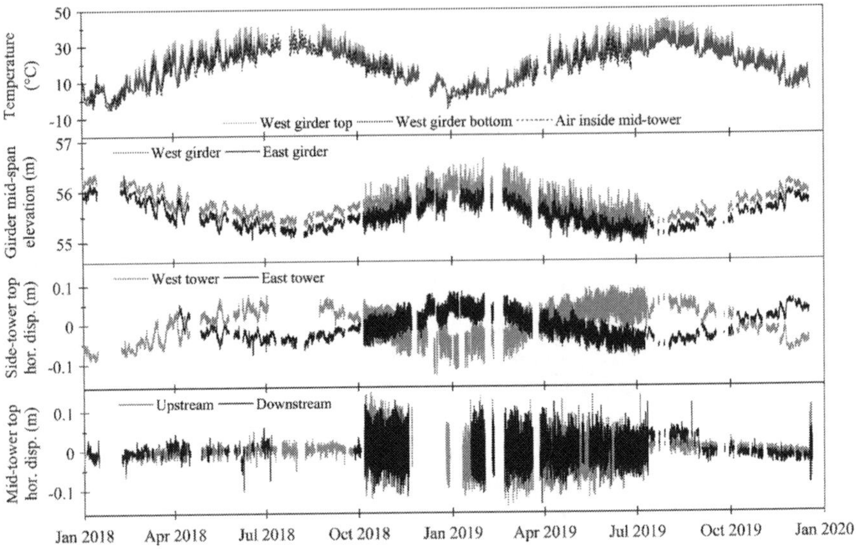

FIGURE 4.38 The measured temperature and displacements over 2 years.

west girder and the air temperature inside the mid-tower. The girder mid-span elevation in Figure 4.38 is the average of the upstream and downstream measurement data. A rising temperature caused a downward movement in the middle of the main-span girders. The tower-top horizontal displacement is positive eastward. The third subplot of Figure 4.38 illustrates that the west and east side towers moved toward the mid-tower when the temperature rose. As the bridge has a symmetrical arrangement about the mid-tower, the horizontal movement of the mid-tower top was slight most of the time, except for the period between October 5, 2018 and July 12, 2019, when the GPS data was very noisy. The measurement data during this period are discarded in the following analysis.

4.5.2.2 Verification of Formulas

As the side towers of the Maanshan Yangtze River Bridge are flexible, their flexural stiffness can be ignored, i.e., $k_{P1}=k_{P3}=0$. Meanwhile, the bridge has a symmetrical configuration, and the cable and tower temperature changes can be assumed to be symmetrical with respect to the mid-tower. Under the conditions of $l_i=l_{5-i}$, $P_i=P_{5-i}$, $M_i=M_{5-i}$, $E_i=E_{5-i}$, $f_i=f_{5-i}$ and $Q_i=Q_{5-i}$, ($i=1$ and 2), δf_i and δl_i ($i=1$, 2, 3, and 4) can be solved from Equations (C40) and (C41). In particular,

$$\delta f_2=-\frac{f_2\left(E_2l_1+l_1l_2M_2\right)}{\sigma_1}Q_1-\frac{f_2\left(l_1l_2M_1-E_1l_1+l_1P_1f_1+l_2P_1f_1\right)}{\sigma_1}Q_2 \tag{4.67}$$

$$\delta l_1=l_1l_2\left(E_2-P_2f_2\right)/\sigma_1.Q_1-l_1l_2\left(E_1-P_1f_1\right)/\sigma_1.Q_2 \tag{4.68}$$

where

$$\sigma_1 = l_1 l_2 \left(E_1 M_2 + E_2 M_1 - M_1 P_2 f_2 - M_2 P_1 f_1\right) + E_1 l_1 P_2 f_2 + E_2 l_2 P_1 f_1 - (l_1 + l_2) P_1 P_2 f_1 f_2 \tag{4.69}$$

Equations (4.67) and (4.68) show that the sag change of the main-span cable and the tower-top horizontal displacement are independent of the mid-tower stiffness k_{P2}. This is valid for symmetric four-span bridges only. The mid-tower's stiffness cannot be canceled out for unsymmetric bridges, which is not studied here. The elevation change in the middle of the main-span girder is

$$\delta D_2 = -\delta f_2 + \left(\delta h_{P1} + \delta h_{P2}\right)/2 \tag{4.70}$$

Figure 4.39 shows the correlation between the structural displacements and temperature, with the average temperature of the top and bottom plates of the west main girder being the abscissa. The data measured during 2:00–6:00 are used here.

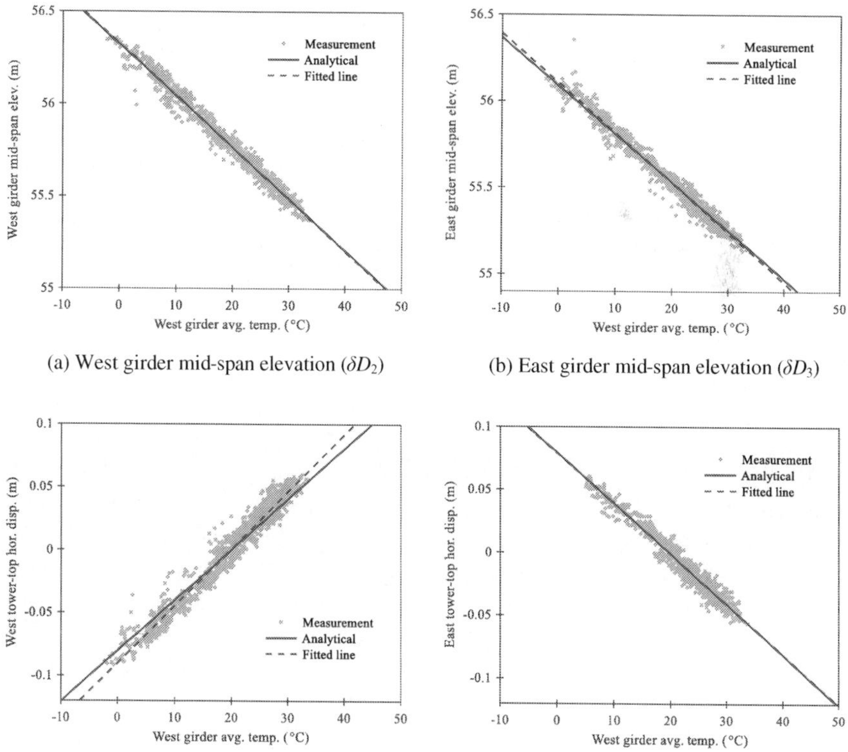

(a) West girder mid-span elevation (δD_2)

(b) East girder mid-span elevation (δD_3)

(c) West tower-top horizontal displacement (δl_1)

(d) East tower-top horizontal displacement ($-\delta l_4$)

FIGURE 4.39 Correlation between structural displacements and temperature: (a) west girder mid-span elevation (δD_2), (b) East girder mid-span elevation (δD_3), (c) West tower-top horizontal displacement (δl_1), and (d) East tower-top horizontal displacement ($-\delta l_4$).

This is because the structural temperature distribution tends to be uniform in the early morning. Thus, the girder temperature can represent the unknown cable and tower temperatures well. The slopes of the linear fitted lines are –2.81 cm/°C and –2.88 cm/°C for the mid-span elevation changes of the west and east main-span girders (δD_2 and δD_3), respectively; and 0.45 cm/°C and –0.40 cm/°C for the west and east tower-top horizontal displacements (δl_1 and $\delta l_1+\delta l_2+\delta l_3=-\delta l_4$), respectively.

The line denoted "Analytical" represents the calculated displacements over the temperature range based on the assumption of $\delta T_C=\delta T_P$. They pass the centroid of the data points, with the slope being determined by Equations (4.68) and (4.70). When the temperature increases by 1°C, the mid-span elevation of the two main-span girders decreases by –2.79 cm, and both side tower tops move 0.40 cm toward the mid-tower. Therefore, the analytical results derived from the formulas herein agree well with those of the field measurement data. The discrepancy between the fitted and calculated displacements is attributed to assumptions and uncertainties. For example, the cable and tower temperatures were not measured, and the girder temperature was used. In addition, the GPS data was inaccurate, mainly when the displacement was small.

The following parameters of the bridge are used for the calculation:

1. $l_1=l_4=360\,\text{m}$, $l_2=l_3=1080\,\text{m}$;
2. $f_1=f_4=2.61\,\text{m}$, $f_2=f_3=120\,\text{m}$;
3. $\alpha_1=22.39°$, $\alpha_2=\alpha_3=0$, $\alpha_4=-22.39°$;
4. $h_{P0}=h_{P4}=0$, $h_{P1}=h_{P2}=h_{P3}=171.30\,\text{m}$;
5. $A_1=A_2=A_3=A_4=0.2976\,\text{m}^2$;
6. $\theta_C=1.2\times10^{-5}°\text{C}^{-1}$, $\theta_P=1.0\times10^{-5}°\text{C}^{-1}$.

4.5.3 Simplified Formulation

Similar to the three-span case, this section simplifies the formulas derived in Section 4.5.1.

4.5.3.1 Ignorance of Tower Flexural Stiffness

The constructed long-span suspension bridges are mainly of three- or four-span types. Most conventional three-span suspension bridges employ a flexible tower design to balance the cable horizontal tensions at the tower tops, and the towers are supposed to have negligible flexural stiffness. The four-span suspension bridges, such as the Taizhou, Maanshan, and Yingwuzhou Yangtze River bridges, adopt flexible side towers and stiffer mid-towers with a symmetric configuration. The above equations (Equations 4.67 and 4.68) show that the temperature-induced sag and span changes of the main cable are independent of the mid-tower stiffness. Therefore, the flexural stiffness of towers can be neglected in calculating the thermal deformation of symmetrical four-span suspension bridges.

If the lateral flexural stiffness of all intermediate supports of a general multi-span suspension bridge is negligible, i.e., $\kappa_i=0$ ($i=1, 2,\ldots, u-1$), which means the horizontal components of the cable tension at the tower tops are always in equilibrium, then the solution to Equation (4.59) can be significantly simplified, as derived in Section C.1 in Appendix C. When the high-order terms of n in c_{ni}, c_{li}, and $c_{\alpha i}$ ($i=1$, 2,..., and u, see Equations (4.44)–(4.46)) are omitted, we have:

$$\delta f_i = n_i/\cos\alpha_i . Q_i + n_i(l_i - z_i)\Big/\sum_{k=1}^{u} z_k \cdot \sum_{j=1}^{u}(Q_j/\cos\alpha_j) \tag{4.71}$$

$$\delta l_i = Q_i/\cos\alpha_i - z_i\Big/\sum_{k=1}^{u} z_k \cdot \sum_{j=1}^{u}(Q_j/\cos\alpha_j) \tag{4.72}$$

$$\delta H_i/H_0 = -\left(1\Big/\sum_{k=1}^{u} z_k\right) . \sum_{j=1}^{u}(Q_j/\cos\alpha_j) \tag{4.73}$$

where

$$z_i = \left(16 l_i n_i^2 \cos^2\alpha_i\right)/3 - E_i/\cos\alpha_i \tag{4.74}$$

$$Q_i = \delta S_i^{\text{Free}} - \sin\alpha_i \cdot \left(\delta h_{Pi} - \delta h_{P(i-1)}\right) \tag{4.75}$$

Equations (4.71)–(4.73) are non-recurrent solutions. They apply to most practical long-span suspension bridges.

4.5.3.2 Simplification of Equations (4.71)–(4.73)

Equations (4.71)–(4.73) will be further simplified to reveal the influence of the cable tension changes on thermal deformation, as discussed in Section 4.5.3.3. It also facilitates the estimation of **δF**, **δL**, and **δH** for the constructed suspension bridges.

Consider the item $l_i - z_i$ in Equation (4.71) first. Suppose the outmost span cables (referred to as the anchor cables hereafter) hang the girder, such as the Yingwuzhou Yangtze River Bridge (Zhang et al., 2017). In that case, all cables are assumed to be subjected to the same vertical distributed load q. According to Equation (4.39), the sag-to-span ratio $n_i = f_i/l_i$ is proportional to the span l_i. Let l_{mx} denote the longest span length, and let mx represent the corresponding span index. The span ratio of the ith span cable is defined as:

$$\zeta_i = l_i/l_{mx} = n_i/n_{mx} \tag{4.76}$$

For a three-span suspension bridge, $n_2 \approx 1/10$ and ζ_i ($i=1$ and 3) is usually less than 0.5. If these conditions are extended to a multi-span suspension bridge, we have $n_{mx} \approx 1/10$ and $n_i = \zeta_i \cdot n_{mx} \leq 1/20$ ($i=1$ and u).

The parameter z_i in Equation (4.74) consists of two items. The ratio of the first item $\left(16 l_i n_i^2 \cos^2\alpha_i\right)/3$ to the span l_i can be estimated as ($i=1, 2,\ldots, u$)

$$16 l_i n_i^2 \cos^2\alpha_i/(3l_i) = 16/3 \cdot n_{mx}^2 \zeta_i^2 \cos^2\alpha_i \leq 16/3 \cdot (1/10)^2 = 5.3\% \tag{4.77}$$

By defining $\varepsilon_i = H_0/(E_c A_i)$, the second item of z_i can be expressed as:

$$-\frac{E_i}{\cos\alpha_i} = \frac{H_0 l_i\left(\sec^2\alpha_i + 16/3n_i^2\right)}{E_C A_i \cos\alpha_i} = \frac{\varepsilon_i l_i\left(\sec^2\alpha_i + 16/3n_i^2\right)}{\cos\alpha_i} \approx \varepsilon_i l_i \sec^3\alpha_i \quad (4.78)$$

n_i^2 is small and thus neglected in the above equation. Note that ε_i is associated with the strength of the cable steel wires divided by the safety factor. It is a relatively stable quantity (about 3×10^{-3}) for different bridges. For example, it is 2.72×10^{-3} for the Maanshan Yangtze River Bridge and 3.45×10^{-3} for the Akashi Kaikyo Bridge. Most suspension bridges have $|\alpha_i| < 30°$, so the second item of z_i is also negligible compared with l_i. Therefore, $l_i - z_i \approx l_i$. This approximation also applies to anchor cables without suspenders, e.g., the Maanshan Yangtze River Bridge, where the anchor cables can be approximated as straight lines without sags and $n_1 = n_u \approx 0$.

By substituting $l_i - z_i \approx l_i$, $\delta S_i^{\text{Free}} = l_i \sec\alpha_i \cdot \theta_C \cdot \delta T_C$, $\sec\alpha_i = \sqrt{l_i^2 + h_i^2}/l_i$, $\tan\alpha_i = h_i/l_i$ $(i = 1, 2,\ldots, u)$, and $\delta h_{Pj} = h_{Pj}\theta_P \cdot \delta T_P$ $(j = 1, 2,\ldots, u)$ into Equation (4.71), one has:

$$\delta f_i = n_i\left[\left(l_i^2 + h_i^2\right)/l_i \cdot \theta_C\delta T_C - h_i\left(h_{Pi} - h_{P(i-1)}\right)/l_i \cdot \theta_P\delta T_P\right]$$
$$+\left(l_i n_i \Big/ \sum_{k=1}^{u} z_k\right)\sum_{j=1}^{u}\left[\left(l_j^2 + h_j^2\right)/l_j \cdot \theta_C\delta T_C - h_j\left(h_{Pj} - h_{P(j-1)}\right)/l_j \cdot \theta_P\delta T_P\right] \quad (4.79)$$

If $h_i^2 = h_i\left(h_{Pi} - h_{P(i-1)}\right)$ and $\theta_C \cdot \delta T_C \approx \theta_P \cdot \delta T_P$ hold, Equation (4.79) becomes

$$\delta f_i = n_i l_i \theta_C \cdot \delta T_C + n_i l_i \Big/ \left(\sum_{k=1}^{u} z_k\right) \cdot \theta_C \delta T_C \cdot \sum_{j=1}^{u} l_j \quad (4.80)$$

The ratio of the first to the second items in Equation (4.80) is

$$\sum_{i=1}^{u} z_i \Big/ \sum_{i=1}^{u} l_i = 16\sum_{i=1}^{u}\left(l_i n_i^2 \cos^2\alpha_i\right) \Big/ \left(3\sum_{i=1}^{u} l_i\right) \quad (4.81)$$

As $n_i \leq 1/20$ ($i = 1$ and u) and $n_i \approx 1/10$ ($i = 2, 3,\ldots, u-1$), Equation (4.81) is less than 5%, which implies that the first item in Equation (4.80) is negligible compared with the second. Consequently, Equation (4.80) can be simplified as

$$\delta f_i = f_i \Big/ \left(\sum_{j=1}^{u} z_j\right) \cdot L\theta_C \cdot \delta T_C \quad (4.82)$$

Likewise, the tower-top horizontal displacement δl_i and the relative cable tension change $\delta H_i/H_0$ $(i = 1, 2,\ldots, u)$ can be simplified as:

$$\delta l_i = \left(l_i - z_i \Big/ \left(\sum_{j=1}^{u} z_j\right) \cdot L\right) \cdot \theta_C \cdot \delta T_C \quad (4.83)$$

$$\delta H_i/H_0 = -L\Big/\left(\sum_{j=1}^{u} z_j\right)\cdot\theta_C\cdot\delta T_C \tag{4.84}$$

It is interesting that Equations (4.82) and (4.83) have a unified form of $L_E\theta_C{\cdot}\delta T_C$, and L_E can be defined as the temperature equivalent length, which has the following form ($i=1, 2,\ldots, u$):

$$L_E = \begin{cases} \left(f_i\Big/\sum_{j=1}^{u} z_j\right)\cdot L & \text{for } \delta f_i \\ l_i - \left(z_i\Big/\sum_{j=1}^{u} z_j\right)\cdot L & \text{for } \delta l_i \end{cases} \tag{4.85}$$

For a u-span suspension bridge with identical main-span cables and short anchor cables, the following two conditions can be assumed: (1) The cable elastic deformation induced by tension changes is ignored, that is $E_i=0$ ($i=1, 2,\ldots, u$); (2) The anchor cable's sag is ignored ($n_1=n_u=0$) and z_i ($i=2, 3,\ldots, u-1$) is a constant $z_{c0}=\left(16l_cn_c^2\cos^2\alpha_c\right)/3$, and then Equation (4.85) can be further simplified as:

$$L_{E0} = \begin{cases} 0 & \text{for } \delta f_i \ (i=1 \text{ and } u) \\ 3L\big/\left[16n_c(u-2)\cos^2\alpha_c\right] & \text{for } \delta f_i \ (i=2,3,\cdots,u-1) \\ l_i & \text{for } \delta l_i \ (i=1 \text{ and } u) \\ l_i - L/(u-2) & \text{for } \delta l_i \ (i=2,3,\cdots,u-1) \end{cases} \tag{4.86}$$

The Maanshan Yangtze River Bridge is used as an example to investigate the accuracy of the above simplified equations. The results are compared in Table 4.5. Case 0 is based on the measurement results as reported in Section 4.5.2.2. Case 1 is obtained from the original solutions without simplifications and is thus regarded as the most accurate result. The ignorance of the cable's elastic deformation (Cases 4–6) causes about a 7% increase in the magnitude of the concerned quantities compared to Cases 1–3. On the other hand, the omission of the high-order terms of n in c_{ni}, c_{li}, and $c_{\alpha i}$ (see Equations (4.44)–(4.46)) tends to decrease the magnitude of δf_2 and δD_2. The combination of the two simplifications (Cases 6 and 7) presents a good estimate of the displacement.

4.5.3.3 Influence of Anchor Cable Sags and Cable's Elastic Deformation

Equation (4.86) is a very simple equation that can be easily applied to practical bridges. It ignores both the anchor cable sags (Effect 1) and the cable's elastic deformation induced by tension changes (Effect 2), which will be studied in this section.

Consider a u-span suspension bridge with the following configurations: (1) The tower flexural stiffness is ignored, i.e., $\kappa_i=0$ ($i=1, 2,\ldots, u-1$); (2) All main-span cables are

TABLE 4.5
Measured and Analytical Displacement per Unit Temperature Change (Unit: cm/°C)

Case	δD_2	δf_2	δl_1	Equations	Remarks	
0	−2.84	–	0.43	–	Measurement results. δD_2 and δl_1 are the averages of the west and east main-span girders and towers, respectively.	
1	−2.79	2.96	0.40	(4.67), (4.68)	Considering the cable elastic deformation, $\varepsilon_i = 2.72 \times 10^{-3}$	–
2	−2.60	2.77	0.40	(4.71), (4.72)		Omitting the high-order terms of n in calculating c_{ni}, $c_{li,}$ and $c_{\alpha i}$
3	−2.57	2.74	0.40	(4.85)		Simplified formulas
4	−2.98	3.15	0.43	(4.67), (4.68)	Ignoring the cable elastic deformation, $\varepsilon_i = 0$	–
5	−2.77	2.94	0.43	(4.71), (4.72)		Omitting the high-order terms of n in calculating c_{ni}, c_{li}, and $c_{\alpha i}$
6	−2.74	2.91	0.43	(4.85)		Simplified formulas
7	−2.74	2.92	0.43	(4.86)		The simplest formula ($n_1 = n_u = 0$)

identical with the parameters $l_i = l_c$, $n_i = n_c$ and $z_i = z_c = \left(16 l_c n_c^2 \cos^2 \alpha_c\right)/3 + \varepsilon_c l_c \sec^3 \alpha_c$ ($i = 2, 3, \ldots, u-1$); (3) The anchor cables are identical except the opposite inclination angles ($\alpha_1 = -\alpha_u$), and they have the sag-to-span ratio $n_1 = n_u = \varphi \cdot \zeta_1 n_c$, where φ is a two-value parameter indicating whether the anchor cable sags are considered. When the anchor cables hang the girder, n_1 and n_u can be reasonably assumed to be proportional to the span ratio ζ_1 (Equation 4.76), and thus $\varphi = 1$. When the anchor cables are straight ($n_1 = n_u = 0$), $\varphi = 0$.

Effects 1 and 2 can be quantified by a modification coefficient, which is defined as the ratio of the temperature equivalent lengths in Equations (4.85) and (4.86), i.e., L_E/L_{E0}. For δf_i ($i = 2, 3, \ldots, u-1$), one has

$$\beta_f = f_c L \Big/ \left(\sum_{j=1}^{u} z_j\right) \times z_{c0}(u-2)/(f_c L) = (u-2) z_{c0} \Big/ \left(2 z_1 + (u-2) z_c\right) \quad (4.87)$$

where $z_{c0} = \left(16 l_c n_c^2 \cos^2 \alpha_c\right)/3$, $z_1 = \left(16 l_1 n_1^2 \cos^2 \alpha_1\right)/3 + \varepsilon_1 l_1 \sec^3 \alpha_1$, and $u \geq 3$. Assuming that the suspension bridge has the same tower-top elevation, i.e., $\alpha_i = 0$ ($i = 2, 3, \ldots, u-1$), and the elevation difference $|h_i|$ ($i = 1$ and u) of the anchor cable is equal to the sag of the main-span cable f_c. Therefore, $\cos^2 \alpha_i$ ($i = 1$ and u) can be approximated as

$$\cos^2 \alpha_1 = \cos^2 \alpha_u = \frac{l_1^2}{l_1^2 + h_1^2} = \frac{l_1^2}{l_1^2 + f_c^2} = \frac{\zeta_1^2 \cdot l_c^2}{\zeta_1^2 \cdot l_c^2 + n_c^2 \cdot l_c^2} = \frac{\zeta_1^2}{\zeta_1^2 + n_c^2} \quad (4.88)$$

By substituting Equation (4.88), $l_1 = \zeta_1 l_c$, $n_1 = \varphi \cdot \zeta_1 n_c$, $\alpha_c = 0$, and $\varepsilon_1 = \varepsilon_c$ into Equation (4.87), one has

$$\beta_f = \frac{16(u-2)n_c^2\zeta_1^2\left(\zeta_1^2+n_c^2\right)}{(u-2)\left(16n_c^2+3\varepsilon_c\right)\zeta_1^2\left(\zeta_1^2+n_c^2\right)+6\varepsilon_c\left(\zeta_1^2+n_c^2\right)^{5/2}+32\varphi^2\zeta_1^7 n_c^2} \tag{4.89}$$

Similarly, the modification coefficient for δl_i ($i = 1, u$) is

$$\beta_l = 1 - \frac{\left[3\varepsilon_c\left(\zeta_1^2+n_c^2\right)^{5/2}+16\varphi^2 n_c^2\zeta_1^7\right]\left(u-2+2\zeta_1\right)}{\zeta_1^3\left(\zeta_1^2+n_c^2\right)\left(16n_c^2+3\varepsilon_c\right)(u-2)+6\varepsilon_c\zeta_1\left(\zeta_1^2+n_c^2\right)^{5/2}+32\varphi^2\zeta_1^8 n_c^2} \tag{4.90}$$

As $2\times\delta l_1 + (u-2)\times\delta l_c = 0$, the modification coefficient for δl_i ($i = 2, 3, \ldots, u-1$) is equal to β_1. When the anchor cable sags and the cable's elastic deformation δS_i^{Elas} are neglected, i.e., $\varphi = 0$ and $\varepsilon_c = 0$, Equations (4.89) and (4.90) produce $\beta_f = 1$ and $\beta_l = 1$.

As mentioned before, ε_c is relatively stable for different bridges. $\varepsilon_c = 3\times10^{-3}$ (considering Effect 2) or $\varepsilon_c = 0$ (neglecting Effect 2) are taken in the following parametric analysis. Also, the span ratio ζ_1 is constrained between 0.15 and 0.50, which applies to almost all suspension bridges. Figure 4.40 presents the relations of β_f and β_l versus ζ_1 in Equations (4.89) and (4.90) with different sag-to-span ratios n_c (for the main-span cables) and span number u. It gives an overall picture of the influence of Effect 1 (the anchor cable sags) and Effect 2 (the temperature-induced cable elastic deformation δS_i^{Elas}) for three- and four-span suspension bridges.

The figure has the following findings: (1) Both β_f and β_l are less than 1.0, indicating that both Effects 1 and 2 reduce the magnitude of the main-span cable sag change and tower-top horizontal movement. (2) For a given span number u, β_f is generally larger than β_l, which means δl_i ($i = 1, 2, \ldots, u$) is more sensitive to Effects 1 and 2 than δf_i ($i = 2, 3, \ldots, u-1$). (3) Effect 1 becomes more significant with the increase of anchor span (ζ_1), while Effect 2 is almost constant for $0.15 \le \zeta_1 \le 0.50$. For bridges with long side spans (e.g., the Akashi Kaikyo Bridge with $\zeta_1 = 0.48$), Effect 1 is more significant than Effect 2. (4) By comparing the case of $u = 3$ with $u = 4$, Effects 1 and 2 influence the three-span bridges more than four-span bridges, as the modification coefficients become closer to 1.0 with a larger u. (5) Compared with Effect 1, Effect 2 is more sensitive to the sag-to-span ratio of the main-span cable (n_c), as the lines for Effect 1 only are almost coincident in Figure 4.40 while the lines for Effect 2 are not. If n_c takes the other values, such as $n_c = 1/12$, this observation will be more obvious.

4.5.4 Discussions

The proposed methodology offers valuable insights into the operational behavior of suspension bridges. According to Equation (4.57), Q_i is linearly dependent on δT_C and δT_P, indicating that a suspension bridge's thermal deformation is the linear superposition of temperature variations in the main cable and towers. On the other hand, the temperature difference between the opposite tower surfaces does not influence the span-wise displacement of the tower top ($\delta l_i = 0$) and the sag change ($\delta f_i = 0$),

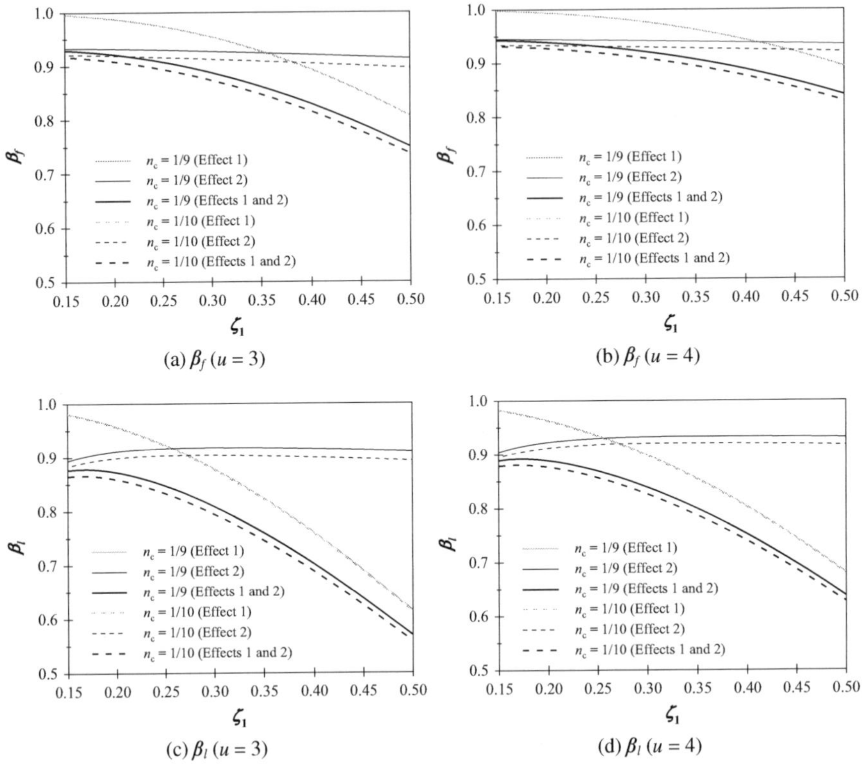

FIGURE 4.40 Variation of β_f and β_l with respect to ζ_1: (a) β_f (u=3), (b) β_f (u=4), (c) β_l (u=3), and (d) β_l (u=4).

because it does not change the tower height, i.e., $\delta h_{Pi}=0$. Also, Equation (4.86) shows that the temperature-induced sag change in the main-span cable is approximately much larger than the tower-top horizontal displacement by one order. This knowledge facilitates the selection of measurement ranges and precision accuracy in a field monitoring system.

The methodology can be easily extended to self-anchored suspension bridges, where the anchorage points of the main cable move due to the changes in the girder temperature. If we neglect the effects of differential movements between the girder and cables and assume the girder is continuous at the girder–tower intersections with the total length of L_G, then the distance change between the anchorage points of the main cable can be estimated by $\delta L_G = L_G\theta_G \cdot \delta T_G$, where θ_G and δT_G are respectively the linear expansion coefficient and temperature change of the girder. Assuming the left end of the girder moves δy, the horizontal equilibrium at tower tops leads to a modified Equation (4.41):

$$\delta H_{j+1} - \delta H_j = k_{Pj}\left(\delta y + \sum_{m=1}^{j} \delta l_m\right) \tag{4.91}$$

Meanwhile, the compatibility condition of all span cables becomes $\sum_{i=1}^{u} \delta l_i = \delta L_G$. As a result, the right-hand side column vector of Equation (4.59) becomes $\begin{bmatrix} \mathbf{Q}_{u\times1}^{\mathrm{T}} & \mathbf{Y}_{u\times1}^{\mathrm{T}} & \mathbf{0}_{1\times u} \end{bmatrix}^{\mathrm{T}}$ with $\mathbf{Y}_{u\times1}$ being $[-\kappa_1\cdot\delta y, -\kappa_2\cdot\delta y, \ldots, -\kappa_{u-1}\cdot\delta y, \delta L_G]$. The general solution to the modified Equation (4.59) is presented in Appendix D. If the tower flexural stiffness is neglected, i.e., $\kappa_i = 0$ ($i = 1, 2, \ldots, u-1$), then the expressions are independent of δy with neat forms:

$$\delta f_i = n_i/\cos\alpha_i \cdot Q_i + n_i(l_i - z_i)\Big/\sum_{k=1}^{u} z_k \cdot \left(-\delta L_G + \sum_{j=1}^{u} \left(Q_j/\cos\alpha_j\right)\right) \quad (4.92)$$

$$\delta l_i = Q_i/\cos\alpha_i - z_i\Big/\sum_{k=1}^{u} z_k \cdot \left(-\delta L_G + \sum_{j=1}^{u} \left(Q_j/\cos\alpha_j\right)\right) \quad (4.93)$$

$$\delta H_i/H_i = -1\Big/\left(\sum_{k=1}^{u} z_k\right)\cdot\left(-\delta L_G + \sum_{j=1}^{u} \left(Q_j/\cos\alpha_j\right)\right) \quad (4.94)$$

where z_i and Q_i refer to Equations (4.74) and (4.75).

Some conventional two-tower suspension bridges have intermediate supports to the main cable, which are usually located at the intersection point of the side-span cable and the girder, such as the Golden Gate Bridge (Figure 4.41). The additional intermediate support divides the side-span cable into two smaller segments: One is below the girder without suspenders with the parameters of l_1, $f_1 = 0$, and α_1, and the other is above the girder with l_2, f_2, and α_2. Under the condition of $\alpha_1 \approx \alpha_2$, Equations (4.71) and (4.72) provide the far-reaching implications that the two smaller cable segments can be combined into a virtual cable with a span of $l_1 + l_2$ and a sag of

FIGURE 4.41 The Golden Gate Bridge in the U.S. with intermediate cable supports in the side span (David Ball, https://commons.wikimedia.org/wiki/File:GG-ftpoint-bridge-2.jpg).

$f_2\sqrt{(l_1+l_2)/l_2}$, which produces almost the same results as the original two-span cable in the calculation of sag and span changes of the main-span cable.

Moreover, the formulas enable us to compare the contribution of different temperature variables to bridge thermal deformation. For the Maanshan Yangtze River Bridge, the rising temperature of the main cable will lower the elevation of the main-span girder and cause the side tower tops to move toward the mid-tower. In contrast, the increasing tower temperature will exert opposite effects on the deformation. Assuming $\delta T_C = \delta T_P$, the amplitude ratio of the cable and tower temperature effects is 10.7:1.0 for the mid-span elevation of the main-span girder and 6.8:1.0 for the side tower top movement. Therefore, the cable temperature is the dominant factor affecting the global deformation of this bridge. For this reason, measuring the main cable temperature is essential in SHM systems.

As mentioned in Section 4.5.3.1, the ignorance of tower flexural stiffness is feasible for existing suspension bridges. However, unsymmetric multi-span suspension bridges are emerging, such as the Chacao Suspension Bridge in Chile, which has two unequal main spans of 1055 m and 1155 m (Peña et al., 2017). For such cases, the tower stiffness comes into play and needs to be considered in the thermal deformation analysis. This effect is not investigated in detail herein due to the unavailability of design parameters. Still, the present study provides an analytical model and formulas for further research on tower stiffness effects. Two extreme cases (tower stiffness is negligible or an infinity) will be presented in Sections C.1 and C.2 in Appendix C.

Last but not least, the coefficient matrix **G** in Equation (4.64) is the inherent characteristic of the thermal deformation of a suspension bridge, which is independent of the temperature changes in structural components. They quantify the influence of various structural parameters on the temperature-deformation relationship (e.g., Equations (4.67) and (4.68)); therefore, they provide a theoretical basis for SHM and evaluation of suspension bridges.

4.5.5 Summary

This section derives the formulas for calculating the temperature-induced global deformation of multi-span suspension bridges. The governing equations are formulated based on three conditions: (1) the horizontal equilibrium at tower tops, (2) the cable's geometric relationship, and (3) the compatibility of all cable spans. The main conclusions can be drawn as follows:

1. Considering the flexural stiffness of towers and the elastic deformation of cables, the derived formulas apply to general suspension bridges with any number of spans. The tower stiffness affects the interaction between individual span cables. For symmetrical four-span suspension bridges, the mid-tower stiffness does not affect the temperature-induced variations in the sag and span of the main cables.
2. Several simplified formulas have been proposed for easy use in practice. By introducing the temperature equivalent length L_E, the tower-top movement and cable sag changes can be unified into a formula $L_E\theta_C{\cdot}\delta T_C$, with L_E component-dependent.

3. The effects of the anchor cable sag and the elastic deformation of cables are quantified by the modification coefficients β_f and β_l. The cable's elastic deformation induced by the cable tension variations tends to reduce the temperature-induced span and sag changes of a suspension bridge by 5%–10%. The anchor cable sag also reduces the magnitude of thermal deformations, and its effect increases rapidly with the span ratio. For a three-span suspension bridge with long side spans, the side-span cable sag affects thermal deformation more significantly than the cable's elastic deformation does.

This analytical model offers engineers a simple and general approach to understanding and calculating the temperature-induced displacement of suspension bridges. The analysis process also applies to transmission lines and other ropeways with certain modifications, such as catenary cables. These will be detailed in Appendix C.

4.6 CHAPTER SUMMARY

This chapter comprehensively studies the temperature behaviors of suspension bridges using field monitoring, numerical simulation, and theoretical analysis. The theoretical formulas, covering the three- and multi-span cases, successfully reveal the underlying mechanisms of temperature-induced bridge deformations. These studies can assist engineers in understanding and quantifying the temperature behaviors of suspension bridges.

In general, the cable temperature (both main- and side-span) contributes most to the deformation of a suspension bridge. As the temperature increases, the mid-span cable and girder deflect downward and the two-tower tops move toward the middle span.

REFERENCES

Abaqus, G., 2011, *Abaqus 6.11*. Providence, RI, Dassault Systemes Simulia Corporation.

ANSYS Inc., 2009, *ANSYS 12.0*. ANSYS Inc., Canonsburg, PA.

Bernstein, D. S., 2009, *Matrix Mathematics: Theory, Facts, and Formulas*. Princeton, NJ, Princeton University Press.

Datta, B. N., 2010, *Numerical Linear Algebra and Applications*, Philadelphia, PA, SIAM.

El-Mikkawy, M. E., 2004, On the inverse of a general tridiagonal matrix. *Applied Mathematics and Computation* **150(3)**: 669–679.

Froli, M., Hariga, N., Nati, G. and Orlandini, M., 1996, Longitudinal thermal behaviour of a concrete box girder bridge. *Structural Engineering International* **6(4)**: 237–242.

Fujino, Y., Siringoringo, D. M., Ikeda, Y., Nagayama, T. and Mizutani, T., 2019, Research and implementations of structural monitoring for bridges and buildings in Japan. *Engineering* **5(6)**: 1093–1119.

Ge, Y. and Xiang, H., 2011, Extension of bridging capacity of cable-supported bridges using double main spans or twin parallel decks solutions. *Structure and Infrastructure Engineering* **7(7–8)**: 551–567.

Gimsing, N. J. and Georgakis, C. T., 2012, *Cable Supported Bridges: Concept and Design*. New York, John Wiley & Sons.

Kitagawa, M., 2004, Technology of the Akashi Kaikyo Bridge. *Structural Control and Health Monitoring* **11(2)**: 75–90.

Mallik, R. K., 2001, The inverse of a tridiagonal matrix. *Linear Algebra and Its Applications* **325(1–3)**: 109–139.
Ministry of Transport of the People's Republic of China, 2015, *Specifications for Design of Highway Suspension Bridge*. JTG/T D65-05-2015. Beijing, China, Communications Press.
Ochsendorf, J. A. and Billington, D. P., 1999, Self-anchored suspension bridges. *Journal of Bridge Engineering* **4(3)**: 151–156.
Ogihara, K., 2015, Design and construction of suspension bridges. In Alampalli, S. and Moreau, W. J. (eds.), *Inspection, Evaluation and Maintenance of Suspension Bridges*. Boca Raton, FL, CRC Press, pp. 70–87.
Peña, Á., Valenzuela, M., Márquez, M. and Pinto, H., 2017, Minimum geotechnical requirements for traditional and singular bridges foundations design: Chacao Suspension Bridge. *Revista de la Construcción. Journal of Construction* **16(3)**: 498–505.
Qiang, Z., He-qiang, T. and Guang-wu, Y., 2012, Selection of a structural system for a three-tower suspension bridge of Maanshan Yangtze River Highway Bridge. *Structural Engineering International* **22(1)**: 139–143.
Timoshenko, S. and Young, D. H., 1965, *Theory of Structures*. New York, McGraw-Hill.
Xu, Y. L., Chen, B., Ng, C. L., Wong, K. Y. and Chan, W. Y., 2010, Monitoring temperature effect on a long suspension bridge. *Structural Control and Health Monitoring* **17(6)**: 632–653.
Xu, Y. L. and Xia, Y., 2011, *Structural Health Monitoring of Long-Span Suspension Bridges*. London, CRC Press.
Zhang, Q., Cheng, Z., Cui, C., Bao, Y., He, J. and Li, Q., 2017, Analytical model for frictional resistance between cable and saddle of suspension bridges equipped with vertical friction plates. *Journal of Bridge Engineering* **22(1)**: 04016103.
Zhang, W.-M., Ge, Y.-J. and Levitan, M. L., 2013, Nonlinear aerostatic stability analysis of new suspension bridges with multiple main spans. *Journal of the Brazilian Society of Mechanical Sciences and Engineering* **35**: 143–151.
Zhou, Y., Xia, Y., Chen, B. and Fujino, Y., 2020, Analytical solution to temperature-induced deformation of suspension bridges. *Mechanical Systems and Signal Processing* **139**: 106568.
Zhou, Y., Xia, Y., Fujino, Y. and Yamaguchi, K., 2021, Analytical formulas of thermal deformation of suspension bridges. *Engineering Structures* **238**: 112228.
Zhou, Y., Xia, Y., Sun, Z. and Fujino, Y., 2022, Analytical formulation of the temperature-induced deformation of multispan suspension bridges. *Structural Control and Health Monitoring* **29(6)**: e2937.

APPENDICES

APPENDIX A

The assumptions presented in Section 4.4.1 are practically reasonable.

When the main cables and suspenders of a suspension bridge are substantially lighter than its suspended girder, the main cables are primarily subjected to the vertical loads uniformly distributed along the span. Thus, the cable curve approximates a parabola. Moreover, the cable elastic deformation caused by the changes in tension force can be ignored. From the formulas of δf_i and δl_i (Equations (4.16) and (4.17)), we obtain

$$\frac{\delta H_i}{H_i} = \frac{\delta l_i}{l_i} - \frac{\delta f_i}{f_i} = -\frac{1}{\sum_{j=1}^{3} z_j} \sum_{j=1}^{3} \left[\frac{\delta S_j}{\cos \alpha_j} - \tan \alpha_j \left(\delta h_{Pj} - \delta h_{P(j-1)} \right) \right] \tag{A1}$$

For the Akashi Kaikyo Bridge, $\delta H_i/H_i = -3.63 \times 10^{-4}$ when $\delta T_C = \delta T_P = 1°\text{C}$. Because the elastic elongation of the main-span cables of this bridge is $S_{e2} = 7.04\,\text{m}$, the cable deformation induced by tension change caused by a unit increase in temperature can be estimated as follows:

$$\delta S_{e2} = S_{e2} \times \delta H_2 / H_2 = -2.57 \text{ mm} \tag{A2}$$

δS_{e2} is nearly one order of magnitude smaller than the unconstrained (or stress-free) thermal elongation of the main-span cables ($\delta S_2 = 24.48\,\text{mm}$):

Second, most long-span suspension bridges employ a flexible tower design to balance the cable horizontal tensions on both sides of a tower. Therefore, towers are not supposed to resist the longitudinal displacement of the main cables (Ogihara, 2015).

Third, main cables dominate the structural global stiffness in the vertical plane, and the contribution of stiffening girders and suspenders as secondary structures is low. Moreover, for an earth-anchored suspension bridge, both the average girder temperature and differential temperature between the top and bottom surfaces of the girder exhibit negligible effects on the main cable's deformation.

Finally, the main cables are firmly anchored at anchorage blocks on both ends. Relative sliding between the main cables and tower-top saddles is not allowed in the operational stage of suspension bridges.

APPENDIX B

Equation (4.15) can be solved as follows.

The coefficient matrix, denoted as **G**, can be partitioned as a 2-by-2 block matrix, each block being a 3-by-3 submatrix:

$$\mathbf{G} = \left[\begin{array}{ccc|ccc} c_{n1} l_1^{-1} & 0 & 0 & M_1 & 0 & 0 \\ 0 & c_{n2} l_2^{-1} & 0 & 0 & M_2 & 0 \\ 0 & 0 & c_{n3} l_3^{-1} & 0 & 0 & M_3 \\ \hline -f_1^{-1} & f_2^{-1} & 0 & l_1^{-1} & -l_2^{-1} & 0 \\ 0 & -f_2^{-1} & f_3^{-1} & 0 & l_2^{-1} & -l_3^{-1} \\ 0 & 0 & 0 & 1 & 1 & 1 \end{array} \right] = \begin{bmatrix} \mathbf{X}_A & \mathbf{X}_B \\ \mathbf{X}_C & \mathbf{X}_D \end{bmatrix} \tag{B1}$$

The inverse of the above block matrix can be expressed as (Bernstein, 2009)

$$\mathbf{G}^{-1}=\begin{bmatrix} \mathbf{G}_A & \mathbf{G}_B \\ \mathbf{G}_C & \mathbf{G}_D \end{bmatrix}$$

$$=\begin{bmatrix} \mathbf{X}_A^{-1}+\mathbf{X}_A^{-1}\mathbf{X}_B\left(\mathbf{X}_D-\mathbf{X}_C\mathbf{X}_A^{-1}\mathbf{X}_B\right)^{-1}\mathbf{X}_C\mathbf{X}_A^{-1} & -\mathbf{X}_A^{-1}\mathbf{X}_B\left(\mathbf{X}_D-\mathbf{X}_C\mathbf{X}_A^{-1}\mathbf{X}_B\right)^{-1} \\ -\left(\mathbf{X}_D-\mathbf{X}_C\mathbf{X}_A^{-1}\mathbf{X}_B\right)^{-1}\mathbf{X}_C\mathbf{X}_A^{-1} & \left(\mathbf{X}_D-\mathbf{X}_C\mathbf{X}_A^{-1}\mathbf{X}_B\right)^{-1} \end{bmatrix} \tag{B2}$$

As $\mathbf{X}_A$ and $\mathbf{X}_B$ are diagonal matrices, $\mathbf{X}_A^{-1}$, $\mathbf{X}_A^{-1}\mathbf{X}_B$, and $\mathbf{X}_C\mathbf{X}_A^{-1}$ are obtained as:

$$\mathbf{X}_A^{-1}=\mathrm{diag}\left(\left[c_{n1}^{-1}l_1,\ c_{n2}^{-1}l_2,\ c_{n3}^{-1}l_3\right]\right) \tag{B3}$$

$$\mathbf{X}_A^{-1}\mathbf{X}_B=\mathrm{diag}\left(\left[c_{n1}^{-1}l_1M_1,\ c_{n2}^{-1}l_2M_2,\ c_{n3}^{-1}l_3M_3\right]\right) \tag{B4}$$

$$\mathbf{X}_C\mathbf{X}_A^{-1}=\begin{bmatrix} -c_{n1}^{-1}n_1^{-1} & c_{n2}^{-1}n_2^{-1} & 0 \\ 0 & -c_{n2}^{-1}n_2^{-1} & c_{n3}^{-1}n_3^{-1} \\ 0 & 0 & 0 \end{bmatrix} \tag{B5}$$

By introducing a variable $\gamma_i=l_i^{-1}+c_{ni}^{-1}n_i^{-1}M_i$ $(i=1,\ 2,\ 3)$, the matrix $\mathbf{G}_D^{-1}=\mathbf{X}_D-\mathbf{X}_C\mathbf{X}_A^{-1}\mathbf{X}_B$ becomes:

$$\mathbf{G}_D^{-1}=\begin{bmatrix} \gamma_1 & -\gamma_2 & 0 \\ 0 & \gamma_2 & -\gamma_3 \\ 1 & 1 & 1 \end{bmatrix} \tag{B6}$$

According to the general formula of 3-by-3 matrix inversion (Bernstein, 2009), $\mathbf{G}_D$ can be expressed as:

$$\mathbf{G}_D=\begin{bmatrix} \gamma_1^{-1}\left[1-\left(\sum_{k=1}^{3}\gamma_k^{-1}\right)^{-1}\sum_{k=1}^{1}\gamma_k^{-1}\right] & \gamma_1^{-1}\left[1-\left(\sum_{k=1}^{3}\gamma_k^{-1}\right)^{-1}\sum_{k=1}^{2}\gamma_k^{-1}\right] & \gamma_1^{-1}\left(\sum_{k=1}^{3}\gamma_k^{-1}\right)^{-1} \\ -\gamma_2^{-1}\left(\sum_{k=1}^{3}\gamma_k^{-1}\right)^{-1}\sum_{k=1}^{1}\gamma_k^{-1} & \gamma_2^{-1}\left[1-\left(\sum_{k=1}^{3}\gamma_k^{-1}\right)^{-1}\sum_{k=1}^{2}\gamma_k^{-1}\right] & \gamma_2^{-1}\left(\sum_{k=1}^{3}\gamma_k^{-1}\right)^{-1} \\ -\gamma_3^{-1}\left(\sum_{k=1}^{3}\gamma_k^{-1}\right)^{-1}\sum_{k=1}^{1}\gamma_k^{-1} & -\gamma_3^{-1}\left(\sum_{k=1}^{3}\gamma_k^{-1}\right)^{-1}\sum_{k=1}^{2}\gamma_k^{-1} & \gamma_3^{-1}\left(\sum_{k=1}^{3}\gamma_k^{-1}\right)^{-1} \end{bmatrix} \tag{B7}$$

By substituting Equations (B3), (B4), (B5), and (B7) into Equation (B2), one has the explicit expressions of $\mathbf{G}_A$, $\mathbf{G}_B$, and $\mathbf{G}_C$:

$$\mathbf{G}_A = \begin{bmatrix} c_{n1}^{-1}\gamma_1^{-1}\left(1+\dfrac{l_1-\gamma_1^{-1}}{\sum_{k=1}^{3}\gamma_k^{-1}}\right) & c_{n2}^{-1}n_2^{-1}\gamma_2^{-1}\dfrac{f_1-n_1\gamma_1^{-1}}{\sum_{k=1}^{3}\gamma_k^{-1}} & c_{n3}^{-1}n_3^{-1}\gamma_3^{-1}\dfrac{f_1-n_1\gamma_1^{-1}}{\sum_{k=1}^{3}\gamma_k^{-1}} \\ c_{n1}^{-1}n_1^{-1}\gamma_1^{-1}\dfrac{f_2-n_2\gamma_2^{-1}}{\sum_{k=1}^{3}\gamma_k^{-1}} & c_{n2}^{-1}\gamma_2^{-1}\left(1+\dfrac{l_2-\gamma_2^{-1}}{\sum_{k=1}^{3}\gamma_k^{-1}}\right) & c_{n3}^{-1}n_3^{-1}\gamma_3^{-1}\dfrac{f_2-n_2\gamma_2^{-1}}{\sum_{k=1}^{3}\gamma_k^{-1}} \\ c_{n1}^{-1}n_1^{-1}\gamma_1^{-1}\dfrac{f_3-n_3\gamma_3^{-1}}{\sum_{k=1}^{3}\gamma_k^{-1}} & c_{n2}^{-1}n_2^{-1}\gamma_2^{-1}\dfrac{f_3-n_3\gamma_3^{-1}}{\sum_{k=1}^{3}\gamma_k^{-1}} & c_{n3}^{-1}\gamma_3^{-1}\left(1+\dfrac{l_3-\gamma_3^{-1}}{\sum_{k=1}^{3}\gamma_k^{-1}}\right) \end{bmatrix} \tag{B8}$$

$$\mathbf{G}_B = \begin{bmatrix} -(f_1-n_1\gamma_1^{-1})\left(1-\dfrac{\sum_{k=1}^{1}\gamma_k^{-1}}{\sum_{k=1}^{3}\gamma_k^{-1}}\right) & -(f_1-n_1\gamma_1^{-1})\left(1-\dfrac{\sum_{k=1}^{2}\gamma_k^{-1}}{\sum_{k=1}^{3}\gamma_k^{-1}}\right) & -\dfrac{f_1-n_1\gamma_1^{-1}}{\sum_{k=1}^{3}\gamma_k^{-1}} \\ (f_2-n_2\gamma_2^{-1})\dfrac{\sum_{k=1}^{1}\gamma_k^{-1}}{\sum_{k=1}^{3}\gamma_k^{-1}} & -(f_2-n_2\gamma_2^{-1})\left(1-\dfrac{\sum_{k=1}^{2}\gamma_k^{-1}}{\sum_{k=1}^{3}\gamma_k^{-1}}\right) & -\dfrac{f_2-n_2\gamma_2^{-1}}{\sum_{k=1}^{3}\gamma_k^{-1}} \\ (f_3-n_3\gamma_3^{-1})\dfrac{\sum_{k=1}^{1}\gamma_k^{-1}}{\sum_{k=1}^{3}\gamma_k^{-1}} & (f_3-n_3\gamma_3^{-1})\dfrac{\sum_{k=1}^{2}\gamma_k^{-1}}{\sum_{k=1}^{3}\gamma_k^{-1}} & -\dfrac{f_3-n_3\gamma_3^{-1}}{\sum_{k=1}^{3}\gamma_k^{-1}} \end{bmatrix} \tag{B9}$$

$$\mathbf{G}_C = \begin{bmatrix} c_{n1}^{-1}n_1^{-1}\gamma_1^{-1}\left(1-\dfrac{\gamma_1^{-1}}{\sum_{k=1}^{3}\gamma_k^{-1}}\right) & -c_{n2}^{-1}n_2^{-1}\gamma_2^{-1}\dfrac{\gamma_1^{-1}}{\sum_{k=1}^{3}\gamma_k^{-1}} & -c_{n3}^{-1}n_3^{-1}\gamma_3^{-1}\dfrac{\gamma_1^{-1}}{\sum_{k=1}^{3}\gamma_k^{-1}} \\ -c_{n1}^{-1}n_1^{-1}\gamma_1^{-1}\dfrac{\gamma_2^{-1}}{\sum_{k=1}^{3}\gamma_k^{-1}} & c_{n2}^{-1}n_2^{-1}\gamma_2^{-1}\left(1-\dfrac{\gamma_2^{-1}}{\sum_{k=1}^{3}\gamma_k^{-1}}\right) & -c_{n3}^{-1}n_3^{-1}\gamma_3^{-1}\dfrac{\gamma_2^{-1}}{\sum_{k=1}^{3}\gamma_k^{-1}} \\ -c_{n1}^{-1}n_1^{-1}\gamma_1^{-1}\dfrac{\gamma_3^{-1}}{\sum_{k=1}^{3}\gamma_k^{-1}} & -c_{n2}^{-1}n_2^{-1}\gamma_2^{-1}\dfrac{\gamma_3^{-1}}{\sum_{k=1}^{3}\gamma_k^{-1}} & c_{n3}^{-1}n_3^{-1}\gamma_3^{-1}\left(1-\dfrac{\gamma_3^{-1}}{\sum_{k=1}^{3}\gamma_k^{-1}}\right) \end{bmatrix} \tag{B10}$$

If the second-order terms of n in c_{li} and $c_{\alpha i}$ ($i=1, 2, 3$) (Equations (4.7) and (4.8)) are omitted, one has:

$$\gamma_i = 3/\left(16 l_i n_i^2 \cos^2 \alpha_i\right) \quad \text{(B11)}$$

$$Q_i = \delta S_i - \sin \alpha_i \cdot \left(\delta h_{Pi} - \delta h_{P(i-1)}\right) \quad \text{(B12)}$$

By introducing a new parameter $z_i = \gamma_i^{-1}$, $\mathbf{G}_A$, $\mathbf{G}_B$, $\mathbf{G}_C$, and $\mathbf{G}_D$ can be rewritten as:

$$\mathbf{G}_A = \begin{bmatrix} \frac{n_1}{\cos\alpha_1} + \frac{n_1(l_1 - z_1)}{\cos\alpha_1 \sum_{k=1}^{3} z_k} & \frac{n_1(l_1 - z_1)}{\cos\alpha_2 \sum_{k=1}^{3} z_k} & \frac{n_1(l_1 - z_1)}{\cos\alpha_3 \sum_{k=1}^{3} z_k} \\ \frac{n_2(l_2 - z_2)}{\cos\alpha_1 \sum_{k=1}^{3} z_k} & \frac{n_2}{\cos\alpha_2} + \frac{n_2(l_2 - z_2)}{\cos\alpha_2 \sum_{k=1}^{3} z_k} & \frac{n_2(l_2 - z_2)}{\cos\alpha_3 \sum_{k=1}^{3} z_k} \\ \frac{n_3(l_3 - z_3)}{\cos\alpha_1 \sum_{k=1}^{3} z_k} & \frac{n_3(l_3 - z_3)}{\cos\alpha_2 \sum_{k=1}^{3} z_k} & \frac{n_3}{\cos\alpha_3} + \frac{n_3(l_3 - z_3)}{\cos\alpha_3 \sum_{k=1}^{3} z_k} \end{bmatrix} \quad \text{(B13)}$$

$$\mathbf{G}_B = \begin{bmatrix} -n_1(l_1 - z_1)\left(1 - \frac{\sum_{k=1}^{1} z_k}{\sum_{k=1}^{3} z_k}\right) & -n_1(l_1 - z_1)\left(1 - \frac{\sum_{k=1}^{2} z_k}{\sum_{k=1}^{3} z_k}\right) & -\frac{n_1(l_1 - z_1)}{\sum_{k=1}^{3} z_k} \\ n_2(l_2 - z_2)\frac{\sum_{k=1}^{1} z_k}{\sum_{k=1}^{3} z_k} & -n_2(l_2 - z_2)\left(1 - \frac{\sum_{k=1}^{2} z_k}{\sum_{k=1}^{3} z_k}\right) & -\frac{n_2(l_2 - z_2)}{\sum_{k=1}^{3} z_k} \\ n_3(l_3 - z_3)\frac{\sum_{k=1}^{1} z_k}{\sum_{k=1}^{3} z_k} & n_3(l_3 - z_3)\frac{\sum_{k=1}^{2} z_k}{\sum_{k=1}^{3} z_k} & -\frac{n_3(l_3 - z_3)}{\sum_{k=1}^{3} z_k} \end{bmatrix} \quad \text{(B14)}$$

$$\mathbf{G}_C = \begin{bmatrix} \frac{1}{\cos\alpha_1}\left(1-\frac{z_1}{\sum_{k=1}^{3} z_k}\right) & -\frac{z_1}{\cos\alpha_2\sum_{k=1}^{3} z_k} & -\frac{z_1}{\cos\alpha_3\sum_{k=1}^{3} z_k} \\ -\frac{z_2}{\cos\alpha_1\sum_{k=1}^{3} z_k} & \frac{1}{\cos\alpha_2}\left(1-\frac{z_2}{\sum_{k=1}^{3} z_k}\right) & -\frac{z_2}{\cos\alpha_3\sum_{k=1}^{3} z_k} \\ -\frac{z_3}{\cos\alpha_1\sum_{k=1}^{3} z_k} & -\frac{z_3}{\cos\alpha_2\sum_{k=1}^{3} z_k} & \frac{1}{\cos\alpha_3}\left(1-\frac{z_3}{\sum_{k=1}^{3} z_k}\right) \end{bmatrix} \tag{B15}$$

$$\mathbf{G}_D = \begin{bmatrix} z_1 - z_1\frac{\sum_{k=1}^{1} z_k}{\sum_{k=1}^{3} z_k} & z_1 - z_1\frac{\sum_{k=1}^{2} z_k}{\sum_{k=1}^{3} z_k} & \frac{z_1}{\sum_{k=1}^{3} z_k} \\ -z_2\frac{\sum_{k=1}^{1} z_k}{\sum_{k=1}^{3} z_k} & z_2 - z_2\frac{\sum_{k=1}^{2} z_k}{\sum_{k=1}^{3} z_k} & \frac{z_2}{\sum_{k=1}^{3} z_k} \\ -z_3\frac{\sum_{k=1}^{1} z_k}{\sum_{k=1}^{3} z_k} & -z_3\frac{\sum_{k=1}^{2} z_k}{\sum_{k=1}^{3} z_k} & \frac{z_3}{\sum_{k=1}^{3} z_k} \end{bmatrix} \tag{B16}$$

The solution to Equation (4.15) is:

$$\begin{bmatrix} \delta f_1 \\ \delta f_2 \\ \delta f_3 \\ \hline \delta l_1 \\ \delta l_2 \\ \delta l_3 \end{bmatrix} = \begin{bmatrix} \mathbf{G}_A & \mathbf{G}_B \\ \mathbf{G}_C & \mathbf{G}_D \end{bmatrix} \begin{bmatrix} Q_1 \\ Q_2 \\ Q_3 \\ \hline 0 \\ 0 \\ 0 \end{bmatrix} = \begin{bmatrix} \mathbf{G}_A \\ \mathbf{G}_C \end{bmatrix} \begin{bmatrix} Q_1 \\ Q_2 \\ Q_3 \end{bmatrix} \tag{B17}$$

Substituting Equations (B12), (B13), and (B15) into Equation (B17) leads to Equations (4.16) and (4.17).

APPENDIX C

This part serves to solve Equation (4.59). Equation (4.59) is rewritten as

$$\begin{bmatrix} \mathbf{P}_{u\times u} & \mathbf{M}_{u\times u} & \mathbf{E}_{u\times u} \\ \mathbf{0}_{u\times u} & \mathbf{K}_{u\times u} & \mathbf{J}_{u\times u} \\ \mathbf{U}_{u\times u} & \mathbf{V}_{u\times u} & \mathbf{I}_u \end{bmatrix} \cdot \begin{bmatrix} \delta\mathbf{F}_{u\times 1} \\ \delta\mathbf{L}_{u\times 1} \\ \delta\mathbf{H}_{u\times 1} \end{bmatrix} = \begin{bmatrix} \mathbf{Q}_{u\times 1} \\ \mathbf{0}_{u\times 1} \\ \mathbf{0}_{u\times 1} \end{bmatrix} \tag{C1}$$

where

$$\mathbf{P}_{u\times u} = diag\left(\left[P_1, P_2, \cdots, P_u\right]\right) \tag{C2}$$

$$\mathbf{M}_{u\times u} = diag\left(\left[M_1, M_2, \cdots, M_u\right]\right) \tag{C3}$$

$$\mathbf{E}_{u\times u} = diag\left(\left[E_1, E_2, \cdots, E_u\right]\right) \tag{C4}$$

$$\mathbf{K}_{u\times u} = (K)_{ij} = \begin{cases} \kappa_i & \text{when } 1 \le j \le i \le u-1 \\ 1 & \text{when } i = u \\ 0 & \text{others} \end{cases} \tag{C5}$$

$$\mathbf{J}_{u\times u} = (J)_{ij} = \begin{cases} 1 & \text{when } 1 \le i = j \le u-1 \\ -1 & \text{when } 2 \le i+1 = j \le u \\ 0 & \text{others} \end{cases} \tag{C6}$$

$$\mathbf{U}_{u\times u} = \text{diag}\left(\left[f_1^{-1}, f_2^{-1}, \cdots, f_u^{-1}\right]\right) \tag{C7}$$

$$\mathbf{V}_{u\times u} = \text{diag}\left(\left[-l_1^{-1}, -l_2^{-1}, \cdots, -l_u^{-1}\right]\right) \tag{C8}$$

$$\delta\mathbf{F}_{u\times 1} = \left[\delta f_1, \delta f_2, \cdots, \delta f_u\right]^{\mathrm{T}} \tag{C9}$$

$$\delta\mathbf{L}_{u\times 1} = \left[\delta l_1, \delta l_2, \cdots, \delta l_u\right]^{\mathrm{T}} \tag{C10}$$

$$\delta\mathbf{H}_{u\times 1} = \left[\delta H_1/H_0, \delta H_2/H_0, \cdots, \delta H_u/H_0\right]^{\mathrm{T}} \tag{C11}$$

$$\mathbf{Q}_{u\times 1} = \left[Q_1, Q_2, \cdots, Q_u\right]^{\mathrm{T}} \tag{C12}$$

and $\mathbf{0}_{u\times u}$ is a $u\times u$ zero matrix, $\mathbf{0}_{u\times 1}$ is a $u\times 1$ zero vector, and $\mathbf{I}_u$ is a $u\times u$ identity matrix.

If $\kappa_i \neq 0$ ($i = 1, 2, \ldots, u-1$), then the inverse of **K** exists. For convenience, let $\kappa_u = 1$, then **K** can be written as

$$\mathbf{K} = \begin{bmatrix} \kappa_1 & & & & \\ \kappa_2 & \kappa_2 & & \mathbf{0} & \\ \vdots & \vdots & \ddots & & \\ \kappa_{u-1} & \kappa_{u-1} & \cdots & \kappa_{u-1} & \\ \kappa_u & \kappa_u & \cdots & \kappa_u & \kappa_u \end{bmatrix} \tag{C13}$$

The inverse of **K** is a special bi-diagonal matrix, which is derived by the mathematical induction and can be validated by the matrix multiplication rule:

$$\mathbf{K}^{-1} = \begin{bmatrix} \kappa_1^{-1} & & & & \\ -\kappa_1^{-1} & \kappa_2^{-1} & & \mathbf{0} & \\ & -\kappa_2^{-1} & \kappa_3^{-1} & & \\ & & \ddots & \ddots & \\ \mathbf{0} & & & & \\ & & & -\kappa_{u-1}^{-1} & \kappa_u^{-1} \end{bmatrix} \tag{C14}$$

From the second equation of Equation (C1), we have

$$\delta\mathbf{L} = -\mathbf{K}^{-1}\mathbf{J} \cdot \delta\mathbf{H} \tag{C15}$$

Substituting Equation (C15) into the third equation of Equation (C1) yields:

$$\delta\mathbf{F} = \mathbf{U}^{-1}\left(\mathbf{V}\mathbf{K}^{-1}\mathbf{J} - \mathbf{I}\right) \cdot \delta\mathbf{H} \tag{C16}$$

By substituting Equations (C15) and (C16) into the first equation of Equation (C1), one has

$$\left[\left(\mathbf{P}\mathbf{U}^{-1}\mathbf{V} - \mathbf{M}\right)\mathbf{K}^{-1}\mathbf{J} - \mathbf{P}\mathbf{U}^{-1} + \mathbf{E}\right] \cdot \delta\mathbf{H} = \mathbf{Q} \tag{C17}$$

From Equations (C6) and (C14), $\mathbf{K}^{-1}\mathbf{J}$ is a tridiagonal matrix:

$$\mathbf{K}^{-1}\mathbf{J} = \begin{bmatrix} \kappa_1^{-1} & -\kappa_1^{-1} & & & & \\ -\kappa_1^{-1} & \kappa_1^{-1} + \kappa_2^{-1} & -\kappa_2^{-1} & & \mathbf{0} & \\ & -\kappa_2^{-1} & \kappa_2^{-1} + \kappa_3^{-1} & -\kappa_3^{-1} & & \\ & & \ddots & \ddots & \ddots & \\ & \mathbf{0} & & -\kappa_{u-2}^{-1} & \kappa_{u-2}^{-1} + \kappa_{u-1}^{-1} & -\kappa_{u-1}^{-1} \\ & & & & -\kappa_{u-1}^{-1} & \kappa_{u-1}^{-1} \end{bmatrix} \tag{C18}$$

Note that $\mathbf{K}^{-1}\mathbf{J}$ is irrelevant to κ_u. As **P**, **U**, **V**, **M**, and **E** are all diagonal matrices, the coefficient matrix $\mathbf{G}_H = (\mathbf{PU}^{-1}\mathbf{V} - \mathbf{M})\mathbf{K}^{-1}\mathbf{J} - \mathbf{PU}^{-1} + \mathbf{E}$ is also tridiagonal:

$$\mathbf{G}_H = \begin{bmatrix} b_1 & c_1 & & & \mathbf{0} \\ a_2 & b_2 & c_2 & & \\ & \ddots & \ddots & \ddots & \\ & \mathbf{0} & a_{u-1} & b_{u-1} & c_{u-1} \\ & & & a_u & b_u \end{bmatrix} \tag{C19}$$

where

$$a_i = -N_i \kappa_{i-1}^{-1}, \quad i = 2, 3, \cdots, u \tag{C20}$$

$$b_i = \begin{cases} N_1 \kappa_1^{-1} + R_1 & \text{when } i = 1 \\ N_i \left(\kappa_{i-1}^{-1} + \kappa_i^{-1}\right) + R_i & \text{when } i = 2, 3, \cdots, u-1 \\ N_u \kappa_{u-1}^{-1} + R_u & \text{when } i = u \end{cases} \tag{C21}$$

$$c_i = -N_i \kappa_i^{-1}, \quad i = 1, 2, \cdots, u-1 \tag{C22}$$

$$N_i = -P_i f_i l_i^{-1} - M_i, \quad i = 1, 2, \cdots, u \tag{C23}$$

$$R_i = -P_i f_i + E_i, \quad i = 1, 2, \cdots, u \tag{C24}$$

The explicit formula for the inverse of a tridiagonal matrix has complicated forms, as studied by Mallik (2001) and El-Mikkawy (2004). The Thomas algorithm (Datta, 2010) is employed here to obtain the recurrent solution to Equation (C17). First, the parameter c_i is modified as:

$$c_i' = \begin{cases} c_1 / b_1 & \text{when } i = 1 \\ c_i / \left(b_i - a_i c_{i-1}'\right) & \text{when } i = 2, 3, \cdots, u-1 \end{cases} \tag{C25}$$

Second, Q_i is modified as

$$Q_i' = \begin{cases} Q_1 / b_1 & \text{when } i = 1 \\ \left(Q_i - a_i \cdot Q_{i-1}'\right) / \left(b_i - a_i c_{i-1}'\right) & \text{when } i = 2, 3, \cdots, u \end{cases} \tag{C26}$$

By combining Equations (C25) and (C26), the solution to Equation (C27) is obtained:

$$\frac{\delta H_i}{H_0} = \begin{cases} Q_u' & \text{when } i = u \\ Q_i' - c_i' \cdot \delta H_{i+1} / H_0 & \text{when } i = u-1, u-2, \cdots, 1 \end{cases} \tag{C27}$$

Equations (C25)–(C27) are based on the condition of $\kappa_i \neq 0$ ($i = 1, 2, \ldots, u-1$): However, by changing the expression forms, we can derive new expressions of c_i' and Q_i' that apply to any number of κ_i being 0. For the sake of simplicity, a new sequence g_i ($i = 1, 2, \ldots, u$) is introduced in the recurrent form:

$$g_i = R_i/N_i + g_{i-1}/\left(1 + \kappa_{i-1} \cdot g_{i-1}\right) \tag{C28}$$

and $g_0 = 0$. Then c_1' can be written as

$$c_1' = \frac{c_1}{b_1} = \frac{-N_1\kappa_1^{-1}}{N_1\kappa_1^{-1} + R_1} = -\frac{N_1}{N_1 + R_1\kappa_1} = -\frac{1}{1 + R_1/N_1 \cdot \kappa_1} = -\frac{1}{1 + \kappa_1 g_1} \tag{C29}$$

and c_i' ($i = 2, 3, \ldots, u-1$) becomes

$$\begin{aligned} c_i' &= \frac{c_i}{b_i - a_i c_{i-1}'} = \frac{-N_i\kappa_i^{-1}}{N_i\left(\kappa_{i-1}^{-1} + \kappa_i^{-1}\right) + R_i + N_i\kappa_{i-1}^{-1} \cdot c_{i-1}'} \\ &= \frac{-N_i\kappa_i^{-1}}{N_i\kappa_i^{-1} + R_i + N_i\kappa_{i-1}^{-1}\left(1 + c_{i-1}'\right)} = \frac{-N_i\kappa_i^{-1}}{N_i\kappa_i^{-1} + R_i + N_i\, g_{i-1}/\left(1 + \kappa_{i-1} g_{i-1}\right)} \\ &= \frac{-1}{1 + \kappa_i\, R_i/N_i + \kappa_i\, g_{i-1}/\left(1 + \kappa_{i-1} g_{i-1}\right)} \\ &= -\frac{1}{1 + \kappa_i\left(R_i/N_i + g_{i-1}/\left(1 + \kappa_{i-1} g_{i-1}\right)\right)} = -\frac{1}{1 + \kappa_i g_i} \end{aligned} \tag{C30}$$

Consequently, we have a unified formula for c_i' ($i = 1, 2, \cdots, u-1$):

$$c_i' = -1/\left(1 + \kappa_i g_i\right) \tag{C31}$$

where any one of κ_i can be 0.

Similarly, by introducing a sequence w_i ($i = 1, 2, \ldots, u$, $w_0 = 0$) in the recurrent form of

$$w_i = Q_i/N_i + w_{i-1}/\left(1 + \kappa_{i-1} g_{i-1}\right) \tag{C32}$$

we have

$$Q_1' = \frac{Q_1}{b_1} = \frac{Q_1}{N_1\kappa_1^{-1} + R_1} = \frac{Q_1\kappa_1}{N_1 + R_1\kappa_1} = \frac{Q_1/N_1 \cdot \kappa_1}{1 + R_1/N_1 \cdot \kappa_1} = \frac{\kappa_1 w_1}{1 + \kappa_1 g_1} \tag{C33}$$

$$Q_i' = \frac{Q_i - a_i \cdot Q_{i-1}'}{b_i - a_i c_{i-1}'} = \frac{Q_i + N_i \kappa_{i-1}^{-1} \cdot Q_{i-1}'}{N_i \left(\kappa_{i-1}^{-1} + \kappa_i^{-1}\right) + R_i + N_i \kappa_{i-1}^{-1} \cdot c_{i-1}'}$$

$$= \frac{Q_i + N_i \kappa_{i-1}^{-1} \cdot \dfrac{\kappa_{i-1} w_{i-1}}{1+\kappa_{i-1} g_{i-1}}}{N_i \kappa_i^{-1} + R_i + N_i \kappa_{i-1}^{-1}\left(1+c_{i-1}'\right)} = \frac{Q_i + N_i \dfrac{w_{i-1}}{1+\kappa_{i-1} g_{i-1}}}{N_i \kappa_i^{-1} + R_i + N_i \dfrac{g_{i-1}}{1+\kappa_{i-1} g_{i-1}}}$$

$$= \frac{\left(\dfrac{Q_i}{N_i} + \dfrac{w_{i-1}}{1+\kappa_{i-1} g_{i-1}}\right)\kappa_i}{1+\kappa_i \dfrac{R_i}{N_i} + \kappa_i \dfrac{g_{i-1}}{1+\kappa_{i-1} g_{i-1}}} = \frac{\left(\dfrac{Q_i}{N_i} + \dfrac{w_{i-1}}{1+\kappa_{i-1} g_{i-1}}\right)\kappa_i}{1+\kappa_i \left(\dfrac{R_i}{N_i} + \dfrac{g_{i-1}}{1+\kappa_{i-1} g_{i-1}}\right)} = \frac{\kappa_i w_i}{1+\kappa_i g_i} (i = 2, 3, \ldots, u-1)$$

(C34)

$$Q_u' = \frac{Q_u - a_u \cdot Q_{u-1}'}{b_u - a_u c_{u-1}'} = \frac{Q_u + N_u \kappa_{u-1}^{-1} \cdot Q_{u-1}'}{N_u \kappa_{u-1}^{-1} + R_u + N_u \kappa_{u-1}^{-1} \cdot c_{u-1}'}$$

$$= \frac{Q_u + N_u w_{u-1} / \left(1+\kappa_{u-1} g_{u-1}\right)}{R_u + N_u \kappa_{u-1}^{-1}\left(1+c_{u-1}'\right)} = \frac{Q_u / N_u + w_{u-1} / \left(1+\kappa_{u-1} g_{u-1}\right)}{R_u / N_u + g_{u-1} / \left(1+\kappa_{u-1} g_{u-1}\right)} = \frac{w_u}{g_u} \quad \text{(C35)}$$

Consequently, Q_i' can be expressed as

$$Q_i' = \begin{cases} \kappa_i w_i / \left(1+\kappa_i g_i\right) & \text{when } i = 1, 2, \cdots, u-1 \\ w_u / g_u & \text{when } i = u \end{cases} \quad \text{(C36)}$$

By combining Equations (C31) and (C36), the relative cable tension change $\boldsymbol{\delta}\mathbf{H}$ can be obtained from Equation (C27) as

$$\frac{\delta H_i}{H_0} = \begin{cases} w_u / g_u & \text{when } i = u \\ \left(\delta H_{i+1} / H_0 + \kappa_i w_i\right) / \left(1+\kappa_i g_i\right) & \text{when } i = u-1, u-2, \cdots, 1 \end{cases} \quad \text{(C37)}$$

The span change $\boldsymbol{\delta}\mathbf{L}$ and sag change $\boldsymbol{\delta}\mathbf{F}$ can be solved as follows by combining Equations (C15)–(C17):

$$\delta\mathbf{L} = -\left(\mathbf{P}\mathbf{U}^{-1}\mathbf{V} - \mathbf{M}\right)^{-1}\left[\mathbf{Q} + \left(\mathbf{P}\mathbf{U}^{-1} - \mathbf{E}\right) \cdot \delta\mathbf{H}\right] \quad \text{(C38)}$$

$$\delta\mathbf{F} = -\mathbf{U}^{-1}\left(\mathbf{V} \cdot \delta\mathbf{L} + \delta\mathbf{H}\right) \quad \text{(C39)}$$

As **P**, **U**, **V**, **V**, and **E** are all diagonal matrices, the above two equations can also be expressed as ($i = 1, 2, \ldots, u$):

$$\delta l_i = -N_i^{-1}\left(Q_i - R_i \cdot \delta H_i / H_0\right) \tag{C40}$$

$$\delta f_i = f_i\left(l_i^{-1} \cdot \delta l_i - \delta H_i / H_0\right) \tag{C41}$$

Equations (C37), (C40), and (C41) do not rely on the invertibility of **K**. Two special cases ($\kappa_i = 0$ and $\kappa_i \to +\infty$, $i = 1, 2, \ldots, u-1$) will be discussed as follows.

C.1 Case 1: Ignorance of Tower Stiffness

When all κ_i ($i = 1, 2, \ldots, u-1$) equal 0, Equations (C31) and (C36) produce $c_i' = -1$ and $Q_i' = 0$ ($i = 1, 2, \cdots, u-1$), and Equations (C28) and (C32) yield:

$$g_u = \sum_{i=1}^{u} R_i / N_i \tag{C42}$$

$$w_u = \sum_{i=1}^{u} Q_i / N_i \tag{C43}$$

From Equation (C37), all $\delta H_i / H_0$ ($i = 1, 2, \ldots, u$) are identical, i.e.,

$$\delta H_i / H_0 = w_u / g_u = \sum_{j=1}^{u} Q_j / N_j \Big/ \sum_{j=1}^{u} R_j / N_j \tag{C44}$$

Substituting Equation (C44) into Equations (C40) and (C41) leads to the sag and span changes.

Equation (C44) shows that the thermal deformation of the ith span cable depends on the structural properties (N_i and R_i) and deformation (Q_i) of all u span cables ($j = 1, 2, \ldots, u$) when the flexural stiffness of all the intermediate supports is negligible. The physical example of this case is the transmission lines with suspension insulators (Figure C1a):

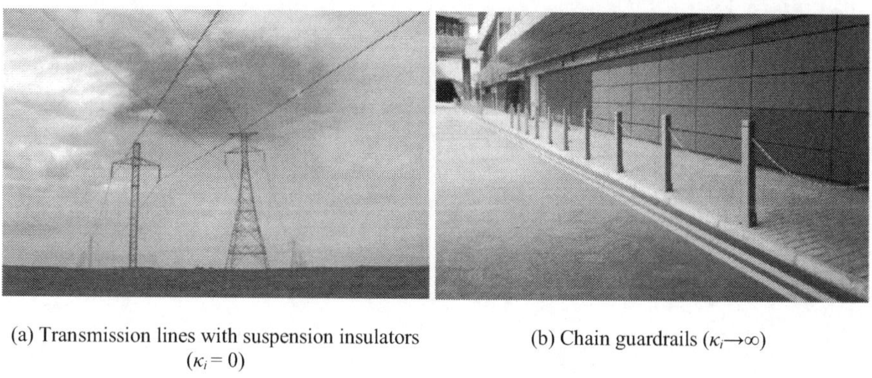

(a) Transmission lines with suspension insulators ($\kappa_i = 0$)

(b) Chain guardrails ($\kappa_i \to \infty$)

FIGURE C1 Physical examples of two special multi-span cable systems. (a) Transmission lines with suspension insulators ($\kappa_i = 0$) and (b) chain guardrails ($\kappa_i \to \infty$) (a—Wtshymanski, https://commons.wikimedia.org/wiki/File:HVDC_Crossover_North-Dakota.jpg)

C.2 Case 2: Rigid Towers

Another extreme case is all κ_i ($i=1, 2,\ldots, u-1$) approaching infinity. From Equations (C31) and (C36), we have $c_i'=0$ ($i=1,2,\cdots,u-1$) and $Q_i'=w_i/g_i$ ($i=1, 2,\ldots, u$): Combining Equations (C28), (C32), and (C37) leads to:

$$\frac{\delta H_i}{H_0}=\frac{w_i}{g_i}=\frac{Q_i}{N_i}\bigg/\frac{R_i}{N_i}=\frac{Q_i}{R_i}=\frac{S_i\theta_C\cdot\delta T_C-c_{\alpha i}\cos^2\alpha_i/l_i\cdot\left(h_{Pi}-h_{P(i-1)}\right)\theta_P\cdot\delta T_P}{-P_i f_i+E_i} \tag{C45}$$

Again, by substituting Equation (C45) into Equations (C40) and (C41), we obtain the sag and span changes:

$$\delta l_i=-N_i^{-1}\left(Q_i-R_i\cdot\delta H_i/H_0\right)=0 \tag{C46}$$

$$\delta f_i=f_i\left(l_i^{-1}\cdot\delta l_i-\delta H_i/H_0\right)=-f_i\cdot\delta H_i/H_0 \tag{C47}$$

Equations (C45)–(C47) indicate that when all the intermediate towers are rigid, i.e., $\kappa_i\to+\infty$ ($i=1, 2,\ldots, u-1$), each span cable deforms independently without interaction with adjacent cables. Chain guardrails (Figure C1b) belong to this case.

If the tower temperature has no change ($\delta T_P=0$), the elastic deformation of the cable is ignored ($E_i=0$), and the cable length is estimated by the first two items in Equation (4.42) ($S_i=l_i\left(\sec\alpha_i+8n_i^2\cos^3\alpha_i/3\right)$), then the temperature-induced sag change of a horizontal single-span cable ($\alpha_i=0$) is

$$\delta f_i=\left(1+8/3\cdot n_i^2\right)/\left(16/3\cdot n_i\right)\cdot l_i\theta_C\cdot\delta T_C=\left(3l_i^2+8f_i^2\right)/\left(16f_i\right)\cdot\theta_C\cdot\delta T_C \tag{C48}$$

Equation (C48) is the same as the result of Timoshenko and Young (1965).

APPENDIX D

This part presents the solution to the modified Equation (4.59), which applies to self-anchored suspension bridges mentioned in Section 4.5.4, i.e.,

$$\begin{bmatrix}\mathbf{P}_{u\times u} & \mathbf{M}_{u\times u} & \mathbf{E}_{u\times u}\\ \mathbf{0}_{u\times u} & \mathbf{K}_{u\times u} & \mathbf{J}_{u\times u}\\ \mathbf{U}_{u\times u} & \mathbf{V}_{u\times u} & \mathbf{I}_u\end{bmatrix}\cdot\begin{bmatrix}\delta\mathbf{F}_{u\times1}\\ \delta\mathbf{L}_{u\times1}\\ \delta\mathbf{H}_{u\times1}\end{bmatrix}=\begin{bmatrix}\mathbf{Q}_{u\times1}\\ \mathbf{Y}_{u\times1}\\ \mathbf{0}_{u\times1}\end{bmatrix} \tag{D1}$$

where

$$\mathbf{Y}_{u\times1}=\left[-\kappa_1\cdot\delta y,\ -\kappa_2\cdot\delta y,\ \cdots,\ -\kappa_{u-1}\cdot\delta y,\ \delta L_G\right]^{\mathrm{T}} \tag{D2}$$

The other parameters in Equation (D1) can be found in Equations (C2)–(C12).

According to the second set of equations in Equation (D1), one has

$$\boldsymbol{\delta}\mathbf{L} = -\mathbf{K}^{-1}\mathbf{J}\cdot\boldsymbol{\delta}\mathbf{H} + \mathbf{K}^{-1}\begin{bmatrix} -\kappa_1\cdot\delta y & -\kappa_2\cdot\delta y & \cdots & -\kappa_{u-1}\cdot\delta y & \delta L_G \end{bmatrix}^{\mathrm{T}} \tag{D3}$$

$\mathbf{K}^{-1}$ refers to Equation (C14) and $\kappa_u = 1$. Therefore, Equations (C15)–(C17) become

$$\boldsymbol{\delta}\mathbf{L} = -\mathbf{K}^{-1}\mathbf{J}\cdot\boldsymbol{\delta}\mathbf{H} + \begin{bmatrix} -\delta y & \mathbf{0}_{1\times(u-2)} & \delta L_G + \delta y \end{bmatrix}^{T} \tag{D4}$$

$$\boldsymbol{\delta}\mathbf{F} = \mathbf{U}^{-1}\left(\mathbf{V}\mathbf{K}^{-1}\mathbf{J} - \mathbf{I}\right)\cdot\boldsymbol{\delta}\mathbf{H} + \begin{bmatrix} -f_1 l_1^{-1}\cdot\delta y & \mathbf{0}_{1\times(u-2)} & f_u l_u^{-1}\cdot\left(\delta L_G + \delta y\right) \end{bmatrix} \tag{D5}$$

$$\left[\left(\mathbf{P}\mathbf{U}^{-1}\mathbf{V} - \mathbf{M}\right)\mathbf{K}^{-1}\mathbf{J} - \mathbf{P}\mathbf{U}^{-1} + \mathbf{E}\right]\cdot\boldsymbol{\delta}\mathbf{H} = \mathbf{Q} - \begin{bmatrix} -M_1\cdot\delta y \\ \mathbf{0}_{(u-2)\times 1} \\ M_u\left(\delta L_G + \delta y\right) \end{bmatrix} - \begin{bmatrix} -P_1 f_1 l_1^{-1}\cdot\delta y \\ \mathbf{0}_{(u-2)\times 1} \\ P_u f_u l_u^{-1}\left(\delta L_G + \delta y\right) \end{bmatrix} \tag{D6}$$

Equation (D6) is the same as Equation (C17) except for the first and last elements of the right-hand side column vector. If we introduce two new quantities $Q_{l\_} = Q_1 - N_1\cdot\delta y$ and $Q_{u\_} = Q_u + N_u(\delta L_G + \delta y)$, where N_1 and N_u are determined by Equation (C23), then the solution to Equation (D6) is identical to that to Equation (C17) by replacing Q_i with $Q_{i\_}$ ($i = 1$ and u): After the cable tension change $\boldsymbol{\delta}\mathbf{H}$ is obtained, the span change $\boldsymbol{\delta}\mathbf{L}$ and sag change $\boldsymbol{\delta}\mathbf{F}$ can be successively solved by Equations (C38) and (C39) or Equations (C40) and (C41). Note that in the application of Equations (C40) and (C41), Q_i cannot be replaced by $Q_{i\_}$ ($i = 1$ and u):

When the tower flexural stiffness is neglected, i.e., $\kappa_i = 0$ ($i = 1, 2, \ldots, u-1$), Equation (C44) with Q_i replaced by $Q_{i\_}$ ($i = 1$ and u) produces Equation (4.94) in the main text, which is independent of δy. Substituting $\delta H_i/H_0$ ($i = 1, 2, \ldots, u$) into Equations (C40) and (C41) leads to Equations (4.93) and (4.92).

5 Temperature Behaviors of Cable-Stayed Bridges

NOTATIONS

$\text{Avg}(\cdots)$	Averaging
E	Young's modulus
H_0	Tower height above the girder
I	Flexural moment of inertia of the girder
I_i	Effect of temperature variable T_i on the mid-span deflection
$\lvert I_i \rvert$	Magnitude of the effect
K_1	Equivalent spring rigidity of stay cables
L_0	Middle span length
L_T, D_T	Thermally induced changes of the girder length and mid-span vertical displacement
S	Stay cable stress
S_0, S_1	Length of cables at the center and side spans
T_G, T_C, T_T	Average girder temperature, cable temperature, average tower temperature
T_{GD}, T_{TD}	Girder vertical differential temperature and tower temperature difference between the opposite external surfaces
$T_{\max}$, $\Delta T_{\max}$, $T_{\text{diff\_max}}$	Maximum temperature, maximum daily temperature variation, maximum temperature differences between top and bottom plates of girder
U_{RMSE}, V_{RMSE}	RMSEs between calculated and measured longitudinal and vertical displacements
X_{-a}, X_{-m}	Calculated or measured variable X, X denotes L_T or D_T
c	Specific heat capacity (J/kg/°C)
ds_0, ds_1	Changes in the lengths of cables due to the increases in their forces
f_1	First-order frequency of the cable
h	Girder depth
$k_i^{D_T}$, $k_i^{L_T}$	Sensitivity coefficients of D_T and L_T due to different temperature variable T_i, $i = \{C, G, GD, T\}$
l	Stay cable length
Δ_{10}, Δ_{11}	Changes in cable lengths due to the effects of T_C
Δ_{12}, Δ_{13}	Horizontal displacement at tower top and vertical displacement at girder mid-span due to the effect of T_C

DOI: 10.1201/9781003480136-5

Δ_{20}, Δ_{21}, Δ_{22}	Horizontal displacement at girder end, horizontal displacement at tower top, and vertical displacement at girder mid-span due to the effect of T_G
Δ_{31}	Vertical displacement of the two-span beam due to the effect of T_{GD}
Δ_{40}, Δ_{41}, Δ_{42}	Vertical displacement at tower top, horizontal displacement at tower top, vertical displacement at girder mid-span due to the effect of T_T
ΔT	Temperature change
ΔT_G, ΔT_C, ΔT_T	Changes in average girder temperature, cable temperature, average tower temperature
ΔT_{GD}, ΔT_{TD}	Changes in girder vertical differential temperature and in tower temperature difference between the opposite external surfaces
$\Delta U_{\max}$, $\Delta V_{\max}$	Maximum daily variation in the total longitudinal displacement and mid-span vertical deflection
$\Delta\sigma_{\max}$, $\sigma_{\max}$	Measured maximum daily stress, and calculated maximum stress in terms of magnitude
ψ	Solar altitude angle
Σ_1, Σ_2, Σ_3	Algebraic sum of the effects of the four mechanisms
α	Absorptivity coefficient
α, α_s, α_c	Thermal expansion coefficient, thermal expansion coefficients of steel and concrete
α^C, α^G	Thermal expansion coefficients of the cable and girder
ε	Emissivity coefficient
$\varepsilon_{\text{total}}$, σ_{total}	Total strain and stress
$\varepsilon_{\text{el}}^{d}$, $\varepsilon_{\text{el}}^{\text{th}}$, ε_{th}	Dead load-induced elastic strain, temperature-induced elastic strain, and free thermal strain
ε^{M}	Measured strain
λ_0	Ratio of tower heights below and above the deck
γ_0	Ratio of side span length to middle span length
ρ	Density (kg/m^3)
σ_{el}^{d}, $\sigma_{\text{el}}^{\text{th}}$	Dead load-induced elastic stress and temperature-induced elastic stress
ξ	ratio of the tower height above the deck and the center span length of the girder

5.1 PREFACE

Cables are widely used in modern structures. They allow for lightweight solutions with high material efficiency and contribute to aesthetically appealing designs. A cable-stayed bridge is a structural system with a bridge girder supported by stay cables from towers. The stay cables can be viewed as alternatives to the piers, which reduce the girder span and thus the moments. Multiple towers may be arranged, and continuous spans can be realized, such as the Millau Viaduct in France (Cachot et al., 2015), which has seven towers and eight continuous spans (204 + 342 × 6 + 204 m span arrangement). The bridge girders may also have multiple choices, such as the slab type, box type, and double deck made of trusses. Multiple cables significantly increase the redundancy of the bridge, thus increasing the mechanical complexity of the structure. Consequently, the temperature behaviors may be more complicated. Comprehensive SHM systems have been installed on cable-stayed bridges to evaluate the bridge performance (Wong, 2004, 2007; Cao et al., 2011).

Given the difficulties in fabricating physical models of cable-stayed bridges in the laboratory, field monitoring, numerical investigations, and theoretical analysis have been conducted to study the temperature characteristics and temperature-induced structural responses of cable-stayed bridges. In this chapter, field monitoring and numerical simulation are carried out on the Qingzhou Bridge, a cable-stayed navigation bridge of the Hong Kong–Zhuhai–Macau Bridge (HZMB). Analytical formulations are derived and applied to the Shanghai Yangtze River Bridge. Different underlying mechanisms will also be compared and discussed.

5.2 FIELD MONITORING STUDIES OF A CABLE-STAYED BRIDGE

5.2.1 SHM System on the Qingzhou Bridge

The 55-km long HZMB is the world's longest sea-crossing bridge connecting three administrative regions of Hong Kong, Macao, and Zhuhai. The Qingzhou Bridge, one of the three main navigation channel bridges of HZMB, located at 22.28°N and 113.73°E, is an east-west cable-stayed bridge with a total length of 1150 m and a main span of 458 m, as shown in Figure 5.1. The steel box girder, covered by 70 mm-thick asphalt, carries a dual three-lane highway and is continuously supported by two H-shaped towers, two auxiliary piers, and two transitional piers. As high as 163 m, the H-shaped tower is composed of two concrete box legs, a concrete lower transom, and a steel upper transom, resembling a Chinese knot.

An SHM system was designed for HZMB and has been operating since 2018 when the bridge was first opened to the public. The SHM system consists of 262 sensors, including hygro-thermometers, thermometers, anemometers, GPS rovers, accelerometers, liquid leveling systems, displacement transducers, cable tensiometers, strain

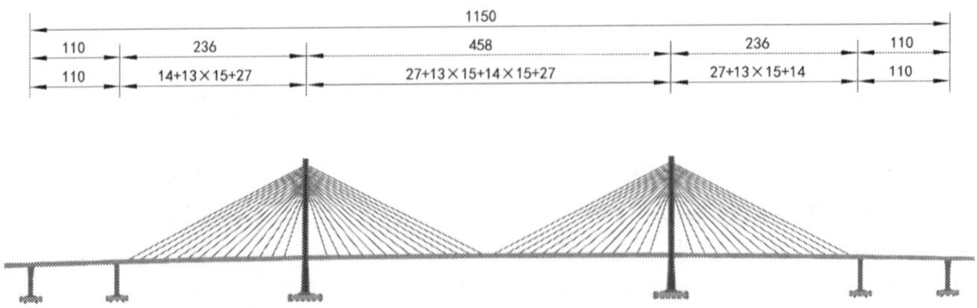

FIGURE 5.1 Configuration of the Qingzhou Bridge.

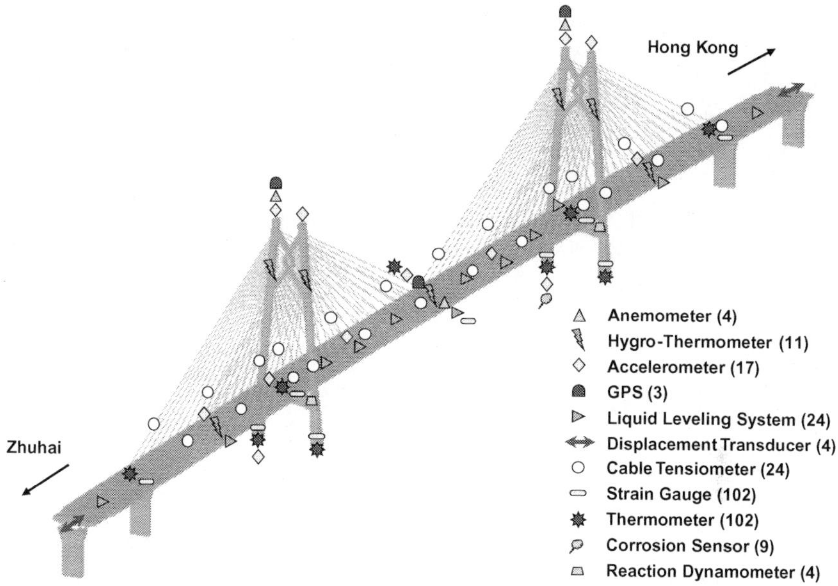

FIGURE 5.2 3D layout of the sensors of the Qingzhou Bridge.

gauges, corrosion sensors, and reaction dynamometers, to monitor the environmental factors, external loads, and structural responses of the bridge. The 3D layout of the sensors is shown in Figure 5.2.

5.2.1.1 Temperature Monitoring

The locations of the sections with thermometers are shown in Figure 5.3. The internal ambient temperatures of the girder and tower are measured by hygro-thermometers, which are located inside the box girder (G04, G09, and G14) and the tower legs (T02 and T06). Each section has two hygro-thermometers, one on the south side and the other on the north side. The structural temperatures are recorded by fiber optic thermometers, which are located on the girder sections of the mid-span (G09), on the auxiliary piers (G03 and G15), on the towers (G05 and G13), and on the lower columns of the towers (T03 and T07). Sections G05 and G13 are supported on the towers and thus have no wind fairings due to the spatial limitation of the tower column. The layouts of the thermometers used to measure the structural temperature on

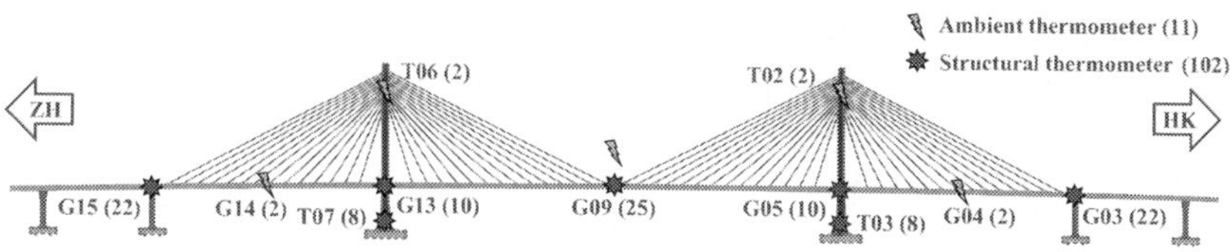

FIGURE 5.3 Location of sections with thermometers.

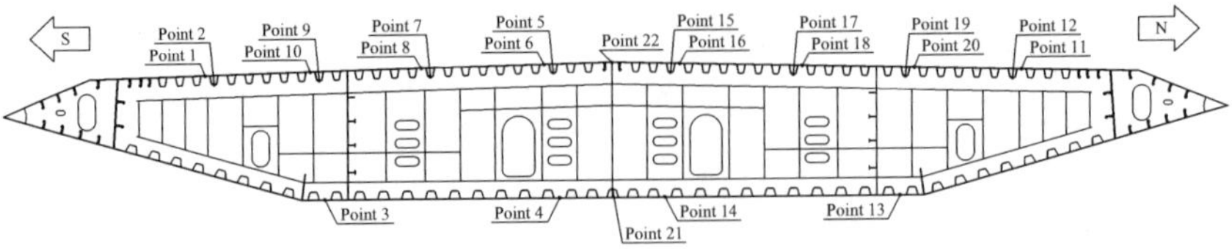

FIGURE 5.4 Layout of thermometers on girder section G09 (with wind fairings).

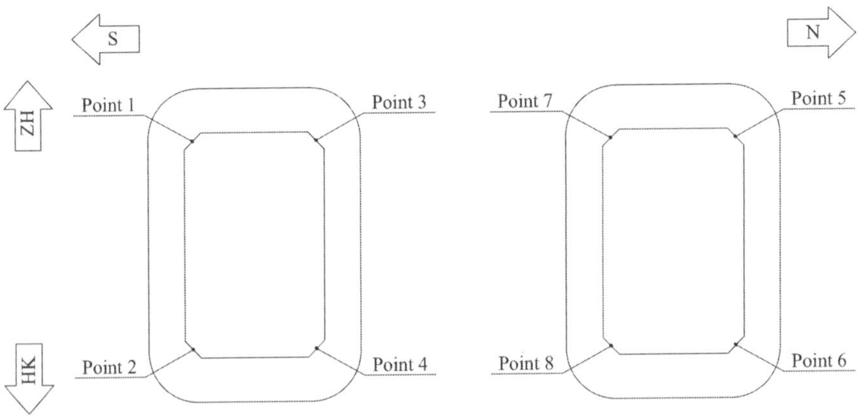

FIGURE 5.5 Layout of thermometers and strain gauges on tower sections (T03 and T07).

girder section G09 (with wind fairings) and tower sections (T03 and T07) are shown in Figures 5.4 and 5.5, respectively.

Two propeller anemometers measure the wind velocity and wind direction on top of the towers, and two ultrasonic anemometers at the mid-span of the girder measure the wind velocity in three orthogonal directions.

A weather station installed at the mid-span of the girder records the ambient temperature of the external air, wind conditions, air pressure, humidity, and rainfall by integrating certain types of sensors. Among them, the ambient temperature of the external air is considered. As the SHM system of the bridge does not have a pyranometer, the solar radiation data collected at the Hong Kong Observatory, approximately 45 km away from the bridge, are adopted in this study.

5.2.1.2 Structural Performance Monitoring

The displacement of the Qingzhou Bridge is measured using GPS rovers, linear variable differential transformers (LVDTs), and liquid leveling systems (LLSs). The locations of the sections with displacement sensors are illustrated in Figure 5.6. Two GPS rovers on the tower top and one on the mid-span of the girder measure the displacement in three directions. Four LVDTs on two ends of the girder (G01 and G17) measure the longitudinal displacement. Each end has two sensors, one on the north and the other on the south. Twenty-four LLSs installed on twelve sections of the girder measure the vertical displacement. These sections include the mid-sections

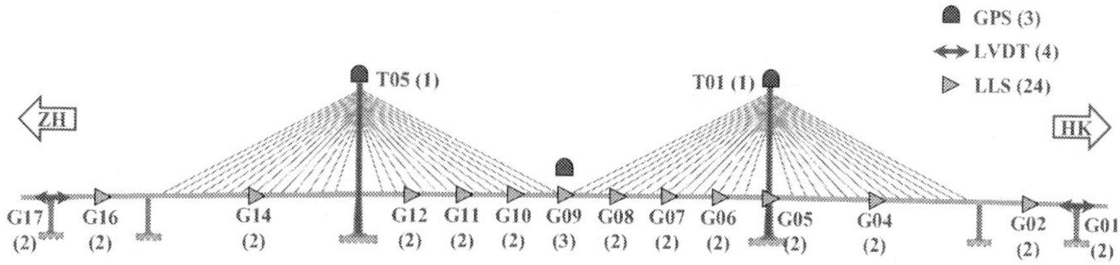

FIGURE 5.6 Location of sections with displacement sensors.

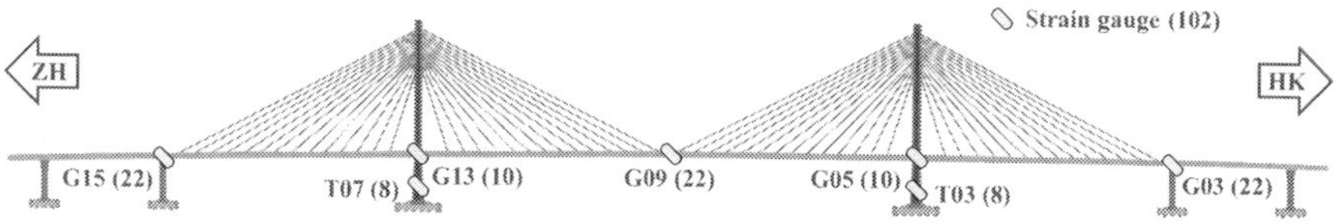

FIGURE 5.7 Locations of sections with strain gauges.

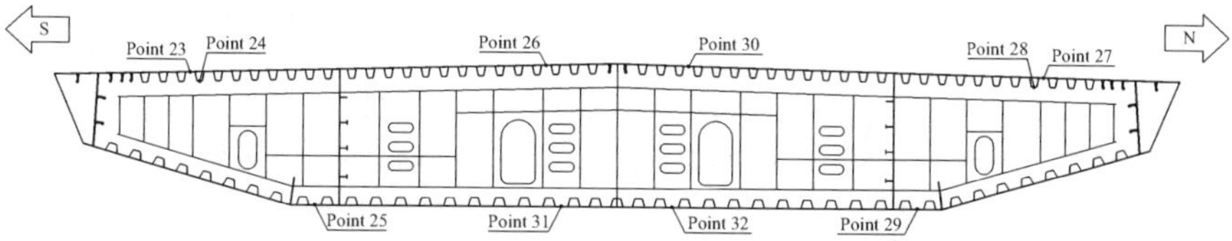

FIGURE 5.8 Layout of strain gauges on girder section G13 (without wind fairings).

of the side spans (G02 and G16), mid-sections of the auxiliary spans (G04 and G14), eight evenly spaced sections of the middle span (G06 to G12), and the section supported on the tower on the Hong Kong side (G05). Each section is equipped with two sensors on the north and south sides.

Strain gauges were installed at the same points as the fiber optic temperature sensors. The locations of the sections with strain gauges are shown in Figure 5.7. A total of 102 fiber optic strain gauges were distributed on seven sections, with 22 positioned on the mid-section of the middle span (G09), 22 × 2 at the two girder sections supported on the auxiliary piers (G03 and G15), 10 × 2 at the two girder sections supported on the towers (G05 and G13), and 8 × 2 at the lower column sections of the tower (T03 and T07). The layout of the strain gauges on girder section G13 (without wind fairings) is shown in Figure 5.8. The strain gauge locations on the tower sections (T03 and T07) are the same as the thermometers on the towers, as shown in Figure 5.5. That is, thermometers and strain gauges were installed at each point.

Altogether, 24 cable tensiometers are installed on the cables of the longest (S14 and M14), the shortest (S1 and M1), and the middle length (S7 and M7) on the side span and the middle span of the north and south sides to measure the accelerations of the cables. The locations of the cable tensiometers are shown in Figure 5.9.

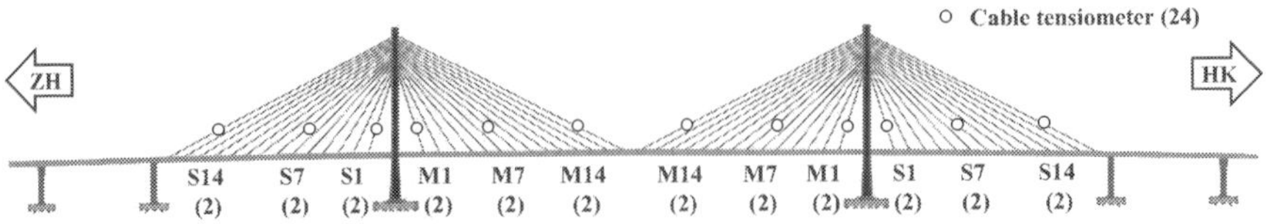

FIGURE 5.9 Location of cable tensiometers.

5.2.2 Measured Temperature Distribution

Based on the meteorological conditions, January 4–6, April 5–7, July 5–7, and October 2–3, 2019, are selected as the representative periods of the four seasons. The measured environmental conditions are briefed first. Then, the findings on the temperature distribution patterns in the bridge during July 5–7, 2019, are thoroughly discussed.

5.2.2.1 Measured Environmental Conditions

The measured ambient temperature, wind, and solar radiation data of the first weeks of January, April, July, and October of 2019 are studied. The hourly averaged wind and ambient temperature data obtained from the SHM system and the total solar radiation data received by a horizontal surface on July 1–7, 2019, are shown in Figure 5.10. During the week, the wind speed varied greatly, mainly from the south and east directions. The solar radiation conditions of consecutive days were not identical. The internal ambient temperature of the girder was higher than that of the external air and had a similar variation pattern as the solar radiation. In contrast, the internal temperature of the tower changed only slightly owing to the good insulation of the concrete tower wall.

5.2.2.2 Temperature Distribution of Main Girder

The temperature distribution in a typical middle span girder section, Section G09, is examined here. The location of the section and the sensor layout are shown in Figures 5.3 and 5.4, respectively.

These thermometers measured the temperatures of the top plate (TP), top U-rib (TU), bottom U-rib (BU), and bottom plate (BP). The initial data processing and data integrity checking show that data from Point 22 exhibit abnormality and are thus discarded. Then, the characteristics of the temperature distribution are explored by selecting the following seven points as the representative measurement points of the section: Points 8 and 18 for TP on the south and north sides, respectively; Points 7 and 17 for TU on the south and north sides, respectively; Points 3 and 13 for BP on the south and north sides, respectively; and Point 21 for BU. The temperatures of the representative points on July 5–7, 2019, are presented in Figure 5.11 for demonstration. Throughout this chapter, the curves with a legend ending with "M" denote the measured ones, while those with a legend ending with "C" are the calculated ones from the FE simulations. The simulation results will be obtained in the following sections.

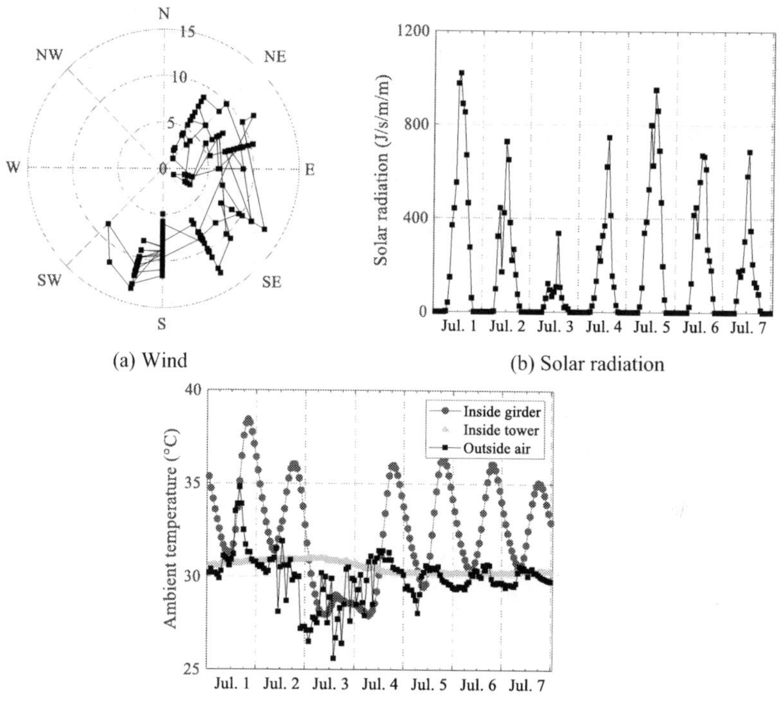

(a) Wind (b) Solar radiation

(c) Ambient temperature

FIGURE 5.10 Environmental conditions on July 1–7, 2019: (a) wind, (b) solar radiation, and (c) ambient temperature.

The following results can be derived from the figure:

1. The temperature of TP in summer could reach as high as 53°C with a daily variation of approximately 23°C. Meanwhile, the maximum temperature of BP was approximately 32°C with a daily variation of 3°C. TP had a much higher temperature and more diurnal variation than BP due to the contribution of solar radiation.
2. On the upper plate, the temperature of TU was lower than that of TP with a lag effect. The vertical temperature difference between TP and TU could reach approximately 9°C. On the lower plate, the temperature of BU was higher than that of BP with a lag effect, and the maximum vertical temperature difference between them was approximately 4°C. The temperature difference between TP and BP could reach as high as 22°C.
3. The temperature on the north side was higher than that on the south side for two reasons: i) the subsolar point was on the north of the bridge during the prescribed period, and ii) the wind was mainly blowing from the south, leading to larger heat dissipation in the south. The measured transverse temperature difference was 4°C according to the representative points on the TP.

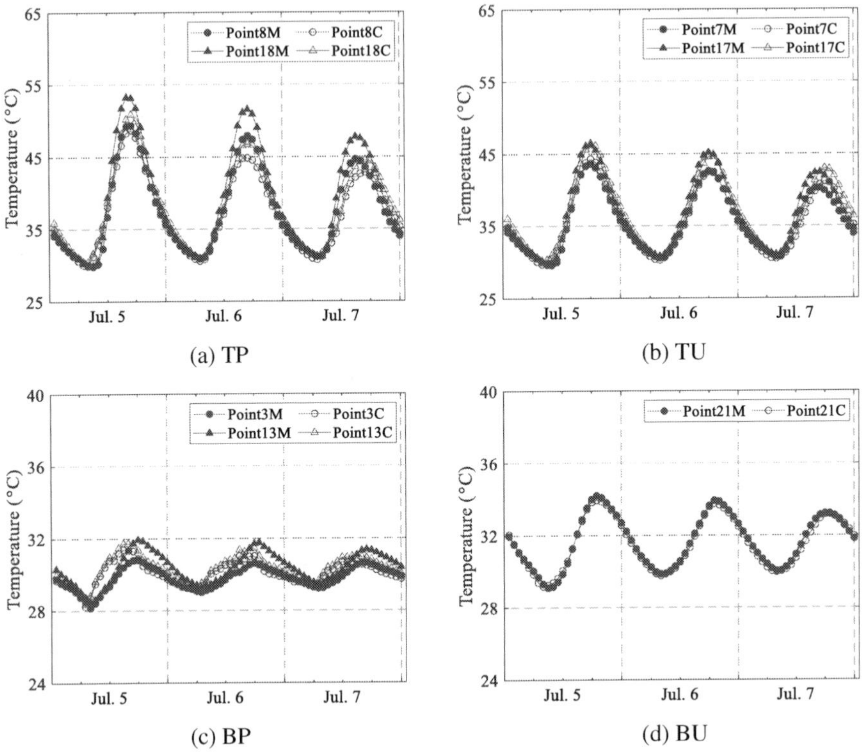

FIGURE 5.11 Measured and simulated temperatures of G09 on July 5–7, 2019: (a) TP, (b) TU, (c) BP, and (d) BU.

The temperature measurement results for the representative days in the other three seasons are shown in Figures 5.12–5.14. The dates with unfavorable weather are not shown in the figures. High absolute temperatures and daily temperature variations for most of the year, with significant vertical temperature differences, are observed in Figures 5.12–5.14. Table 5.1 summarizes the characteristic values of the measured temperature of the main girder, including the maximum temperatures T_{max} of TP and BP, the maximum daily temperature variations ΔT_{max} of TP and BP, and the maximum temperature differences between TP and BP (i.e., T_{diff_max}) for different seasons in 2019.

For TP, the maximum temperature was approximately 50°C, with a daily variation larger than 20°C except in winter when the corresponding values were 27°C and 9°C, respectively. For BP, the maximum temperature was approximately 30°C in the last three seasons, which was 10°C higher than that in winter, with a daily variation of approximately 4°C for the whole year. The vertical temperature differences

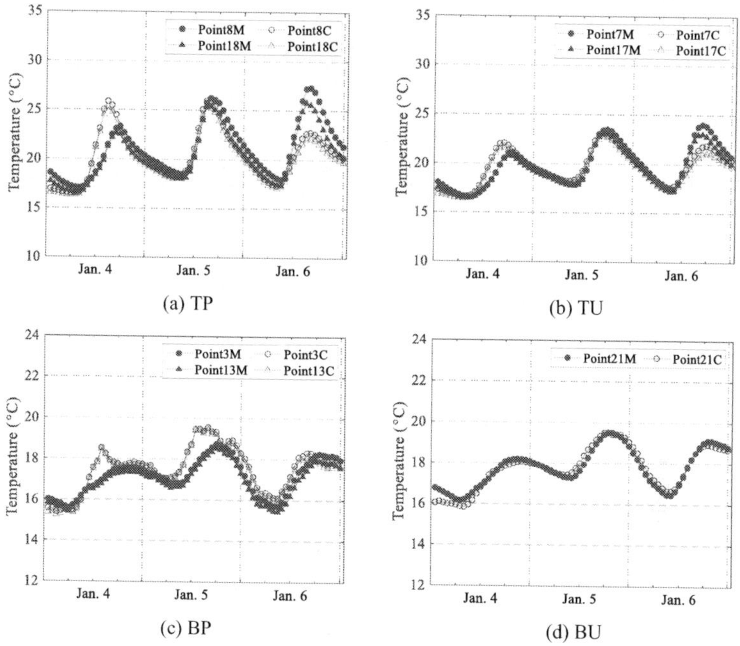

FIGURE 5.12 Measured and simulated temperatures of G09 on January 4–6, 2019: (a) TP, (b) TU, (c) BP, and (d) BU.

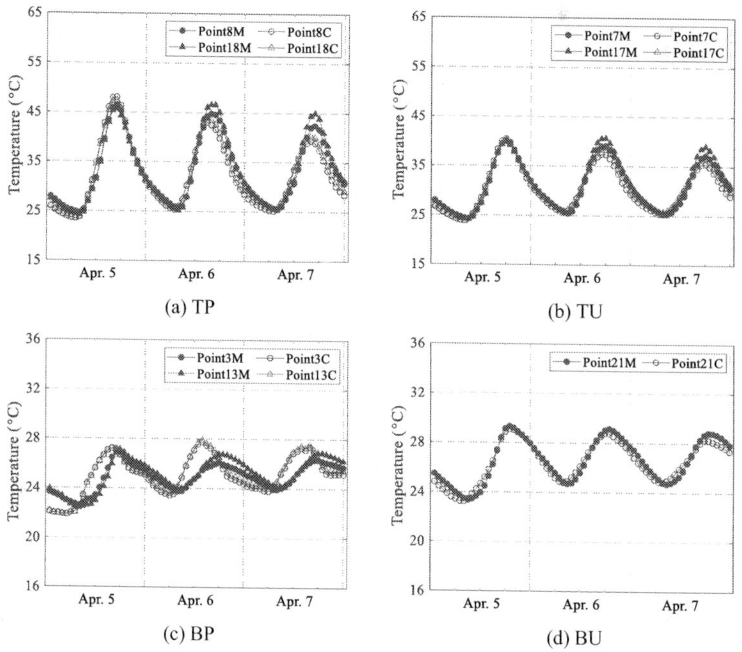

FIGURE 5.13 Measured and simulated temperatures of G09 on April 5–7, 2019: (a) TP, (b) TU, (c) BP, and (d) BU.

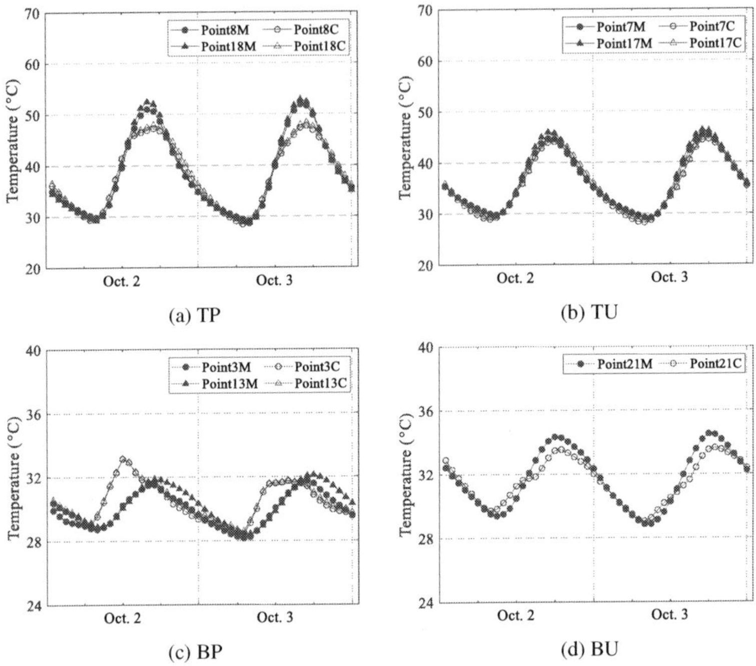

FIGURE 5.14 Measured and simulated temperatures of G09 on October 2–3, 2019: (a) TP, (b) TU, (c) BP, and (d) BU.

TABLE 5.1
Measured Temperature of the Main Girder in Different Seasons in 2019 (Unit: °C)

	TP		BP		
Period	T_{max}	ΔT_{max}	T_{max}	ΔT_{max}	T_{diff_max}
January 4–6	27	9	19	3	10
April 5–7	47	22	27	5	20
July 5–7	53	23	32	3	22
October 2–3	53	24	33	4	21

between TP and BP for most of the year were more than 20°C except in winter (less than 10°C). Apart from the diurnal variation, TP also experienced a higher seasonal temperature variation than BP. However, the maximum temperatures and daily variations did not necessarily occur in summer because the wind speed on July 5–7, 2019, was relatively high, indicating rapid heat dissipation.

The transverse temperature difference was not as significant but still nonnegligible. The maximum temperature of each sensor on the selected days, shown as the transverse temperature profiles of TP, TU, and BP in different seasons, are presented

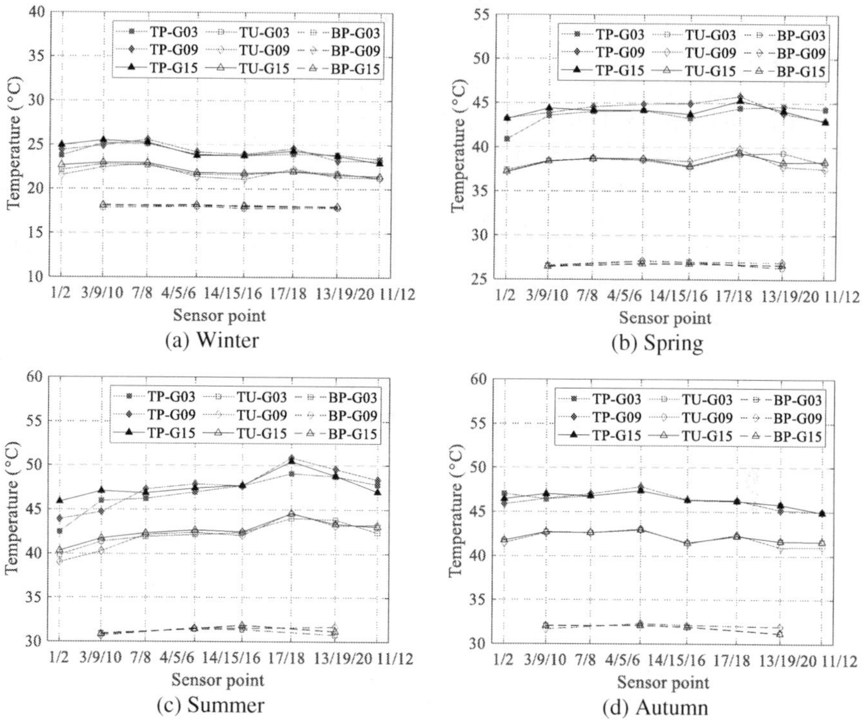

FIGURE 5.15 Transverse temperature profiles of TP, TU, and BP: (a) winter, (b) spring, (c) summer, and (d) autumn.

in Figure 5.15. The transverse temperature profile of BU is unavailable because each section has only one sensor point.

The vertical temperature difference was significant for most of the year, and the difference between TP and TU constituted over one-third of that between TP and BP, indicating a rapid and nonlinear temperature decrease with distance from the top plate. The transverse temperature profiles of TP and TU were similar throughout the year because of the high heat conductivity of steel. In winter, the temperatures of TP and TU on the south side were higher than those on the north side, with a maximum difference of approximately 3°C. The opposite was true in summer, with the maximum difference of approximately 7°C, while the difference was not obvious in spring and autumn. This is because the Qingzhou Bridge is located slightly lower than the Tropic of Cancer. The relative position change in the subsolar point to the bridge site in different seasons causes varying transverse heat accumulation. The transverse temperature difference of BP was trivial because the bottom part had a slight difference in solar radiation and wind on the two sides. In addition, the transverse temperature profiles in the three girder sections of G03, G09, and G15 almost coincided, indicating relatively small temperature differences in the longitudinal direction.

5.2.2.3 Temperature Distribution of Main Tower

Tower section T07 is selected as the representative section to investigate the temperature distribution of the tower. The location of Section T07 and the thermometer layout in the section are shown in Figures 5.3 and 5.5, respectively.

The measured temperatures of Points 1 and 6 for the four seasons are shown in Figure 5.16. The internal temperatures of the tower remained almost unchanged owing to the good insulation effect of the concrete.

The temperatures of the stay cables were not measured as no thermometers were equipped in the cables. However, the temperatures of the cables may have a remarkable influence on the bridge's performance and thus will be investigated through numerical investigations in later sections.

5.2.3 Temperature-Induced Structural Responses

The temperatures in the bridge can influence the structural responses, including the displacement and forces. The structural displacement, strain/stresses, and cable forces during the four prescribed representative periods will be presented in this section.

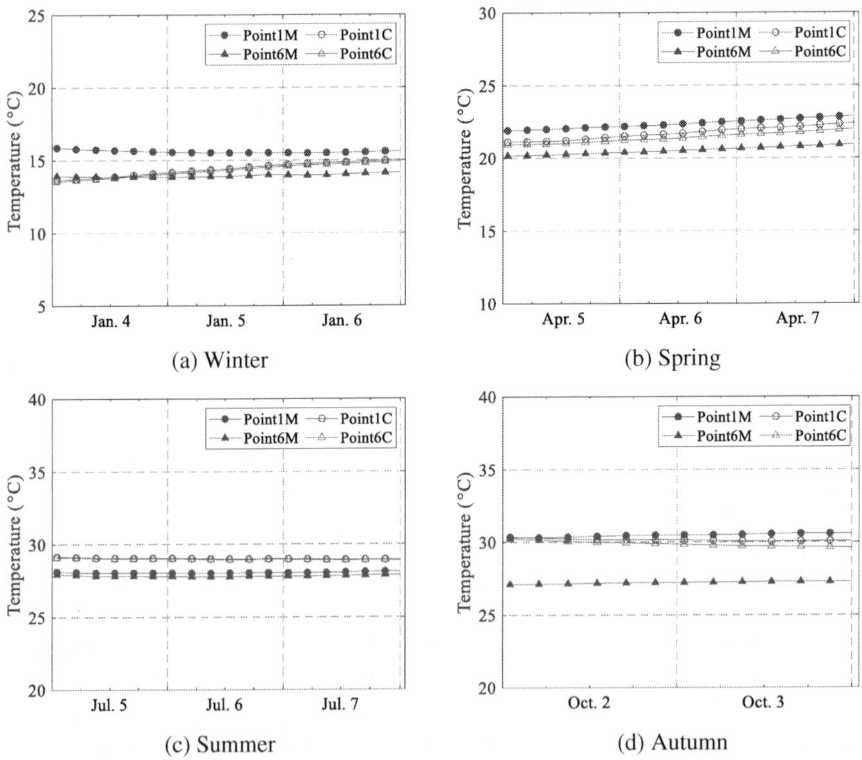

FIGURE 5.16 Measured and simulated temperature of the tower: (a) winter, (b) spring, (c) summer, and (d) autumn.

5.2.3.1 Displacement of the Main Girder

The longitudinal displacement of the girder end sections (G01 and G17) and the vertical deflection of the mid-span girder section (G09) are investigated here. The locations of the sections are shown in Figure 5.6. The measured displacements on July 5–7, 2019, are shown in Figure 5.17. In each section, the results of the north and south sides are averaged. Initial values of the displacements are subtracted to focus on the effect of the thermal variation in a day. The longitudinal displacement is positive along the expansion direction, and the vertical deflection is positive in the upward direction.

As shown in Figure 5.17a, the girder expanded with an increasing temperature and contracted with a decreasing temperature, as the longitudinal displacement of the girder is mainly dominated by the average girder temperature. The longitudinal displacement of girder section G01 is smaller than that of girder section G17. This is because the longitudinal displacement of the girder section is constrained on the tower on the HK side, and thus, the free length of the girder on this side is shorter than that on the Zhuhai side.

Figure 5.17b shows a downward movement of the girder mid-span in the morning and its upward movement in the afternoon, which aligns with most observations on cable-stayed bridges. The mid-span deflection is affected by two opposite factors: (1) when the temperature increased, the elongation of the cable led to the downward deflection of the girder; (2) the expansion of the girder indirectly caused the mid-span to move upward by drawing the tower outward through the cables. For the Qingzhou Bridge, the effect of the cable gives way to that of the average girder temperature. Therefore, the girder deflected upward in the afternoon when the temperature increased. The maximum deflection in one day was measured at approximately 49 mm.

The displacements in other seasons are shown in Figures 5.18–5.20. The variation trends of the measured displacements are similar. The displacement curves in winter were more rugged than those in summer, possibly due to the weak and unstable solar radiation during the period.

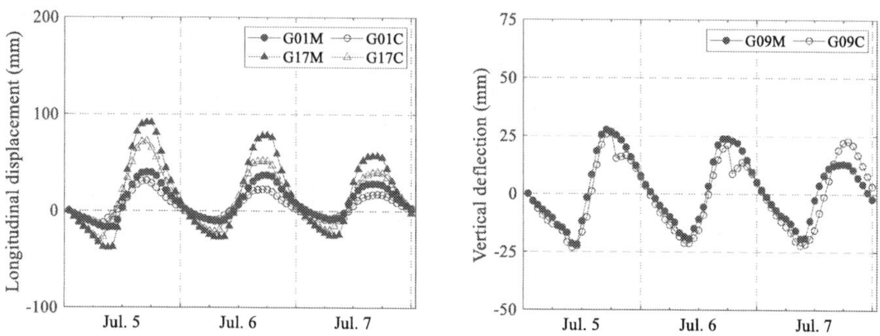

(a) Longitudinal displacement at the end

(b) Vertical deflection at the mid-span

FIGURE 5.17 Displacement of girder on July 5–7, 2019: (a) Longitudinal displacement at the end and (b) vertical deflection at the mid-span.

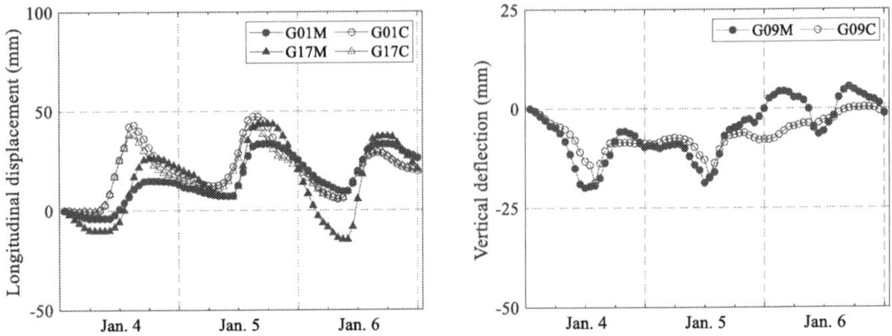

(a) Longitudinal displacement at the end (b) Vertical deflection at the mid-span

FIGURE 5.18 Displacement of girder on January 4–6, 2019: (a) Longitudinal displacement at the end and (b) vertical deflection at the mid-span.

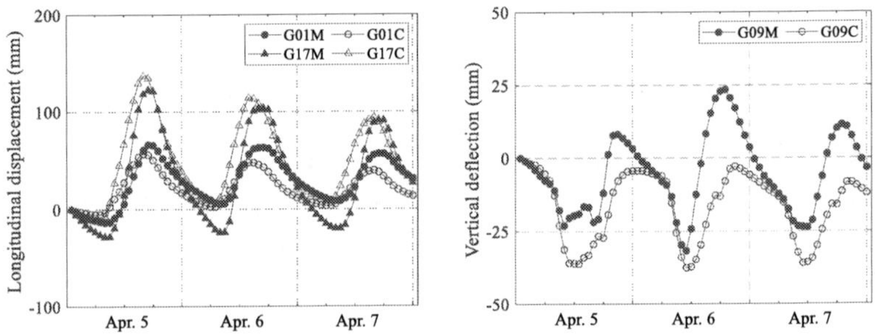

(a) Longitudinal displacement at the end (b) Vertical deflection at the mid-span

FIGURE 5.19 Displacement of girder on April 5–7, 2019: (a) Longitudinal displacement at the end and (b) vertical deflection at the mid-span.

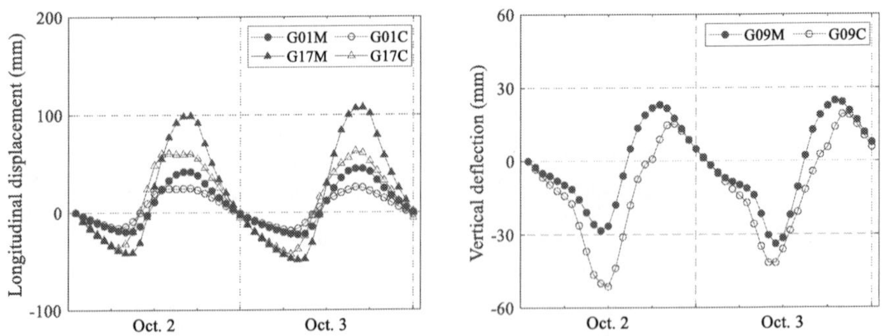

(a) Longitudinal displacement at the end (b) Vertical deflection at the mid-span

FIGURE 5.20 Displacement of girder on October 2–3, 2019: (a) Longitudinal displacement at the end and (b) vertical deflection at the mid-span.

TABLE 5.2
Displacement Results of the Main Girder in Different Seasons in 2019 (Unit: mm)

Period	ΔU_{max}	U_{RMSE}	ΔV_{max}	V_{RMSE}
January 4–6	76	20	20	4
April 5–7	231	32	55	15
July 5–7	187	21	49	5
October 2–3	223	32	59	12

Table 5.2 summarizes the maximum daily variations in the measured displacement of the main girder in 2019. ΔU_{max} and ΔV_{max} represent the maximum daily variation in the total longitudinal displacement and the mid-span vertical deflection, respectively. The daily displacements in the other three seasons were larger than those in winter because the diurnal temperature variations were significant in most of the year except in winter. Moreover, the displacement results in summer were slightly lower than those in spring and autumn. The maximum daily variation of the total longitudinal displacement could reach 231 mm, and the maximum daily mid-span vertical deflection was 59 mm.

5.2.3.2 Displacement of the Main Towers

The longitudinal displacements of the tower top sections (T01 and T05) measured by GPS rovers in the four seasons of 2019 are shown in Figure 5.21. The initial values are also subtracted from the results. The longitudinal displacement is positive along the outward direction.

As shown in the figure, the longitudinal displacements of the two-tower tops varied similarly to those of the two girder ends. As the temperature decreased in the morning, the tower tops moved inward with the contraction of the girder. As the temperature increased in the afternoon, the tower tops moved outward with the expansion of the girder. The movement range of the tower on the Zhuhai side was larger than that on the HK side, corresponding to a larger expansion of the girder end on the same side. This is because the cables draw the tower in the same movement direction as the girder to achieve displacement compatibility. The maximum daily longitudinal displacement could reach 60 mm for the Hong Kong tower top and 93 mm for the Zhuhai tower top in autumn. Thus, the average girder temperature was also the main factor influencing the longitudinal displacement of the tower.

5.2.3.3 Strain/Stress of the Main Girder

For cable-stayed bridges, the girder sections supported on the tower may be subject to high compressive stress. Here, the strain and stress of section G13 are studied. The locations of Section G13 and the strain sensor layout are shown in Figures 5.7 and 5.8, respectively. Given that strain gauges are at the same location as the thermometers, the strain results at TP, TU, BU, and BP are also analyzed. As strain gauges

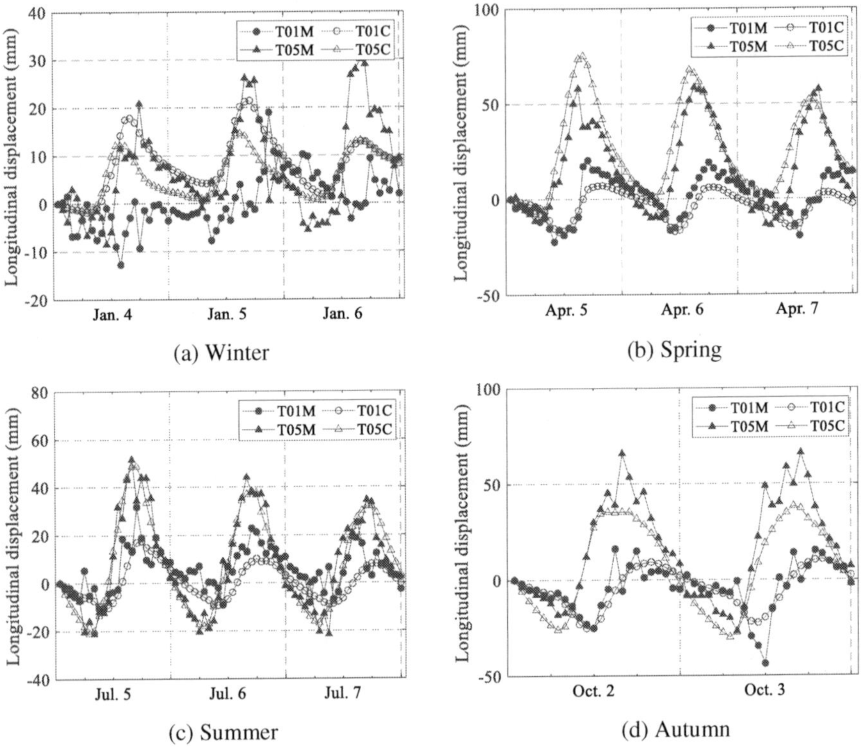

FIGURE 5.21 Longitudinal displacement of the tower top in 2019: (a) winter, (b) spring, (c) summer, and (d) autumn.

are placed in the longitudinal direction of the girder, only longitudinal strains and stresses of the girder are discussed.

The strains on section G13 on July 5–7, 2019 are shown in Figure 5.22. The relative strain with respect to the initial value is studied. The strain of TU is similar to that of TP with a slight lag effect. Only the results of TP and BP are illustrated here as no strain gauge has been installed on the surface of BU. In particular, Points 23 and 27 are selected to show the strain results of TP on the south and north sides, respectively, and points 31 and 32 are chosen for BP on the south and north sides, respectively.

The maximum strain of TP due to the thermal effect was approximately 83 $\mu\varepsilon$ with a maximum daily variation of 119 $\mu\varepsilon$, and that of BP was 55 $\mu\varepsilon$ with a maximum daily variation of 84 $\mu\varepsilon$. The thermal strain of the girder was significant.

The total strain is caused by the dead load and thermal effects. It includes the free thermal strain and the elastic strains. The elastic strain is expected to generate elastic stress, while the free thermal strain would not. The total strain and total stress can be expressed as follows:

$$\varepsilon_{\text{total}} = \varepsilon_{\text{el}}^{d} + \varepsilon_{\text{el}}^{\text{th}} + \varepsilon_{\text{th}} \tag{5.1}$$

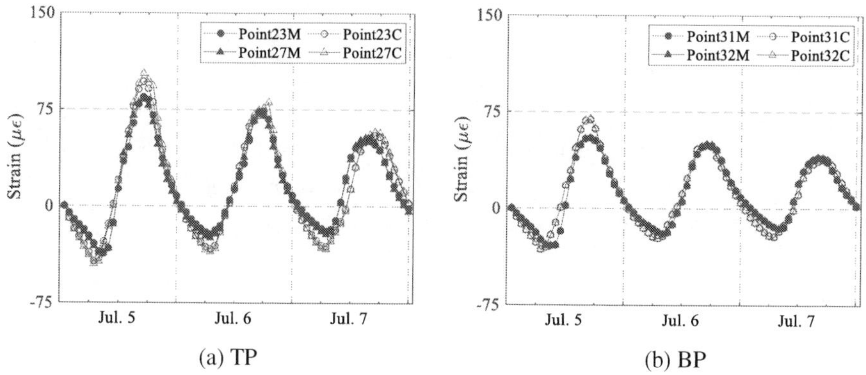

FIGURE 5.22 Strain on G13 on July 5–7, 2019: (a) TP and (b) BP.

$$\sigma_{\text{total}} = \sigma_{\text{el}}^{d} + \sigma_{\text{el}}^{\text{th}} \tag{5.2}$$

where $\varepsilon_{\text{total}}$, σ_{total} are the total strain and stress, respectively; $\varepsilon_{\text{el}}^{d}$, $\varepsilon_{\text{el}}^{\text{th}}$, ε_{th} are the dead load-induced elastic strain, temperature-induced elastic strain, and the temperature-induced free thermal strain, respectively; correspondingly, σ_{el}^{d} and $\sigma_{\text{el}}^{\text{th}}$ are the dead load-induced elastic stress and temperature-induced elastic stress, respectively.

As the sensors were installed after the completion of the bridge construction, the dead load-induced strain could not be measured. The monitored strain thus only consists of the elastic strain due to the temperature change and the free thermal strain. The latter should be removed to calculate the measured thermal stress, that is,

$$\varepsilon_{\text{el}}^{\text{th}} = \varepsilon^{M} - \varepsilon_{\text{th}} = \varepsilon^{M} - \alpha\Delta T \tag{5.3}$$

$$\sigma_{\text{el}}^{\text{th}} = E\varepsilon_{\text{el}}^{\text{th}} \tag{5.4}$$

where ε^{M} is the measured strain, α is the thermal expansion coefficient, E is Young's modulus, and ΔT is the temperature change.

As the initial temperature state of the bridge after construction is complicated and its data are not available, only the stress change with respect to the initial value is studied. Figure 5.23 shows the stresses of section G13 on July 5–7, 2019. The temperature variation caused TP to be in compression and BP to be in tension most of the time.

The measured stress variations of the other three seasons are shown in Figure 5.24. In these seasons, TP was in compression, while BP was in tension, which is similar to the scenario in summer because the vertical temperature difference between TP and BP existed throughout the year. The stress variation was significant except in winter, aligning with the vertical temperature difference in different seasons.

Table 5.3 summarizes the measured maximum daily stress variation of the main girder for different seasons in 2019. The maximum stress variations of TP were all higher than those of BP, indicating that TP was more subject to daily temperature

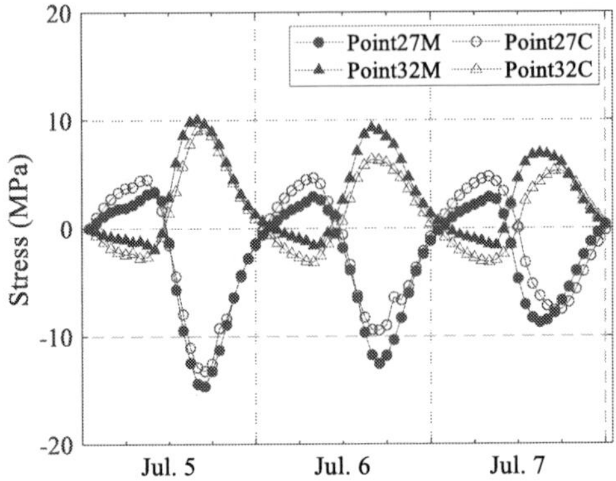

FIGURE 5.23 Stress on G13 on July 5–7, 2019.

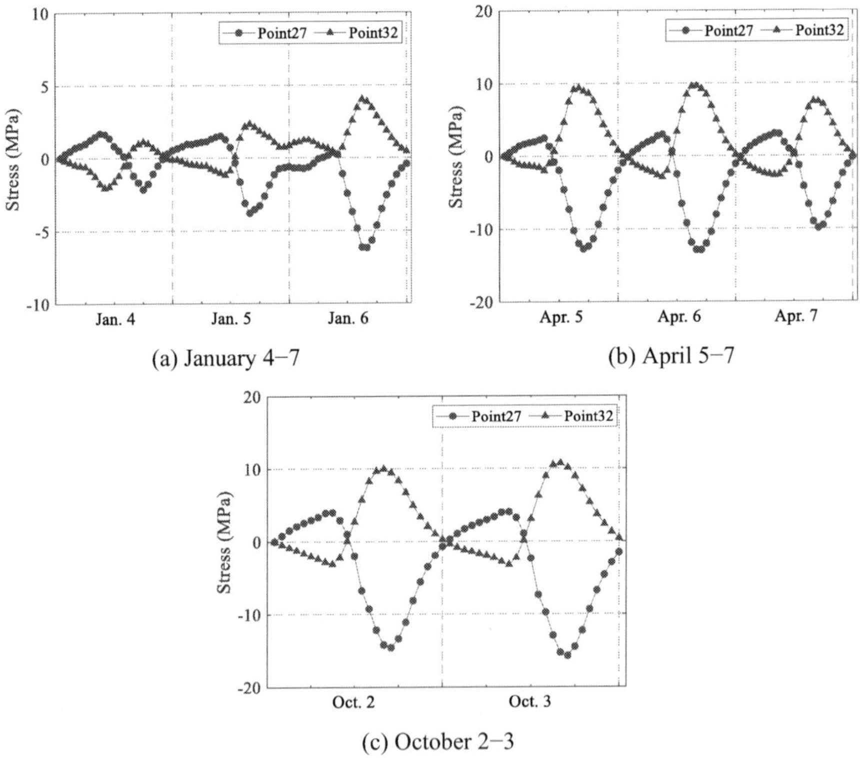

FIGURE 5.24 Stress on G13 in the other three seasons of 2019: (a) January 4–7, (b) April 5–7, and (c) October 2–3.

variations. The maximum daily stress variations in different seasons were close to each other except in winter. The maximum daily variations were approximately 18 MPa for TP and 13 MPa for BP in the other three seasons, while those in winter were 7 and 4 MPa for TP and BP, respectively. Thus, the temperature-induced strain

TABLE 5.3
Stress of the Main Girder in Different Seasons in 2019 (Unit: MPa)

Period	TP		BP	
	$\Delta\sigma_{max}$	σ_{max}	$\Delta\sigma_{max}$	σ_{max}
January 4–6	7	−29	4	−23
April 5–7	16	−42	13	−23
July 5–7	18	−42	12	−24
October 2–3	20	−39	14	−25

and stress were significant, indicating the need to give more attention to them in bridge design and health monitoring.

5.2.3.4 Strain/Stress of the Main Towers

The measured total strains of tower section T03 on the Hong Kong side and tower section T07 on the Zhuhai side are extracted. The corresponding measured thermal stresses are calculated according to Equations (5.3) and (5.4). Figure 5.25 presents the measured strains and stresses of Point 3 on the Zhuhai side and Point 6 on the Hong Kong side. The initial values are subtracted from the results. The strains and stresses of the towers are in the vertical direction (i.e., along the longitudinal direction of the tower).

Figure 5.25 shows that the variations in the strains and stresses of the two towers were similar and small because their temperatures were highly stable, as elaborated previously in Section 5.2.2.3. In particular, as shown in Figure 5.25a and c, Point 3 in section T03 contracted in the morning and elongated in the afternoon relative to 1:00, whereas Point 6 changed slightly. These trends indicate that the tower on the Hong Kong side is bent to the Zhuhai side in the morning and bent conversely in the afternoon. Meanwhile, the opposite results are observed for section T07. In other words, the two towers were bent inward in the morning and outward in the afternoon. In addition, the strain and stress of T03 on the Hong Kong side were smaller than those of T07 on the Zhuhai side. This is because the expansion and contraction of the girder were the main causes of the tower's movement, and the girder was longitudinally constrained on the tower on the Hong Kong side. Thus, the side span girder on the Hong Kong side had a smaller movement than the Zhuhai side span (Section 5.2.3.1), which further induced a smaller tilt of the Hong Kong tower than the Zhuhai tower.

The stresses of T07 in the other three seasons are shown in Figure 5.26. Similarly, when the temperature increased, the two towers bend outward. The overall stress of the tower was less than 1 MPa throughout the year because of the slight temperature variation.

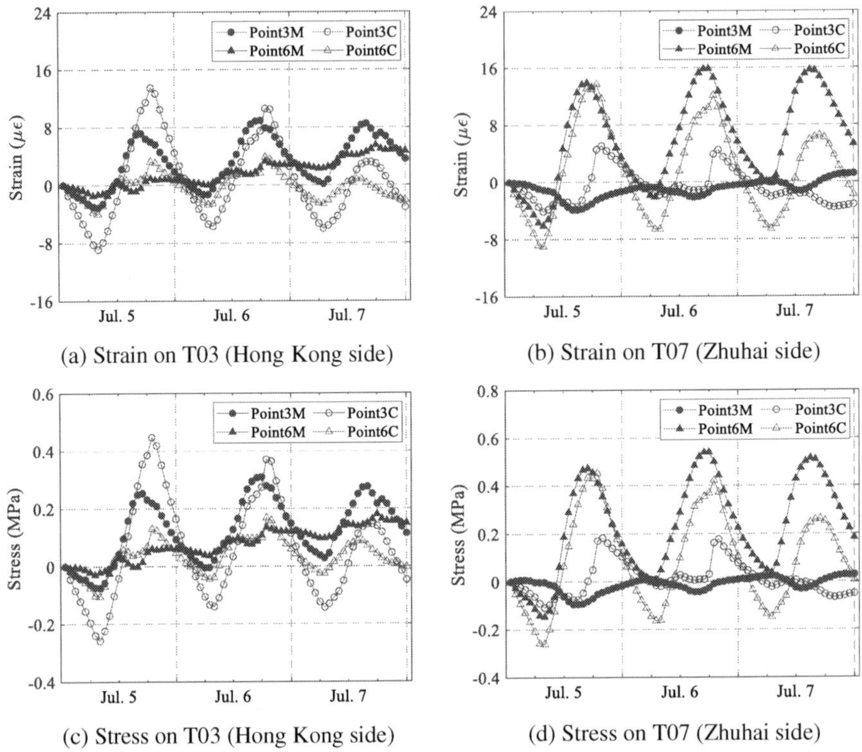

FIGURE 5.25 Strain and stress on the tower on July 5–7, 2019: (a) Strain on T03 (Hong Kong side), (b) Strain on T07 (Zhuhai side), (c) Stress on T03 (Hong Kong side), and (d) Stress on T07 (Zhuhai side).

5.2.3.5 Stresses of Stay Cables

Based on the acceleration data obtained by cable tensiometers, the cable stress is calculated using a frequency-based method as follows (Cho et al., 2013):

$$S = 4\rho l^2 f_1^2 \tag{5.5}$$

where S represents the cable stress, ρ is the cable density, l is the cable length, and f_1 denotes the first-order frequency of the cable.

Figure 5.27 shows the stress of the mid-span shortest cable anchored on the north leg of the tower on the Zhuhai side (M1) in 2019. The initial values are subtracted from the results to reflect the temperature effect.

The cable stress variation was generally within 20 MPa. The cable prestress was slightly released in the morning as the temperature decreased, and increased in the afternoon when the temperature increased. The cable stress variation was highly

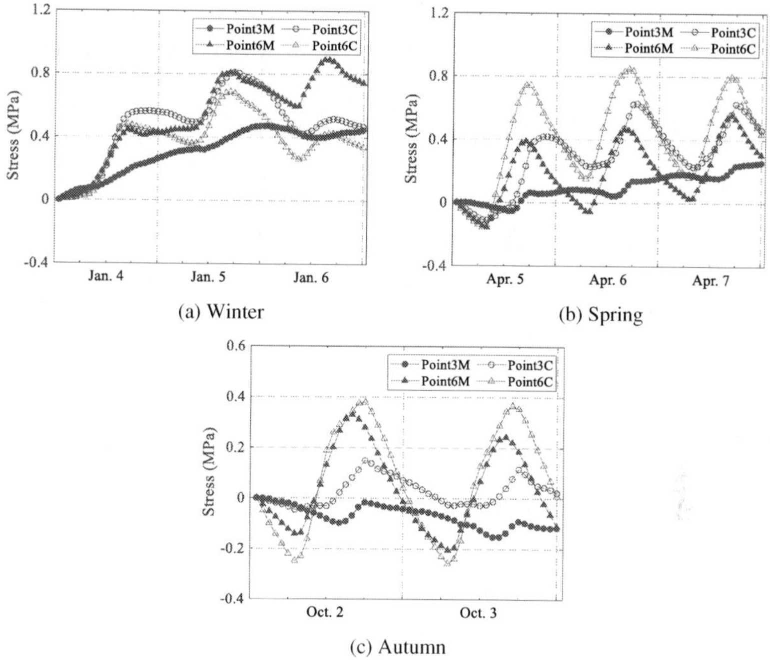

FIGURE 5.26 Stress on tower section T07 in winter, spring, and autumn of 2019: (a) winter, (b) spring, and (c) autumn.

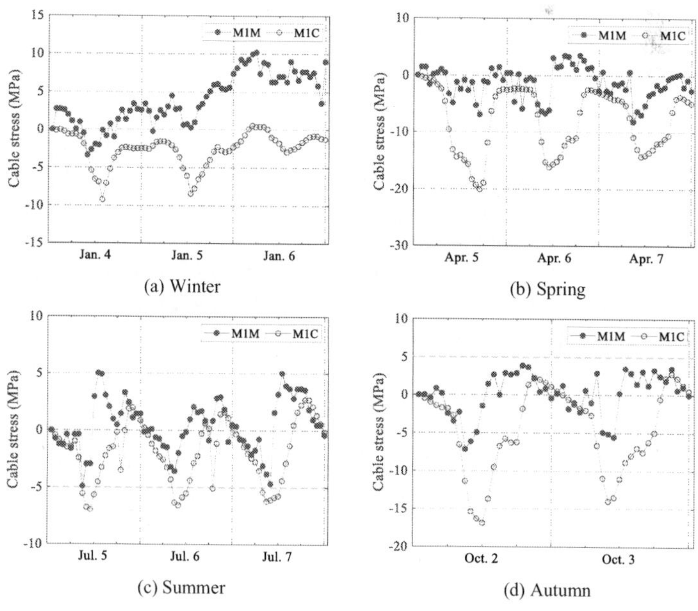

FIGURE 5.27 Stress of cable M1 on the north, Zhuhai side in 2019: (a) winter, (b) spring, (c) summer, and (d) autumn.

related to the distance between its two anchor points on the tower and the girder, so it was influenced by the movements of the tower and the girder, which were mainly governed by the average girder temperature. Therefore, the cable stress was determined simultaneously by both the cable temperature and the average girder temperature.

5.2.4 Summary

In this section, the temperature distribution pattern and characteristics of the temperature-induced structural responses of the Qingzhou Bridge, including the bridge displacement and forces, are comprehensively analyzed from the field monitoring data. The daily and seasonal temperature variations were significant. In particular, the transverse temperature gradients were dominated by the solar radiation conditions. The bridge displacement, strain/stresses, and cable stress were also remarkably influenced by the temperature.

5.3 NUMERICAL INVESTIGATIONS OF A CABLE-STAYED BRIDGE

In this section, the thermal distribution and temperature-induced structural responses of the Qingzhou Bridge are comprehensively and accurately simulated by establishing the global 3D FE model of the bridge using ANSYS, following the unified analysis scheme introduced in Chapter 3.

5.3.1 FE Model of the Qingzhou Bridge

5.3.1.1 Details of FE Model

The refined FE model of the Qingzhou Bridge consists of the following parts: 493,941 nodes and 520,422 elements, including 170,336 solid elements for the steel box girder, the asphalt concrete layer, tower legs, lower tower transoms, and piers; 349,962 shell elements for the diaphragms, U-ribs, and upper tower transoms; and 124 link elements for the cables and bearings. The FE model of the bridge and close-range illustrations of representative components are shown in Figure 5.28.

The webs, diaphragms, and U-ribs of the girder simulated with shell elements contain 305,455 nodes and 344,650 elements, constituting most of the model's nodes and elements. The upper transom of the tower, which is in the shape of a Chinese knot and composed of steel, is also carefully modeled with shell elements. The bearings are modeled with link elements that connect the central nodes of the stiffened surfaces of the girder and the pier or the tower. For the connection between the tower and stay cables, the nodes of the concrete pylon, the steel anchor box, and the cables are joined together. Close views of these parts are also shown in Figure 5.28.

Conventional studies of the temperature behaviors of cable-stayed bridges generally exclude cables from the heat-transfer analysis. In this section, a refined FE model of four cables is established with solid elements for simulating the steel wire inner core and the polyethylene (PE) sheath to calculate the temperature distribution of the cables. The shortest and longest side cables (S1 and S14) and middle cables

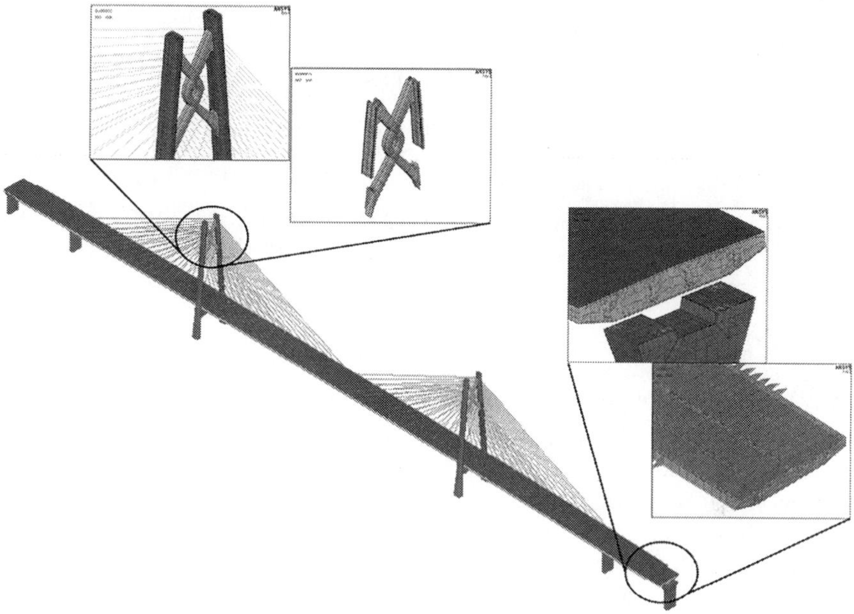

FIGURE 5.28 Global 3D FE model of the Qingzhou Bridge.

(M1 and M14) anchored on the south leg of the tower on the Zhuhai side are studied for simplicity (Figure 5.29).

SOLID70, SHELL57, and LINK33 are assigned to the solid, shell, and link elements, respectively, in the heat-transfer analysis. The FE model contains a total of 493,941 nodes, each of which has one temperature degree of freedom (DOF). The material parameters used in the heat-transfer analysis are listed in Table 5.4.

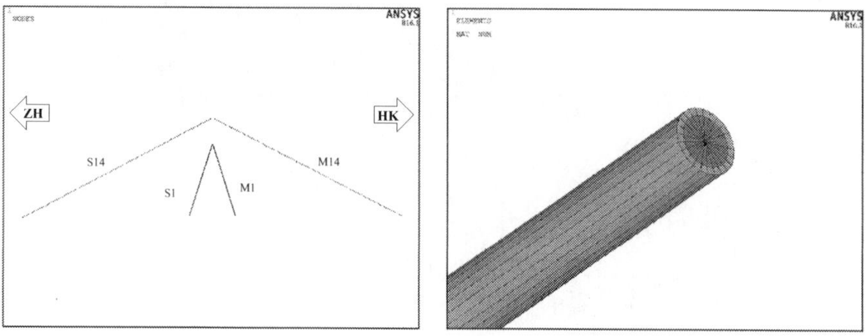

(a) Elevation (b) Close-range view

FIGURE 5.29 Refined FE model of stay cables: (a) elevation and (b) close-range view.

TABLE 5.4
Material Parameters for Heat-Transfer Analysis

Parameter	Unit	Asphalt	Steel	Concrete	PE
Density	kg/m^3	2365	7850	2600	965
Specific heat capacity	J/kg/°C	1075	460	925	2300
Thermal conductivity	W/m/°C	1.8	55	2.71	0.44
Emissivity coefficient	–	0.92	0.8	0.88	0.90
Absorptivity coefficient	–	0.90	0.685	0.65	0.65

5.3.1.2 Convergence Analysis

The mesh size of the FE model and the timestep of the analysis affect the computational efficiency and accuracy (Beckmann et al., 2013). Thus, a convergence analysis is conducted on a 3D segmental model, as shown in Figure 5.30.

The period of July 5–7, 2019, is selected for the convergence analysis. First, the heat-transfer analysis is conducted with one step per hour, totaling 72 load steps. The mesh division is one layer in the thickness direction of the steel girder and the asphalt, and the default mesh size is 2.5 m in the longitudinal direction. The total computational time is the recorded time on a computer with an i9–10900K CPU, a basic frequency of 3.70 GHz, and RAM of 32 GB. The root mean square error (RMSE) of the temperature results is calculated with respect to a baseline result, which is obtained using a 1-minute timestep and a mesh size six times denser in both the thickness and longitudinal directions. The calculation accuracy is measured

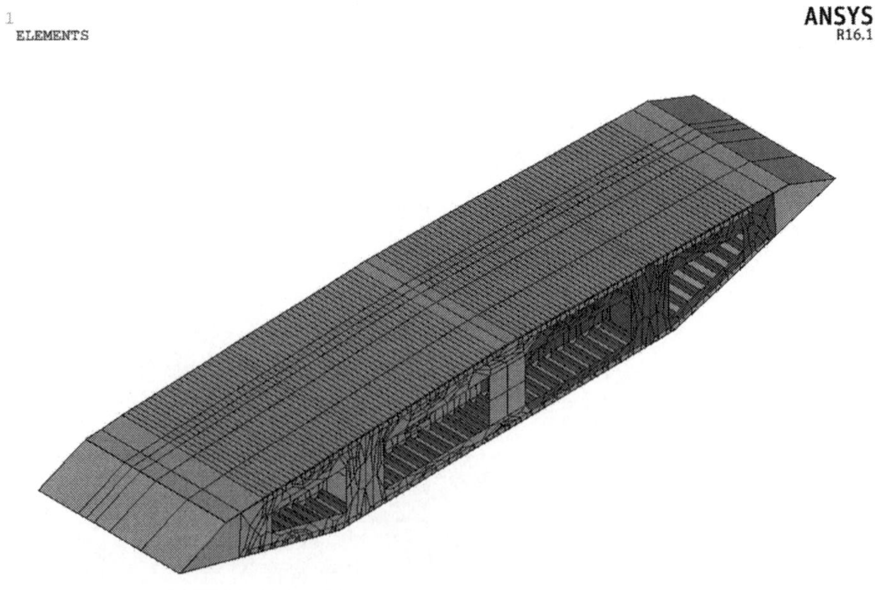

FIGURE 5.30 A 3D segmental model for convergence analysis.

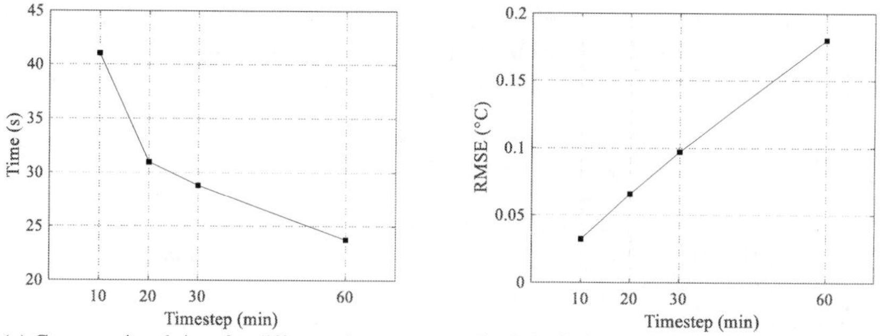

(a) Computational time for different timesteps (b) Calculation accuracy for different timesteps

FIGURE 5.31 Timestep convergence analysis: (a) computational time for different timesteps and (b) calculation accuracy for different timesteps.

based on the average RMSE of all effective sensor data, weighted by area, on the mid-span section of the girder during the three days.

Four different timesteps (60, 30, 20, and 10 minutes/step) are studied. For each case, the total computational time and the RMSE of the temperature results of the mid-span girder section are compared in Figure 5.31. As expected, a longer timestep consumes less computational time but gives a larger error. When the timestep is 30 minutes, the average RMSE is nearly reduced by half, with the total calculation time increasing by only approximately 5 seconds compared with the timestep of 60 minutes. Thus, the timestep of 30 minutes is used hereafter.

Four different mesh sizes are investigated, corresponding to the deck with one to four layers of meshes in the thickness direction with longitudinal sizes of 2.5, 1.25, 0.833, and 0.625 m, resulting in 5,666, 26,726, 74,615, and 159,820 elements, respectively. Figure 5.32 shows the total computational times and the RMSEs of the four mesh schemes. In general, more elements require longer computational times

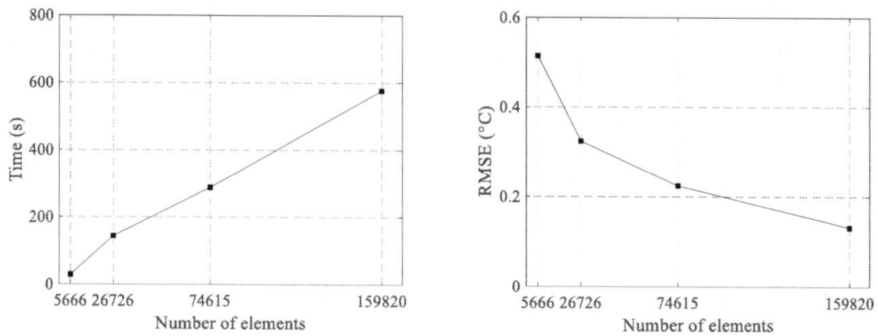

(a) Calculation time for different numbers of elements (b) Calculation accuracy for different numbers of elements

FIGURE 5.32 Mesh convergence analysis: (a) calculation time for different numbers of elements and (b) calculation accuracy for different numbers of elements.

and yield more accurate results. However, when the mesh size is halved, although the RMSE is reduced by over one-third, the computational time increases fivefold. Therefore, the original mesh size is adopted in the subsequent analysis.

5.3.2 Thermal Boundary and Initial Conditions

Given the measured environmental conditions of the representative days in the four seasons, the thermal boundary conditions for the heat-transfer analysis are determined. The boundary conditions in the summer are given here.

As July 3, 2019, was a rainy day, and the rain effect was not considered in the heat-transfer analysis, July 1–4 was then taken as the pre-analysis for the initial conditions, according to the procedures specified in Chapter 3. The results on July 5–7 were examined. After several days of heat-transfer analysis with the measured environmental conditions, the effect of the incorrect initial condition on the temperature distribution is assumed to be negligible.

5.3.2.1 External Thermal Boundary Conditions

For the external thermal boundaries, all outer surfaces of the main components of the whole bridge are categorized into four subsets and listed in Table 5.5, according to their geometric configurations and materials. For instance, the outer surfaces of the box girder include the upper surfaces of the asphalt and the side and bottom surfaces of the steel girder. The configuration is simplified into 7 × 2 surfaces, which are doubled considering the longitudinal slope. The seven surfaces are shown in Figure 5.33. The 8th and 9th surfaces constitute the outer surfaces of the girder segments without wind fairings.

TABLE 5.5
Outer Surfaces of the Main Girder

Component	Material	No. of surfaces
Box girder	Asphalt concrete, steel	14
Pylon and lower transom of tower	Concrete	113
Upper transom of tower	Steel	154
Cable	PE	96

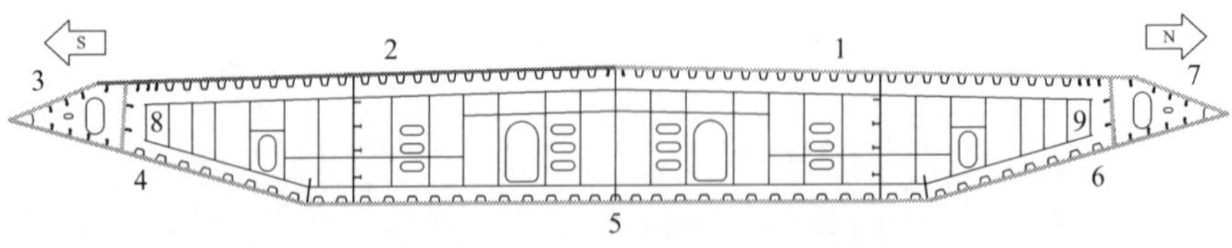

FIGURE 5.33 Outer surfaces of the main girder.

The ambient temperature of all outer surfaces is taken as identical to that of the external air. The wind speed and solar radiation imposed on each surface are taken differently.

Wind speed is a dominant factor in determining the convection heat-transfer coefficient, and thus it significantly affects the heat-transfer results. Most studies typically use a constant wind speed in their simulations; however, real structural surfaces are exposed to varying wind conditions due to different incident angles. Furthermore, the actual wind speed near the structural surface is lower than the measurement of the anemometer, which is generally located several meters from the structure. Therefore, the effective wind speed on different structural surfaces is calculated using empirical equations in Chapter 3, according to the wind incident angle.

As no pyranometer is installed in the bridge, the solar radiation data collected by the Hong Kong Observatory are adopted. The data are the sum of the direct and diffuse solar radiation on a horizontal surface. They are decomposed and transformed to the direct, diffuse, and reflected solar radiations on tilted surfaces of the structure following the methods introduced in Chapter 3.

The external thermal boundaries on all outer surfaces can be determined and expressed as the overall heat-transfer coefficients and the equivalent ambient temperatures. The external thermal boundaries of the asphalt deck surface and the south-side surface, the north-side surface, and the bottom surface of the steel girder, corresponding to surfaces 2, 5, 8, and 9 in Figure 5.33, are illustrated in Figure 5.34.

As shown in Figure 5.34a, the overall heat-transfer coefficients of the four surfaces differed from each other due to different wind incident angles. The overall coefficient of the south-side surface was generally the highest, whereas that of the north-side surface was the lowest because the wind mainly came from the south and the east, as depicted in Figure 5.10a. Meanwhile, as shown in Figure 5.34b, the deck surface had the highest equivalent ambient temperature, whereas the bottom surface had the lowest. This is because the deck surface was subject to the most direct solar radiation, whereas the bottom surface received only a slight reflected solar radiation. The equivalent ambient temperature of the north-side surface was higher than that of the south-side surface and exhibited a sudden peak at sunset. This difference is because

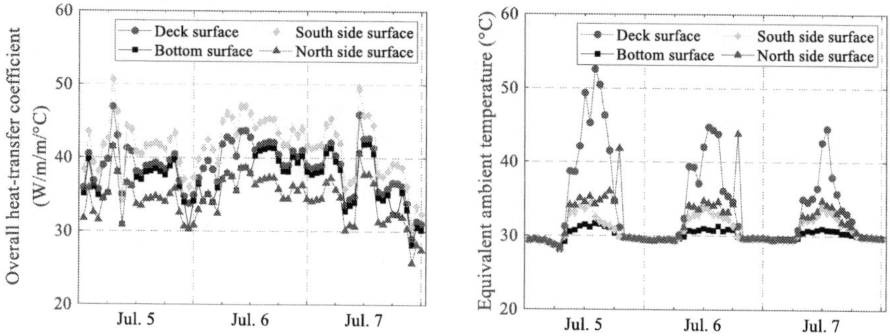

(a) Overall heat-transfer coefficient (b) Equivalent ambient temperature

FIGURE 5.34 Thermal boundary conditions of the four outer surfaces of the girder: (a) overall heat-transfer coefficient and (b) equivalent ambient temperature.

the Qingzhou Bridge is located slightly lower than the Tropic of Cancer. Thus, the northern surface received more solar radiation than the southern surface during the period, particularly at sunset when the solar incident angle reached the maximum.

Obviously, ambient temperature, wind, and solar radiation are highly related to the heat dissipation and accumulation process and thus play significant roles in the heat-transfer analysis of long-span cable-stayed bridges. Therefore, in-situ measurements of environmental conditions are necessary to accurately calculate the temperature distribution of bridges.

5.3.2.2 Internal Thermal Boundary Conditions

The internal thermal boundary conditions of the box girder and the tower must also be determined. As the wind speed and solar radiation inside the box section were zero, the equivalent ambient temperature equaled the ambient temperature inside the box section.

The temperature of the inner surface of the tower was approximately evenly distributed owing to the good insulation effect of the thick concrete tower wall. Therefore, the monitored structural temperature is applied to the inner surfaces of the tower as the equivalent ambient temperature.

For the inner surface of the girder, the vertical temperature gradient is generally significant. However, almost all SHM systems (including the present one) have a single thermometer (even none in others) inside the box to measure the air temperature. Moreover, typical numerical studies did not consider the air temperature difference between the top and bottom plates. In this section, the inner surface of the girder is divided into four components, namely, the bottom surface of the top plate, the surface of the U-ribs on the top plate, the surface of the U-ribs on the bottom plate, and the top surface of the bottom plate, abbreviated as TP, TU, BU, and BP, respectively. Different thermal boundaries are assigned to these surfaces, as discussed below.

As the internal periodic temperature variation of the girder is similar to a sine function, four-piecewise sine functions are constructed to simulate the daily air temperature variations. Figure 5.35 shows a sketch of a constructed equivalent ambient temperature on an inner surface in one day. The magnitude difference and the phase difference between the equivalent ambient temperature of the four surfaces and the measured internal ambient temperature are jointly considered. Four amplification coefficients and four phase indices for indicating two extreme points and two inflection points are determined iteratively according to the monitored structural surface temperature. For the above four surfaces, a total of 16 amplification coefficients and 16 phase indices need to be determined.

A temperature difference also exists in the transverse direction, but the difference in the longitudinal direction inside the girder box is insignificant. Therefore, the internal thermal boundaries on the south- and north-side surfaces of the box girder are based on the longitudinally averaged ambient temperature on the south and north sides, respectively.

The overall heat-transfer coefficients and the equivalent ambient temperatures of the four inner surfaces of the girder (i.e., TP, TU, BU, and BP) on the south and north sides are shown in Figure 5.36.

The vertical temperature gradient is reflected by a descending equivalent ambient temperature of TP, TU, BU, and BP. As the overall heat-transfer coefficient

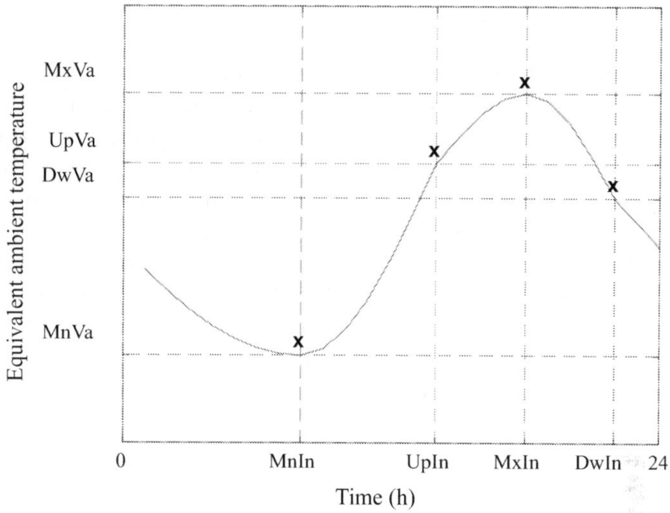

FIGURE 5.35 Constructed equivalent ambient temperature in one day.

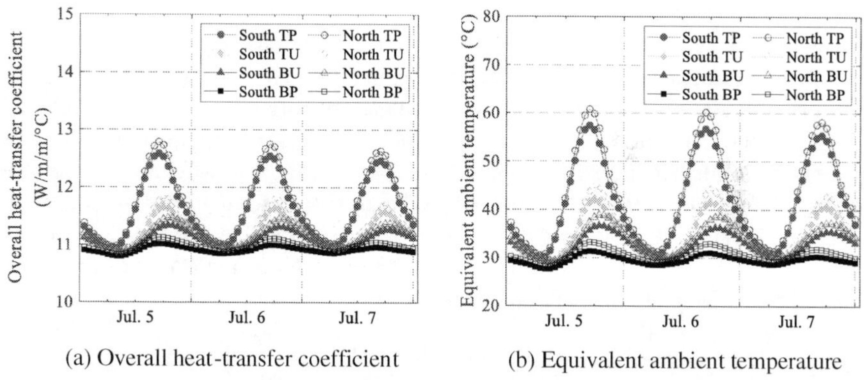

FIGURE 5.36 Thermal boundary conditions on the inner surfaces of the four components of the girder: (a) overall heat-transfer coefficient and (b) equivalent ambient temperature.

was positively related to the ambient temperature, the patterns of the corresponding curves were also similar. The overall heat-transfer coefficient and the equivalent ambient temperature on the north side exceeded those on the south side because the heat accumulation on the former, caused by solar radiation, was slightly greater than that on the south side during the calculation period.

Based on the monitored ambient temperature, wind, and solar radiation data, the thermal boundary conditions, characterized by the overall heat-transfer coefficients and the equivalent ambient temperatures of all outer and inner surfaces, are calculated and input into the FE model. Then, transient heat-transfer analysis is conducted, and the temperature distribution of the entire bridge is obtained. The findings are presented in the subsequent section.

5.3.3 Temperature Distributions

5.3.3.1 Temperature Distribution of Main Girder

The temperature of the entire bridge is obtained from the transient heat-transfer analysis. The calculated temperatures at the sensor points are compared with their measurement counterparts below. For demonstration, the temperatures of the representative points (Points 8 and 18 for TP, Points 7 and 17 for TU, Points 3 and 13 for BP, and Point 21 for BU) are compared with the corresponding measurements in Figures 5.11–5.14. In the figures, the curves with a legend ending with "M" denote the measured temperatures, while the curves ending with "C" are the calculated ones from the FE model.

In the four seasons, the simulated and measured temperatures are consistent on all days. The discrepancies over some periods are due to the solar radiation data obtained from the Hong Kong Observatory, which may differ from the in-situ conditions. The RMSEs of the temperature results of G09 in the four seasons are summarized in Table 5.6. The average RMSEs of the effective temperature do not exceed 1.4°C. TP has the largest difference because the top plate received more heat flow from solar radiation. BU has the smallest difference, as its temperature was mainly controlled by the internal air temperature.

5.3.3.2 Temperature Distribution of the Main Tower

The computed temperatures of the main tower also agree with the corresponding measurement counterparts, as shown in Figure 5.16. The average RMSEs of all measuring points in January, April, July, and October are 0.8°C, 0.9°C, 0.9°C, and 1.7°C, respectively. The large RMSE in October is due to the significant difference in Point 6. A further investigation shows that the significant difference in Point 6 continued after October 2019. This may be due to the sensor drift after that period. As thermometers are only inside the lower column of the concrete tower wall, the calculated temperature results of the upper part and the outer surfaces of the tower cannot be verified and, thus, are not shown here.

5.3.3.3 Temperature Distribution of Stay Cables

The heat-transfer analysis is also conducted on cables to obtain their real-time temperature variation using the refined model of the stay cables, as shown in Figure 5.29.

TABLE 5.6
RMSEs of the Simulated Temperature of G09 in 2019 (Unit: °C)

Period	TP	TU	BU	BP	Average
January 4–6	1.5	0.8	0.2	0.6	1.0
April 5–7	2.0	1.0	0.4	1.2	1.4
July 5–7	1.8	1.4	0.2	0.5	1.1
October 2–3	2.1	1.1	0.6	1.0	1.4

As no thermometers are on the cable, only the calculated results are presented here. The cable temperature was uniformly distributed along the whole steel inner core owing to the high heat conductivity of steel. Then, the calculated temperatures of the center points of the four simulated cables are extracted, as shown in Figure 5.37. The S1C and M1C curves refer to the calculated temperatures of the cables S1 and M1, which are the shortest cables on the side and middle spans, respectively, while S14C and M14C are those of the longest cables (S14 and M14). The locations of the cables are shown in Figures 5.9 and 5.29a.

As presented in Figure 5.37, the temperature of the cable could reach approximately 42°C in autumn, and the maximum variation in a day was 21°C in spring. By contrast, the cable temperature in summer was relatively low, mainly due to the high wind speed and thus the fast heat dissipation during the calculation period. Furthermore, the temperature curves of the cables have spikes and are less smooth than those inside the girder. These features differ from those of the tower and girder because the cables' diameter is only approximately 20 cm and thus are more sensitive to external conditions. Moreover, the two longest cables generally had a single peak temperature at noon, whereas the two shortest cables tended to have one more peak at sunset when they were subject to a large incident angle and received significant

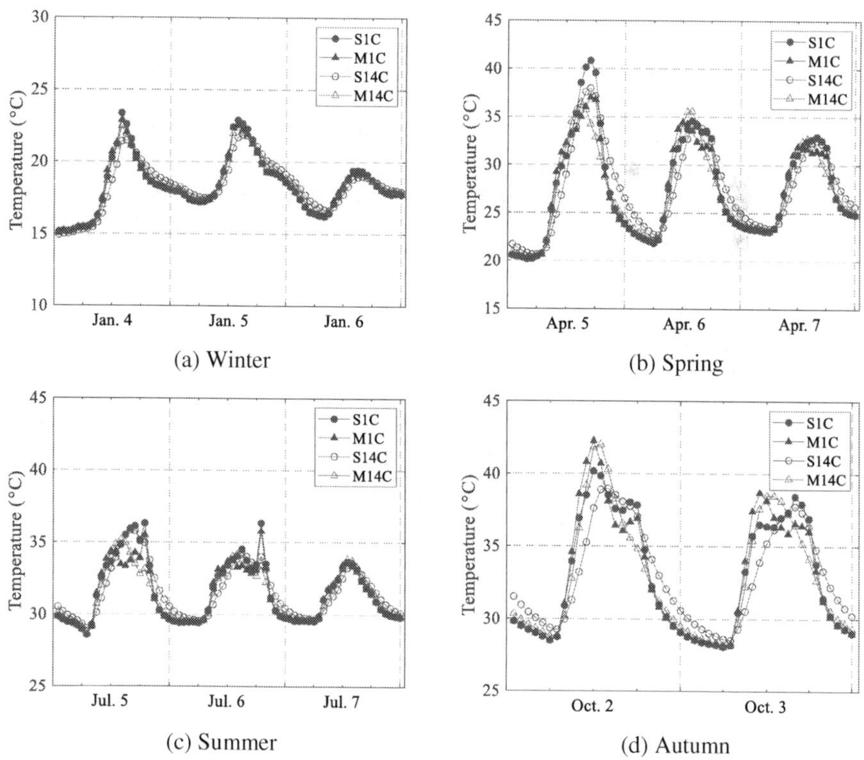

FIGURE 5.37 Calculated temperatures of the center points of four cables: (a) winter, (b) spring, (c) summer, and (d) autumn.

solar radiation. Different cables with varying patterns might cause stress redistribution within the structure.

5.3.3.4 Summary

In this section, the temperature distribution of the entire Qingzhou Bridge is computed from a global 3D FE model. The calculated temperatures at the sensor points matched well with the measurement counterparts in all four seasons, verifying the effectiveness and accuracy of the constructed FE model of the bridge. In the next section, the computed temperature results will be converted to thermal loads and input to the structural FE model of the bridge to calculate the temperature-induced structural responses.

5.3.4 Temperature-Induced Structural Responses

5.3.4.1 Integrated Analysis

As explained in Chapter 3, the conventional separate thermal and structural analyses involve many manual interventions and thus may cause inaccurate results and inefficient calculation. The integrated approach with a unified FE model is adopted here. Specifically, the FE model used in the preceding subsections will remain unchanged, and its element types will be converted to the corresponding structural elements. Then, the calculated temperature results are input into the FE model as the thermal load. The structural analysis is then conducted to calculate the temperature-induced responses. The complete procedure is shown in Figure 5.38.

The thermal FE model of the Qingzhou Bridge is transferred to a structural FE model by redefining the element types. The thermal elements SOLID70, SHELL57, and LINK33 are changed to structural elements SOLID45 or SOLID65, SHELL181, and LINK180, respectively. SOLID45 or SOLID65 has six/eight nodes, each having

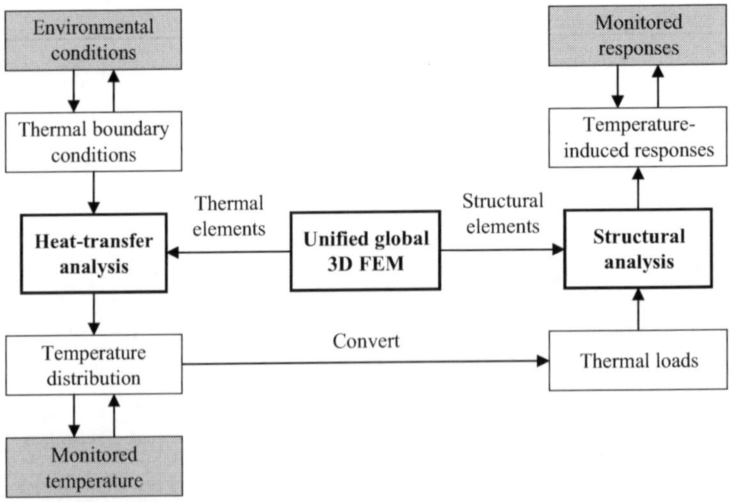

FIGURE 5.38 Framework of the integrated analysis.

three translational DOFs; SHELL181 has four nodes, each having three translational and three rotational DOFs; and LINK180 has two nodes, each having three translational DOFs.

The mechanical parameters of the converted FE model are estimated from the design drawings. They may be inaccurate and thus need to be verified and updated using measurement data. Vibration-based model updating techniques were previously developed to modify the FE model such that the calculated modal parameters optimally match the measured counterparts (Friswell and Mottershead, 1995; Zhang et al., 2008). However, given the large scale of the bridge model, updating the entire model is difficult. In this section, the density of the deck and Young's moduli of the steel and reinforced concrete are adjusted based on the measured global frequencies of the bridge. The density of the deck is increased equivalently to consider the weight of non-structural members, such as the handrails, light poles, bolts, and conduits. This simplified process costs much less computational effort than the global model updating approach. The calculated first four vertical bending frequencies of the FE model are compared with the measured ones in Table 5.7, together with the modal assurance criterion (MAC) values before and after updating. The average discrepancy of the frequencies decreases from 10.85% to 0.87% after updating, so the fine-tuned model is more accurate than the initial one in representing the real bridge. The MAC values remain unchanged at a high value close to 1, indicating that the calculated mode shapes resemble the measured counterparts before and after updating.

The calculated temperatures of all nodes at each time step are input into the FE model for the structural analysis. The temperature change to the initial temperature (or the reference temperature) is regarded as the temperature load in the structural analysis. As only the two shortest and two longest cables are simulated with solid elements, the temperature field of the remaining cables is assumed to vary linearly and interpolated from these four cables. The temperatures are then imposed on the link elements of the cables as body forces. The calculated structural responses, including the displacement, strains, and stresses, are compared with their measured counterparts.

The displacement sensors of the Qingzhou Bridge include GPS rovers, LVDTs, and LLSs, as illustrated in Figure 5.6. The locations of strain sensors are shown in Figures 5.5, 5.7, and 5.8. The locations of cable tensiometers are shown in Figure 5.9.

TABLE 5.7
Natural Frequencies and MAC Values of the First Four Vertical Bending Modes

Mode Order	Mea. (Hz)	Before Updating			After Updating		
		Freq. (Hz)	Diff. (%)	MAC	Freq. (Hz)	Diff. (%)	MAC
First	0.266	0.290	9.02	0.99	0.266	0.00	0.99
Second	0.372	0.417	12.10	0.99	0.379	1.88	0.99
Third	0.577	0.638	10.57	1.00	0.580	0.52	1.00
Fourth	0.641	0.716	11.70	0.97	0.648	1.09	0.97

5.3.4.2 Effect of Cable Temperature Variation

Theoretical analysis (Zhou and Sun, 2019) and numerical simulation (Zhu and Meng, 2017) both identified the significant influence of the temperature variation of cables on structural responses. This influence is quantitatively investigated in this section. The structural analyses were conducted at 8:00 and 18:00 on July 5, 2019, when the structural temperature reached the lowest and highest values, respectively.

The cable stresses at these two instants, with and without considering the cable's thermal loads, are compared in Figure 5.39. The corresponding differences relative to the cable stress under the dead load are also shown in the figure. Here, the stress results are caused by the temperature variation of the day, with the cable stresses at 1:00 being the baseline. Most cable prestresses were slightly released in the morning and increased in the afternoon due to the combined thermal expansions and contractions of different bridge components (Zhou and Sun, 2019). At 8:00, the cable stress did not change considerably (<2 MPa) regardless of whether the cables' thermal loads were considered or not. At 18:00, the cable stress decreased by 10 MPa, accounting for 1.2% of that due to the dead load, when the cables' thermal loads were

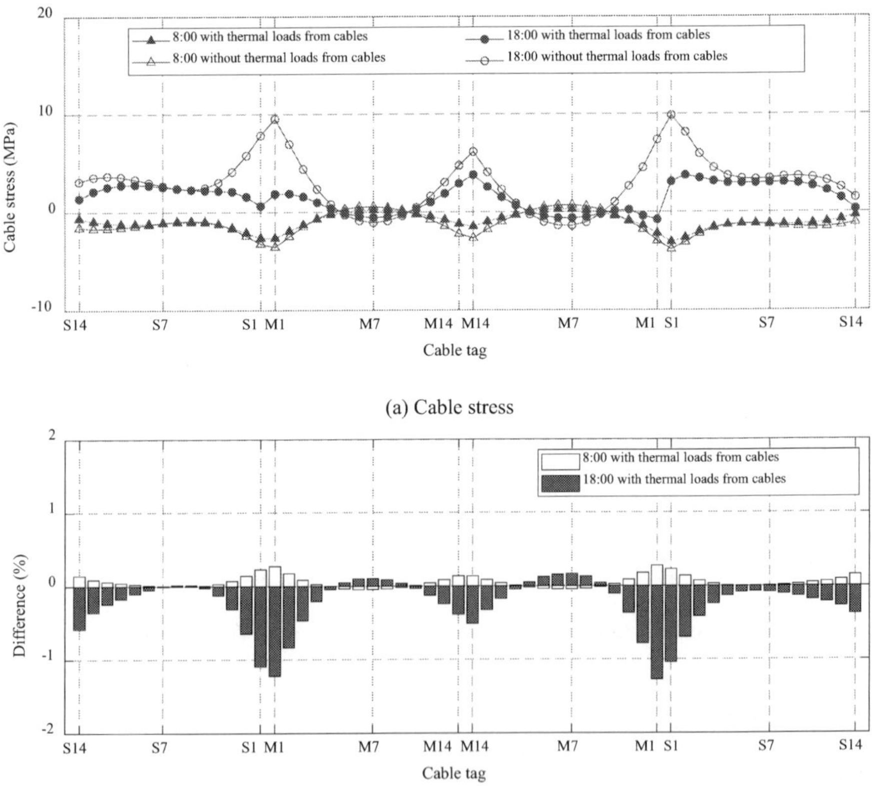

FIGURE 5.39 Influence of the cables' thermal loads on cable stress: (a) cable stress and (b) ratio of cable stress difference to dead load stress.

included. Therefore, integrating the temperature variation of cables can substantially reduce their tensile stress, which results from the stress redistribution induced by the free elongation and contraction of these cables.

Figure 5.40 compares the vertical deflection of the two instants, with and without considering the thermal loads of cables. The corresponding differences relative to the vertical deflection under the dead load are also illustrated. The vertical deflection at 1:00 is also subtracted from the results to investigate the effect of the cable temperature variation.

Figure 5.40a shows that the temperature variation had a more significant vertical deflection on the middle span than on the side and auxiliary spans. Moreover, the cables' thermal loads played an important role. In the middle span, the girder deflected downward at 8:00 and upward at 18:00 compared to 1:00. Considering the cables' temperature, the cables contracted in the morning and pulled the deck up, resulting in a smaller downward deflection. Conversely, the cables elongated in the afternoon and relaxed the deck down, resulting in a smaller upward deflection. Consequently, the cables' thermal load reduced both the downward and upward deflections. As shown

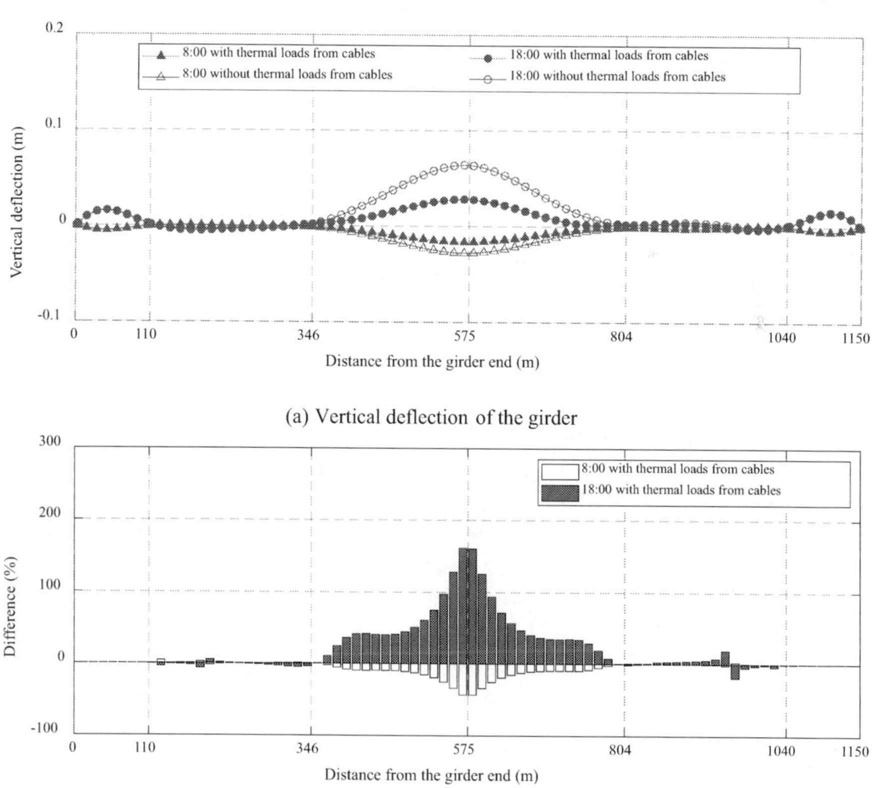

(b) Ratio of vertical deflection difference to dead load deflection

FIGURE 5.40 Cable temperature effects on the vertical deflection of the girder: (a) vertical deflection of the girder and (b) ratio of vertical deflection difference to dead load deflection.

in Figure 5.40b, relative to the dead load action, an approximately 43% difference in the vertical deflection at 8:00 and an approximately 161% difference at 18:00 can be observed for the two cases (i.e., with and without considering the cables' thermal loads). Therefore, the thermal load of the cables significantly affects both the cable stress and the vertical deflection of the girder and should be taken into account during the structural analysis.

Considering the thermal loads on the cables, the structural analysis presents responses such as the displacement of the main girder and the main tower, the strain/stress in the main girder and the main tower, and the stress in the cables. Representative days from each of the four seasons in 2019 are examined. The air temperature was 28°C when the construction was completed; thus, the reference temperature of the bridge is set to 28°C to calculate the temperature-induced responses relative to this baseline.

5.3.4.3 Displacement of the Main Girder

The longitudinal displacement of the girder end sections (G01 and G17) and the vertical deflection of the mid-span girder section (G09) are extracted. The locations of the sections with displacement sensors are shown in Figure 5.6. The calculated displacements on July 5–7, 2019, are compared with the corresponding measurements in Figure 5.17. The calculated total expansion magnitude in one day is close to the measured results. The RMSE between the calculated total longitudinal displacement and the measured data is 21 mm, approximately 11% of the monitored magnitude. According to the previous heat-transfer analysis, the average temperature change of the girder in one day is approximately 14°C. On this basis, the total expansion magnitude can be estimated as $1.2 \times 10^{-5} \times 14 \times 1150 \times 103 = 193$ mm, which is in good agreement with the monitored and calculated results.

Figure 5.17b shows the downward movement of the girder in the morning and its upward movement in the afternoon, which is in line with most observations on cable-stayed bridges. The RMSE between the calculated vertical deflection and the measured data is 5 mm, approximately 10% of the monitored magnitude. The calculated vertical deflection of the mid-span section agrees with the monitored results. The displacements in the other seasons are shown in Figures 5.18–5.20.

The variation trends of the calculated displacement and measured results in the other three seasons are similar. The effects of the vehicle and wind loads are not considered in the numerical analysis, which may induce errors in the results. The RMSEs between the calculated and measured displacements of the main girder in 2019 are also summarized in Table 5.2. U_{RMSE} and V_{RMSE} represent the RMSEs between the corresponding calculated and measured results. Most of them are approximately 15% of the corresponding displacement variation. Thus, good agreement is achieved in all seasons, indicating the accuracy of the proposed method in simulating the temperature-induced displacement of the girder.

5.3.4.4 Displacement of the Main Towers

The calculated longitudinal displacements of the tower top sections (T01 and T05) in the four seasons of 2019 are compared with the corresponding measurements in Figure 5.21. The calculated results are in good agreement with the measurements.

5.3.4.5 Strain/Stress of Main Girder

The calculated strains on section G13 on July 5–7, 2019, agree with their corresponding measurements, as shown in Figure 5.22. The calculated elastic strain and thermal strain at section G13 are extracted, as shown in Figure 5.41. Points 27 and 32 are selected to show the strain of TP and BP, respectively.

The elastic strains of TP and BP were both compressive because the girder was subject to considerable axial compression from stay cables. Compared with the strain state at 1:00, the elastic strain of TP was relatively tensioned in the morning and compressed in the afternoon, whereas that of BP exhibited the opposite behavior. Meanwhile, the thermal strains of both TP and BP were relatively compressed in the morning and tensioned in the afternoon. These phenomena can be explained by the morning temperature decrease in both TP and BP, with TP experiencing a larger decrease, resulting in greater contraction in TP than in BP. This led to TP being in tension and BP in compression to maintain internal deformation compatibility within the section. In the afternoon, the temperature of TP increased more than that of BP, causing TP to elongate more than BP. Consequently, TP was subject to compression while BP was in tension, again to maintain deformation compatibility. The vertical temperature difference between TP and BP significantly contributed to their elastic strains. After summing the elastic and thermal strains, the total strain variation of TP is slightly higher than that of BP, and both exhibit the same pattern of variation.

The stress of the section was calculated using Hook's Law. The results are shown in Figure 5.42. The calculated elastic stresses of TP and BP under the dead load were −22 and −25 MPa, respectively, as indicated by the flat dashed lines in the figure. The stresses of TP (Point 27) and BP (Point 32) were both compressive. The total elastic compressive stress of TP was larger than its dead load stress, indicating that its thermal stress was also compressive (see Equation 5.2). Conversely, BP's thermal stress was tensile. These findings are also attributed to the vertical temperature difference between TP and BP, deformation compatibility, and the internal force equilibrium effect.

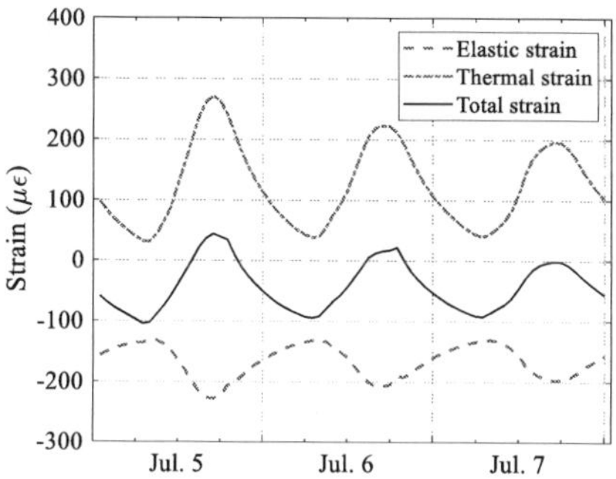

(a) TP (point 27)

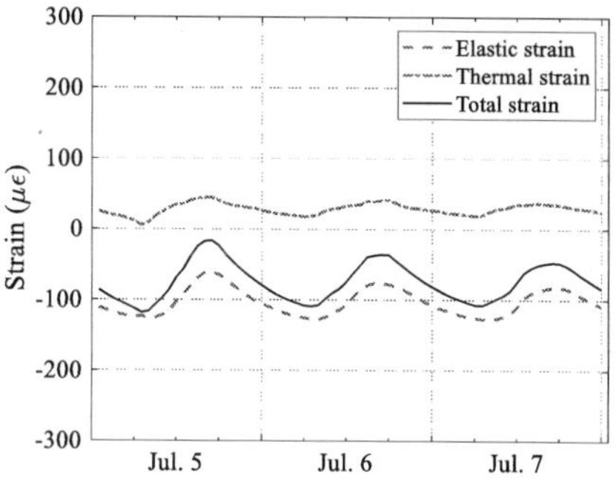

(b) BP (point 32)

FIGURE 5.41 Calculated strain on G13 on July 5–7, 2019: (a) TP (point 27) and (b) BP (point 32).

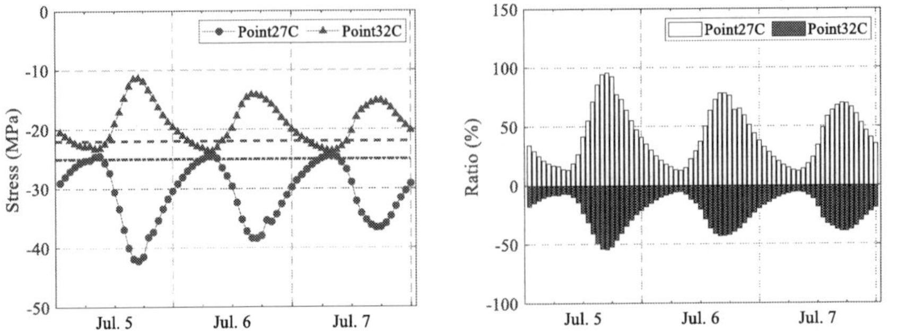

(a) Elastic stress (b) Ratio of thermal stress to dead load stress

FIGURE 5.42 Calculated strain on G13 on July 5–7, 2019: (a) elastic stress and (b) ratio of thermal stress to dead load stress.

As shown in Figure 5.42a, the compressive stress of TP decreased by 5 MPa in the morning and increased by 13 MPa in the afternoon relative to 1:00, whereas that of BP increased by 3 MPa in the morning and decreased by 9 MPa in the afternoon. The maximum and minimum compressive stresses of TP were −42 and −24 MPa, respectively, whereas those of BP were −23 and −11 MPa, respectively. All values are within the elastic range of the steel.

Figure 5.42b shows that the ratio of the thermal stress to the dead load stress could reach 96% for TP and 54% for BP, indicating the dominant role of the thermal effect in the strain and stress responses of the cable-stayed bridge. Figure 5.23 shows the calculated and measured stress results of section G13 on July 5–7, 2019. A generally good agreement between the simulated and monitored results of TP and BP is observed. The temperature variation caused TP to be in compression and BP to be in tension most of the time.

The calculated total stresses are compared with the measured stress variations in the other three seasons in Figures 5.43–5.45. The calculated and measured data are

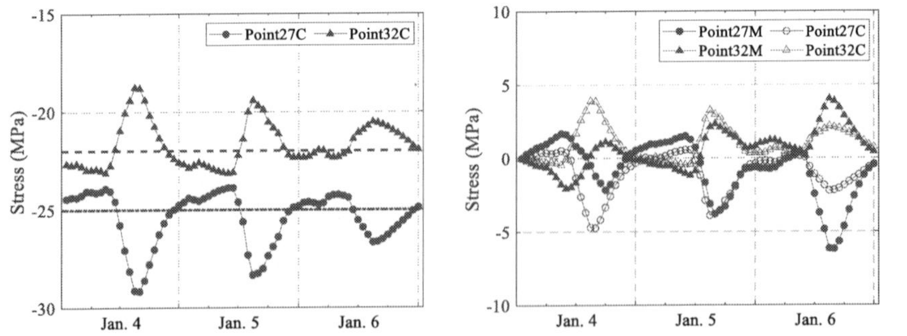

(a) Calculated stress (b) Calculated and measured stress variations

FIGURE 5.43 Stress on G13 on January 4–6, 2019: (a) calculated stress and (b) calculated and measured stress variations.

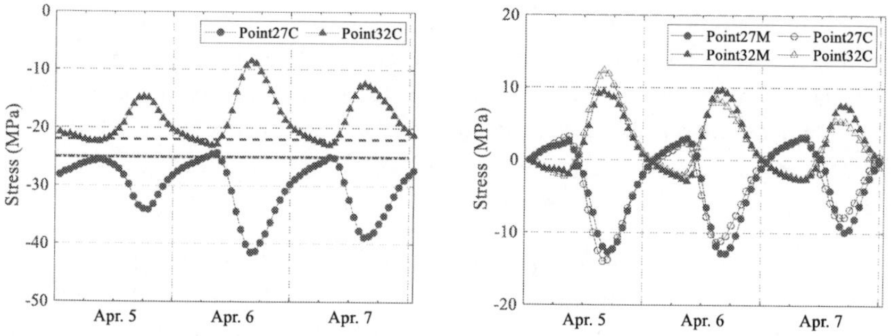

(a) Calculated stress (b) Calculated and measured stress variations

FIGURE 5.44 Stress on G13 on April 5–7, 2019: (a) calculated stress and (b) calculated and measured stress variations.

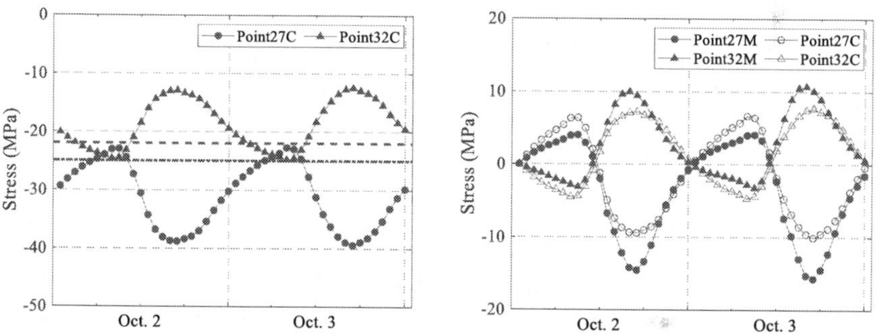

(a) Calculated stress (b) Calculated and measured stress variations

FIGURE 5.45 Stress on G13 on October 2–3, 2019: (a) calculated stress and (b) calculated and measured stress variations.

consistent in all three seasons. A relatively large difference between the calculated and measured stress in winter might be because the solar radiation in winter was weak and unstable, causing an inaccurate simulation temperature and thus a significant error in the stress of the deck. Moreover, the sheltering effect of tower legs and complex support constraints might also lead to a relatively large discrepancy in calculated stresses.

5.3.4.6 Strain/Stress of the Main Towers

The calculated and measured total strains of tower section T03 on the Hong Kong side and tower section T07 on the Zhuhai side are compared in Figure 5.25. The calculated results generally have a similar variation trend as the measurements. The stresses of T07 in the other three seasons are shown in Figure 5.26. A general fit between measurements and simulated counterparts can also be observed. The overall stress of the tower was less than 1 mPa throughout the year because of the slight temperature variation.

5.3.4.7 Stress of Stay Cables

The stress of the mid-span shortest cable anchored on the north leg of the tower on the Zhuhai side (M1) in 2019 is compared with the measurements in Figure 5.27. A good agreement is achieved between the calculated and measured cable stresses.

5.3.5 Summary

This section presents the integrated thermal and structural analysis of a long-span cable-stayed bridge. A unified global 3D FE model of the bridge is built. The measured environmental conditions are processed as the thermal boundary conditions for the heat-transfer analysis. The temperature results of different structural components match well with the monitoring counterparts. The temperature data are then applied to the same FE model as the thermal loads to compute the temperature-induced structural responses, including displacements, strain/stress, and cable stresses. The calculated structural responses generally agree with the measurement counterparts, verifying the accuracy of the integrated analysis approach. With little manual intervention, the integrated approach efficiently analyses the 3D global thermal behaviors of cable-stayed bridges.

5.4 ANALYTICAL FORMULATIONS

The temperature behaviors of the Qingzhou Bridge were investigated through field monitoring and numerical simulations. The underlying mechanisms are yet not revealed. This section will theoretically analyze the temperature-induced displacement of cable-stayed bridges by deriving analytical formulations. The Shanghai Yangtze River Bridge will be used as an example to discuss the mechanisms of the temperature effects of the bridge displacements. The strain/stress responses of the bridges are highly localized structural features and will not be presented.

5.4.1 Temperature-Induced Bridge Deformations

5.4.1.1 Field Measurement of the Shanghai Yangtze River Bridge

Located in the southeast coast of China, the Shanghai Yangtze River Bridge is a cable-stayed bridge with dual towers, dual cable planes, and a separate steel box girder with a main span of 730 m, as shown in Figure 5.46. The girder is a five-span, continuous, fully floating system with a total width of 51.5 m and a height of 4 m. The towers are herringbone-shaped single-column towers with a height of 216.3 m. The bridge contains 192 parallel twisted steel wire stay cables. The SHM system includes 227 sensors of various kinds. The temperature of the girder and towers, as well as the extension at the girder ends, are measured using fiber Bragg grating sensors. The mid-span displacement of the girder is measured using GPS sensors. The sampling interval of the thermometers, GPS sensors, and extensometers is 1 minute, 0.1 seconds (10 Hz), and 0.05 seconds (20 Hz), respectively.

Annual and daily temperature variations are the two dominant cyclic variations. The measurement data in February and July 2012, the coldest and warmest months

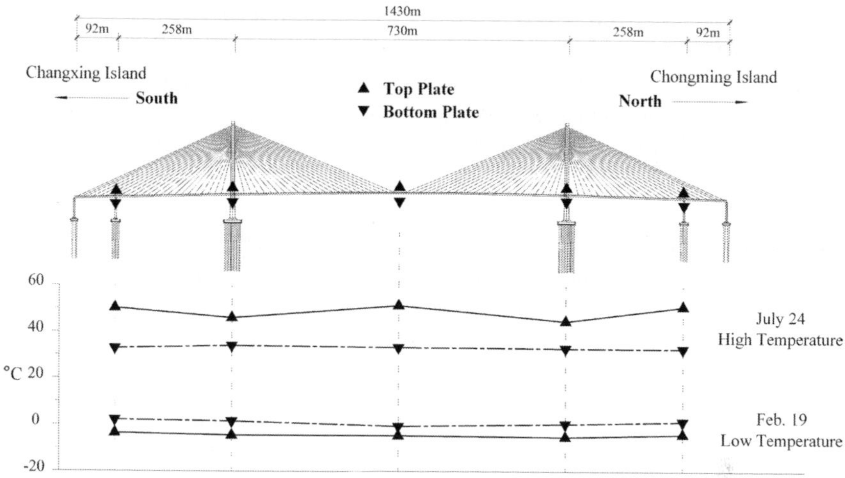

FIGURE 5.46 Temperature distribution in the longitudinal direction.

of the year, are studied. The temperature changes within these 2 months are relatively stable, and the daily variation range is close to the monthly variation range. As a result, the data obtained during each of the 2 months reflect the effect of the daily temperature cycles, while the difference between the 2 months represents the effect of the annual cycles.

5.4.1.2 Temperature Field Simplification

The bridges' temperature varies slowly. Thus, the hourly mean can be used to study temperature change and its effects. The temperature means of the top and bottom plates of the different sections on 19 February (winter) and 24 July (summer) are shown in Figure 5.46. The temperatures along the longitudinal direction of the girder were consistent, except for the tower-girder intersections, which were shaded by the towers during the day and had a slightly lower temperature than other sections.

Also, the vertical temperature difference between the top and bottom plates of the girder was far more significant than the transverse temperature differences of each section. Meanwhile, the temperature fields of both concrete towers were highly similar, with a considerable temperature difference between the south and north external surfaces. The temperature changes of different cables were generally similar (Zhou et al., 2015).

Based on the measured temperatures, the structural temperature field of the investigated cable-stayed bridge can be simplified as follows:

1. Girder: The temperatures of the different sections do not vary in the longitudinal direction. Only the vertical temperature difference between the top and bottom plates must be considered, while the transverse temperature gradient of each section is negligible. The average girder temperature is denoted as T_G, and the temperature difference between the top and bottom plates is denoted as T_{GD}.

2. Tower: The two towers have the same temperature field; only the temperature difference between the sun-facing and shaded external surfaces must be considered. The average tower temperature and the temperature difference are thus denoted as T_T and T_{TD}, respectively.
3. Stay cable: All cables have the same temperature distribution and are denoted as T_C.

Field measurements showed that such simplification of the temperature field was acceptable for structural analyses (Zhou et al., 2015). Consequently, the variations in the structural temperature distribution of a cable-stayed bridge can be decomposed into the following five temperature variables: the average temperature (T_G) of the top and bottom plates in the mid-span section of the girder, the vertical differential temperature (T_{GD}) in the mid-span section of the girder, the cable temperature (T_C), the average temperature (T_T) of the sun-facing and shaded sides of each tower, and the temperature difference (T_{TD}) between the opposite external surfaces of each tower.

5.4.1.3 Temperature-Induced Deformation

The measured structural responses are the combination of diverse environmental or operational actions. As the temperature varies slowly, the temperature-induced quasi-static responses can be extracted from the overall measurement data. Here, the temperature-induced changes in the girder length and the mid-span deflection are concerned. They are closely correlated with the average girder temperature (T_G); therefore, the present subsection focuses on their variations with T_G.

Figure 5.47a presents the longitudinal displacement at the northern end of the girder on July 24, with the girder temperatures shown in Figure 5.47b for reference. A northward displacement is positive. It can be seen that the dynamic component of the movement was small. Therefore, the hourly average can sufficiently represent the temperature variation.

The temperature-induced change in the girder length, L_T, can be determined by subtracting the southern displacement from the northern displacement. Figure 5.48a plots L_T versus T_G. The girder length increased with the rising girder temperature. The slope in the figure approximates the simple linear regression coefficient with

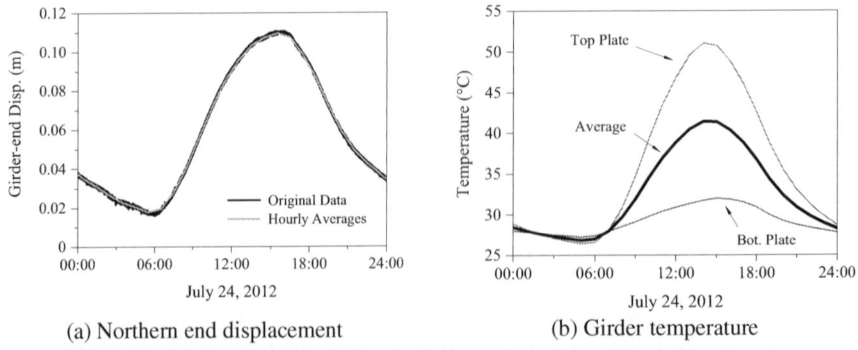

FIGURE 5.47 The girder end horizontal displacement and girder temperature: (a) northern end displacement and (b) girder temperature.

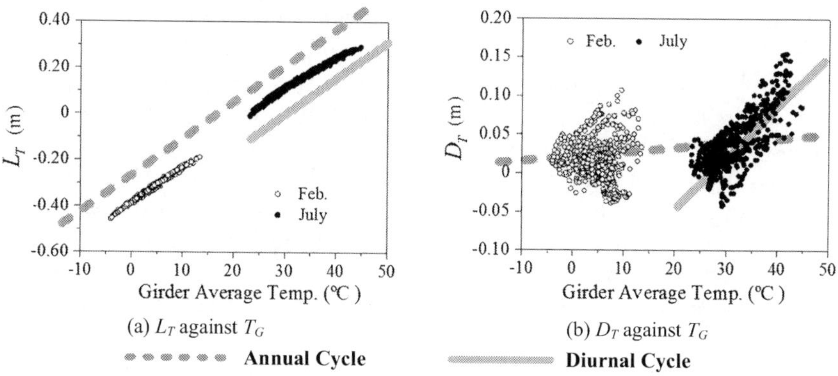

FIGURE 5.48 The deformation versus the average girder temperature: (a) L_T against T_G and (b) D_T against T_G.

respect to T_G. L_T in July minus that in February could represent the annual variation, whereas the difference within a single month, i.e., February or July, could reflect the daily variation. The annual variation in L_T is larger than the daily one.

The origin point of the girder mid-span deflection (D_T) is defined by the elevation of the sensor when it was installed, with subsequent upward movement being positive. The wavelet packet filtering method is used to separate the temperature-induced mid-span deflection. The hourly mean of the mid-span deflection is also plotted in Figure 5.49 for comparison. The wavelet packet filtering leads to a smoother result.

Figure 5.48b shows the scatter plots of D_T versus T_G. They exhibit a roughly linear relationship in summer (correlation coefficient = 0.760), with the mid-span bridge deck moving upwards with increasing temperature. However, they have no clear linear relation in winter, with a small correlation coefficient of −0.050. When considering annual temperature changes, the average elevations of the girder mid-span in February and July 2012 are approximately the same, indicating that the deck level does not significantly rise with increasing temperatures.

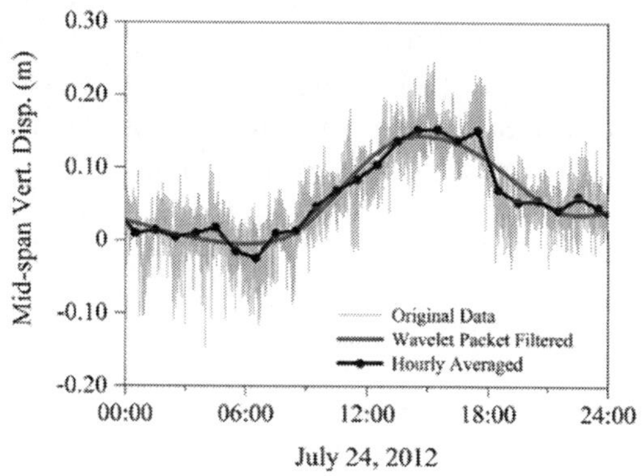

FIGURE 5.49 Extraction of the temperature-induced mid-span deflection.

The distribution pattern of data points concerning girder length in Figure 5.48a differs significantly from that of mid-span deflection in Figure 5.48b. First, the variation in L_T shows less discreteness than in D_T. For a given average girder temperature T_G, the normalized range of variation in L_T, defined as the actual range of variation at T_G divided by the annual range, is much smaller than that of D_T.

Second, in terms of similarity between annual and diurnal temperature dependencies (fitted by a simple linear regression model using T_G as the only explanatory variable), L_T correlates well with T_G. The slope of each fitted linear relationship remains almost constant over both annual and diurnal cycles. Conversely, D_T does not exhibit a consistent linear correlation with T_G; the slope of the fitted line for the annual cycle is much smaller than that for the diurnal cycle, as observed in July 2012.

Finally, for girder length, the amplitude of variations caused by annual temperature changes is significantly larger than that caused by diurnal temperature changes. In contrast, mid-span deflections do not vary monotonically with temperature, with the annual amplitude of variation being approximately equal to the diurnal amplitude.

These observations highlight the complexity of temperature effects on the structural responses of a cable-stayed bridge.

5.4.2 Mechanisms of Temperature Effects

The temperature effects of cable-stayed bridges are usually discussed in terms of the uniform temperature variations in the bridge, temperature differences between the cables and deck (or tower), temperature differences within the girder, and temperature differences between opposite sides of the outer faces of the towers. Most design standards specify the ranges of these temperature variables (European Committee for Standardization CEN, 2003; Ministry of Transport of the People's Republic of China, 2007). However, the effects of these temperature variables on the structural responses have not been intuitively understood. This section will study the effects of five temperature variables (T_C, T_G, T_{GD}, T_T, and T_{TD}) on the deformation of a symmetric two-tower cable-stayed bridge. They will be introduced individually in the following sections.

5.4.2.1 Plane Geometric Model

To better illustrate the mechanisms of the temperature-induced deformation, the changes in L_T and D_T due to a unit temperature increase are estimated using the simplified plane geometric model of the Shanghai Yangtze River Bridge shown in Figure 5.51. The lengths of the center span and side span of the model are respectively defined as $L_0 = 730\,\text{m}$ and $\gamma_0 L_0 = 350\,\text{m}$, whereas the heights of the tower above and below the deck are $H_0 = 150\,\text{m}$ and $\gamma_0 L_0 = 60\,\text{m}$, leading to $S_0 = 150\,\text{m}$ and $S_1 = 380.79\,\text{m}$ for the length of the cables at the center and side spans. The thermal expansion coefficients of the steel and concrete are respectively set as $\alpha_s = 1.2 \times 10^{-5}\,°\text{C}^{-1}$ and $\alpha_c = 1.0 \times 10^{-5}\,°\text{C}^{-1}$ (Figure 5.50).

The girder length varies owing to its thermal expansion and contraction. Hence, L_T is dominated by the change in average girder temperature T_G (Ni et al., 2007) and can be estimated by

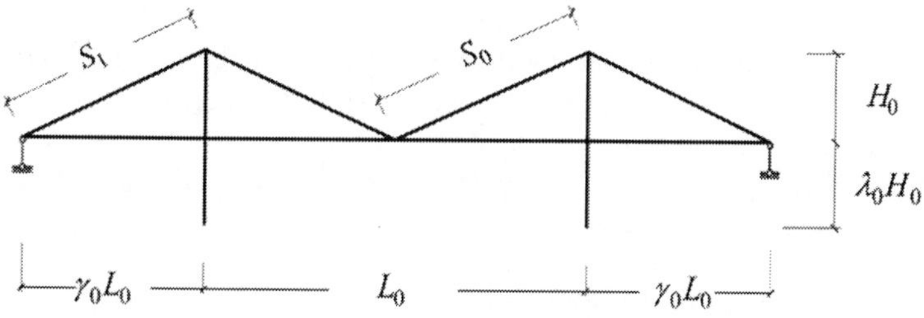

FIGURE 5.50 Simplified plane geometric model of a symmetric two-tower cable-stayed bridge.

$$L_T = \alpha_s \cdot \left(L_0 + 2\gamma_0 L_0\right) \cdot \Delta T_G \tag{5.6}$$

Let $\Delta T_G = 1°C$ and substitute the model parameters into Equation (5.6); then, the sensitivity of L_T with respect to T_G is 0.0172 m/°C. Compared with the girder length, the temperature-induced mid-span deflection is relatively complex. Its mechanism is comprehensively discussed in the following sections.

Figure 5.51 shows the schematic plots of the mid-span movement of the girder with each temperature variable. Although the temperature differences between opposite

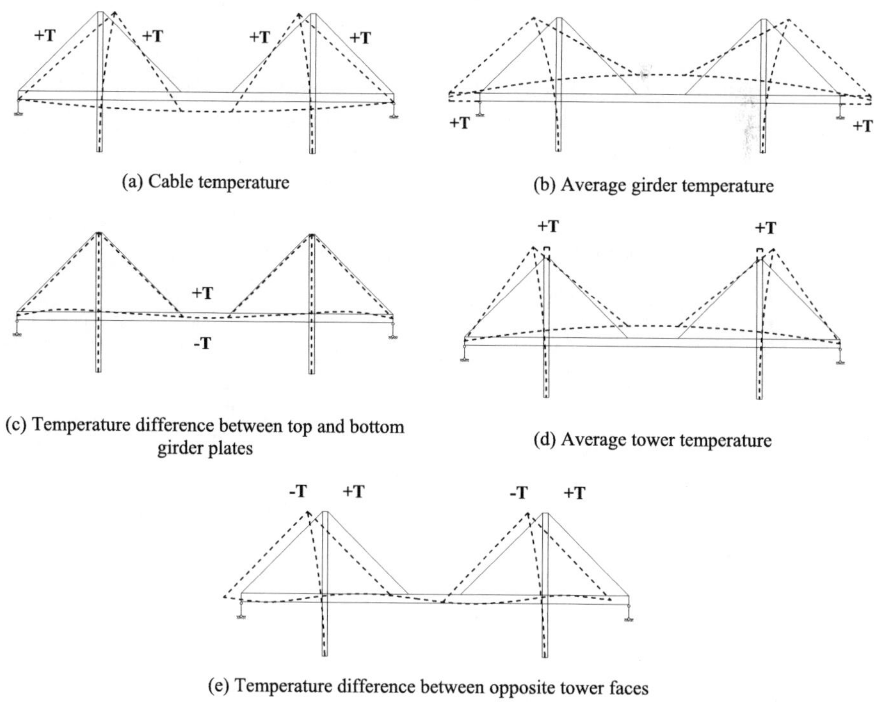

FIGURE 5.51 Schematic diagram of variations in mid-span deflection (D_T).

external surfaces of the tower (T_{TD}) may induce lateral bending deformations of towers, the resultant girder deformations are antisymmetric and, thus, minimally affect the mid-span deflection. Hence, the effect of T_{TD} (Figure 5.51e) is not discussed further. In addition, the temperature-induced changes in the elastic moduli of structural materials have negligible effects on structural deformations (Zhou et al., 2015), so this mechanism is also excluded.

5.4.2.2 Effects of Cable Temperature T_C

With increasing temperature, the thermal expansion of the mid-span cables causes the girder to move downwards (Figure 5.51a). Meanwhile, due to the vertical support at the girder ends, the elongation of the side span cable will cause the towers to move close to each other, further lowering the girder mid-span. In Figure 5.52, the elongation of the stay cables is given by $\Delta_{10} = \alpha_s \cdot T_C \cdot S_1$ and $\Delta_{11} = \alpha_s \cdot T_C \cdot S_0$, and according to the Pythagorean Theorem (an upward Δ_{13} is positive), we get:

$$\begin{cases} (\gamma_0 L_0 + \Delta_{12})^2 + H_0^2 = (S_1 + \Delta_{10})^2 \\ (L_0/2 - \Delta_{12})^2 + (H_0 - \Delta_{13})^2 = (S_0 + \Delta_{11})^2 \end{cases} \tag{5.7}$$

Δ_{13} can be determined by

$$\Delta_{13} = H_0 - \sqrt{(S_0 + \alpha_s T_C S_0)^2 - \left(L_0/2 + \gamma_0 L_0 - \sqrt{(S_1 + \alpha_s T_C S_1)^2 - H_0^2}\right)^2} \tag{5.8}$$

The model geometric parameters and $T_C = 1°C$ lead to $\Delta_{13} = -0.0246\,m/°C$. Here, the minus sign indicates the downward movement of the deck at mid-span. Although Equation (5.8) indicates a nonlinear relationship between Δ_{13} and T_C, the curves of Δ_{13} versus T_C are approximately linear and pass through the origin over the normal temperature range. Therefore, a linearized formula of the form $\Delta_{13} = k_C^{D_T} \cdot T_C$ can be used to approximate Equation (5.8):

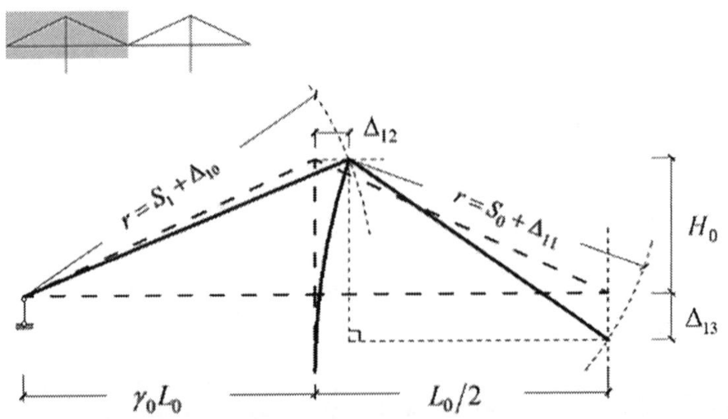

FIGURE 5.52 Analytical model of the effects of T_C.

$$\Delta_{13} \approx k_C^{D_T} \cdot T_C = \frac{\alpha_s(2\gamma_0+1)(2H_0^2+\gamma_0 L_0^2)}{4H_0 \cdot \gamma_0} T_C \quad (5.9)$$

where $k_C^{D_T}$ is the temperature sensitivity coefficient of the mid-span vertical displacement with respect to the temperature change T_C, representing the displacement per unit temperature increase.

5.4.2.3 Effects of Average Girder Temperature T_G

Changes in the average girder temperature T_G may indirectly affect the mid-span deflection of the bridge. The girder expands at both ends with increasing temperature. Considering the constraints of the stay cables, the towers move apart from each other due to the girder expansion, resulting in the rise of the mid-span of the girder, as shown in Figure 5.51b. In Figure 5.53, ds_0 and ds_1 denote the changes in the length of the cables due to the increase in their forces. If $ds_0 = ds_1 = 0$, then $\Delta_{21} = \Delta_{20} = \alpha_s \cdot T_G \cdot (\gamma_0 L_0 + L_0/2)$. The vertical movement at the mid-span corresponding to the unit temperature increase would be

$$\Delta_{22} = H_0 - \sqrt{S_0^2 - \left(L_0/2 + \alpha_s \cdot T_G\left(L_0/2 + \gamma_0 L_0\right)\right)^2} \quad (5.10)$$

Using the given parameters, $\Delta_{22} = 0.0209\,\text{m/°C}$, representing an upward movement of the mid-span deck. The linearized form of Equation (5.10) is:

$$\Delta_{22} \approx k_G^{D_T} \cdot T_G = \frac{\alpha_s L_0^2(2\gamma_0+1)}{4H_0} T_G \quad (5.11)$$

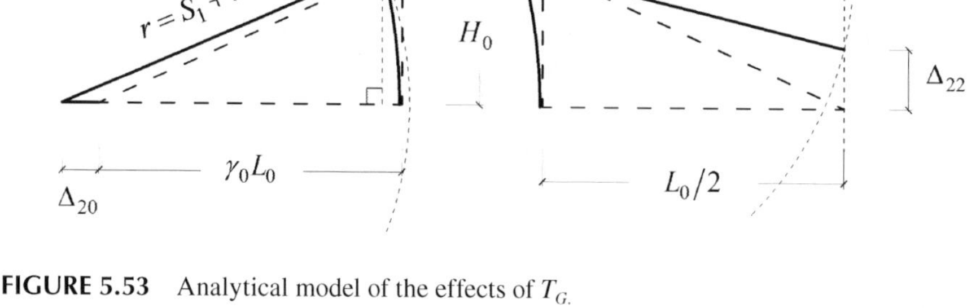

FIGURE 5.53 Analytical model of the effects of T_G.

5.4.2.4 Effects of Girder Vertical Differential Temperature T_{GD}

The temperature difference between the top and bottom plates of the steel box girder is significant and may exceed 20°C under direct sunlight. When the elongation of the TP differs from that of the BP, a curvature is generated in the girder section. Assuming that the temperature varies linearly over the girder depth h and the temperature-generated curvature is constant along the length of the bridge, and considering the elastic constraints of the mid-span cables only, the vertical displacement Δ_{31} of the two-span beam in Figure 5.54b is given by

$$\Delta_{31} = \alpha_s T_{GD}/h \cdot \left(3E_s I/L_1^3\right)/\left(2K_1 + 6E_s I/L_1^3\right) \cdot L_1^2 \tag{5.12}$$

where I is the flexural moment of inertia of the girder, and K_1 is the equivalent spring rigidity of the stay cables (Figure 5.54a). For the Shanghai Yangtze River Bridge, $L_1 = 715\,\text{m}$, $h = 4\,\text{m}$, $I = 4.45\,\text{m}^4$, and $K_1 = 1.98 \times 10^6\,\text{N/m}$ produces $\Delta_{31} = 0.0030\,\text{m/°C}$, indicating upward movement with increasing temperature.

A cable-stayed bridge is analogous to a continuous beam elastically supported on multiple points. Hence, the magnitude and direction of the temperature-induced mid-span deflection are related to the number of spans. Figure 5.54c shows an example of a four-span elastically supported continuous beam, in which a positive T_{GD} causes a downward (negative) Δ_{31}. Therefore, the direction of the mid-span deflection is uncertain. Moreover, as the span number increases, the absolute value of the mid-span deflection would decrease rapidly.

5.4.2.5 Effects of Average Tower Temperature T_T

A change in the temperature of the towers alters the elevation of the anchorage points of the stay cables on the tower side, resulting in an upward movement of the girder mid-span (Figure 5.51d). Moreover, because of the vertical restraints at the girder ends, the side span cables would pull the towers to move outwards, further raising

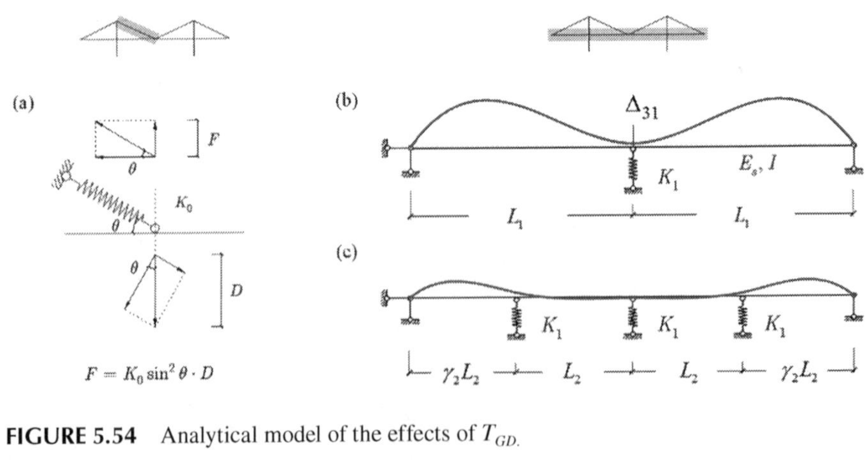

FIGURE 5.54 Analytical model of the effects of T_{GD}.

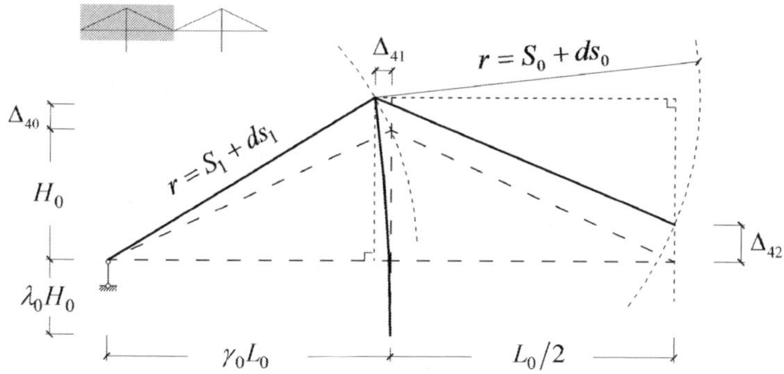

FIGURE 5.55 Analytical model of the effects of T_T.

up the deck at the mid-span. Neglecting the elongation of the cables ($ds_0 = ds_1 = 0$) induced by the cable force variation, the following geometric relationships exist in Figure 5.55:

$$\begin{cases} (H_0 + \Delta_{40})^2 + (\gamma_0 L_0 - \Delta_{41})^2 = S_1^2 \\ (H_0 + \Delta_{40} - \Delta_{42})^2 + (L_0/2 + \Delta_{41})^2 = S_0^2 \end{cases} \tag{5.13}$$

where $\Delta_{40} = \alpha_c \cdot T_T \cdot (\lambda_0 H_{0\,+} H_0)$. As a result, the mid-span vertical displacement Δ_{42} is given by

$$\Delta_{42} = H_0 + \alpha_c T_T H_0 (1 + \lambda_0) - \sqrt{S_0^2 - \left(L_0/2 + \gamma_0 L_0 - \sqrt{S_1^2 - \left(H_0 + \alpha_c T_T H_0 (1 + \lambda_0) \right)^2} \right)^2} \tag{5.14}$$

Equation (5.14) produces an upward movement per unit temperature increase, $\Delta_{42} = 0.0043\,\text{m/°C}$. The linearized form of Equation (5.14) is:

$$\Delta_{42} \approx k_T^{D_T} \cdot T_T = \frac{\alpha_c H_0 (2\gamma_0 + 1)(\lambda_0 + 1)}{2\gamma_0} T_T \tag{5.15}$$

5.4.2.6 FE Model

To calculate the sensitivity of the structural deformation with respect to individual temperature variables, a 3D beam-element model of the Shanghai Yangtze River Bridge with double girders is developed using ANSYS (ANSYS Inc., 2009). The girders, towers, cables, additional masses, and elastic connections at the boundaries are respectively modeled by BEAM44, BEAM4, LINK10, MASS21, and COMBIN14 elements. The model contained 1,673 elements and 1,805 nodes in total.

Based on the characteristics of the structural temperature fields described in Section 5.4.1.2, the FE model considered five kinds of in-situ temperature measurements as inputs, namely, the temperatures of the top and bottom plates of the girder (for T_G and T_{GD}), the temperatures of the external north and south surfaces of the towers (for T_T and T_{TD}), and the cable temperature (for T_C). The thermal expansion/contraction effect is simulated by considering the linear expansion coefficients of the materials.

Owing to the unavailability of the initial temperature fields and structural responses when the bridge construction was completed, calculating the absolute structural response under a particular temperature field is impossible. Hence, the structural response is assumed to be a function of the temperature change. In addition, a virtual reference state is set as the structure having a uniform temperature of 27°C (the girder closure temperature). The difference between the structural response under a given temperature distribution and the reference state is considered the temperature effect. Therefore, the difference between the structural responses in two temperature states is not affected by the reference state.

The changes in the girder length and mid-span deflection with respect to the temperature variables are then calculated. The variables and their ranges are the cable temperature T_C (−10°C to 50°C), average girder temperature T_G (−10°C to 50°C), girder vertical differential temperature T_{GD} (−30°C to 30°C), and average tower temperature T_T (−10°C to 50°C). Although the geometric nonlinearity is considered in the FE calculation, L_T and D_T are observed to vary almost linearly with each temperature variable over the normal temperature ranges.

Figure 5.56 shows the variation in the calculated mid-span deflection with the temperature variables. The slopes of the four curves represent the sensitivity coefficients of D_T, i.e., the change in D_T per unit temperature increase, denoted as $k_i^{D_T}$ in Table 5.8 (subscript i corresponds to the temperature variable T_i, $i = \{C, G, GD, T\}$).

The sensitivities of the mid-span deflection to the four temperature variables are sorted as follows: $T_C > T_G > T_T > T_{GD}$. The first two are larger than the last two by

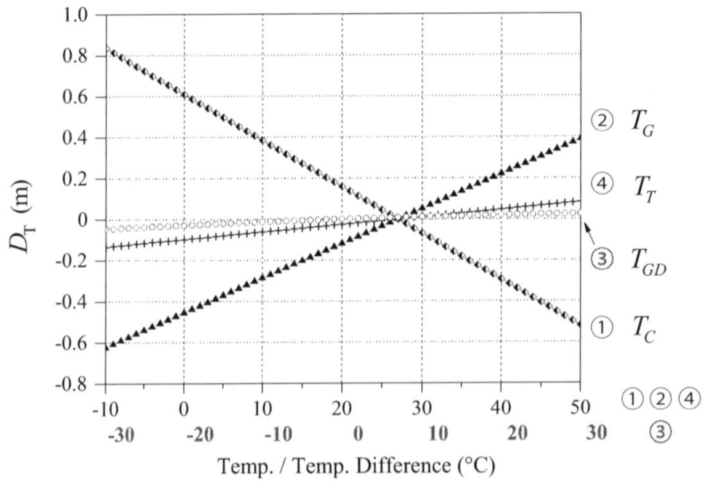

FIGURE 5.56 Sensitivities of D_T to four temperature variables.

TABLE 5.8
Summary of the Sensitivity Coefficients for D_T and L_T (Unit: m/°C)

	$k_i^{D_T}$		$k_i^{L_T}$	
Variables	**Plane Geometrical Analysis**	**FE Analysis**	**Plane Geometrical Analysis**	**FE Analysis**
(1)	(2)	(3)	(4)	(5)
Cable temperature T_C	−0.0246 (Equation 5.8)	−0.0226	0	0.0005
Average girder temperature T_G	0.0209 (Equation 5.10)	0.0169	0.0172 (Equation 5.6)	0.0166
Girder vertical differential temperature T_{GD}	0.0030 (Equation 5.12)	0.0012	0	−0.0015
Average tower temperature T_T	0.0043 (Equation 5.14)	0.0036	0	−0.0002

one order. In addition, an increase in the cable temperature lowers the mid-span deflection, whereas an increase in the other three variables raises the deck. The sensitivities derived from the simplified plane geometric model are close to those calculated by the FE analysis.

Meanwhile, Table 5.8 also lists the sensitivity coefficients of L_T. The average girder temperature is the dominating factor that affects the girder length more significantly than the other three temperatures by one or two orders. The relative difference between the calculated $k_G^{D_T}$ in the FE analysis and that estimated from the simplified plane geometric model is about 3%.

Therefore, the simplified model can reflect the general variation trend of the mid-span deflection and girder length. Despite a slight difference, the simplified plane geometric model provides analytical formulas to calculate the temperature-induced displacement. It is easier to understand and more convenient to apply in situ than the FE model.

5.4.3 Discussion of Mechanisms

This section will use the Shanghai Yangtze River Bridge as an example to discuss and compare the five aforementioned temperature variables, which are considered temperature behavior mechanisms.

5.4.3.1 Comparison of Effects of Different Mechanisms

The contribution of a temperature variable to the structural responses depends on the sensitivity coefficient and the variation amplitude. This subsection further compares the effects of each temperature variable on D_T and L_T by considering the actual temperature variation ranges.

Based on the maximum and minimum temperatures recorded during two typical winter and summer days, Tables 5.9 and 5.10 calculate the actual variation in the temperature variables (or T_i, $i = \{C, G, GD, T\}$) on the daily and annual cycles.

TABLE 5.9
Contribution of Different Mechanisms to D_T

Mechanism	$k_i^{D_T}$ (m/°C)	Daily Cycle: February 19 T_i (°C)	I_i (m)	$\|I_i\|/I_G$ (%)	Daily Cycle: July 24 T_i (°C)	I_i (m)	$\|I_i\|/I_G$ (%)	Annual Cycle: July 24–February 19 T_i (°C)	I_i (m)	$\|I_i\|/I_G$ (%)
(1)	(2)	(3)	(4) (2) × (3)	(5)	(6)	(7) (2) × (6)	(8)	(9)	(10) (2) × (9)	(11)
T_C	−0.0226	8.5	−0.192	74.4	4.5	−0.103	41.2	33.9	−0.765	101.5
T_G	0.0169	15.3	0.258	100.0	14.8	0.250	100.0	44.6	0.754	100.0
T_{GD}	0.0012	18.6	0.022	8.5	20.2	0.024	9.6	23.0	0.028	3.7
T_T	0.0036	6.3	0.023	8.9	2.4	0.009	3.6	29.8	0.107	14.2
ΣI_i (Sum)			$\Sigma_1 = 0.111$			$\Sigma_2 = 0.180$			$\Sigma_3 = 0.123$	

TABLE 5.10
Contribution of Different Mechanisms to L_T

Mechanism	$k_i^{L_T}$ (m/°C)	Daily Cycle: February 19 T_i (°C)	I_i (m)	$\|I_i\|/I_G$ (%)	Daily Cycle: July 24 T_i (°C)	I_i (m)	$\|I_i\|/I_G$ (%)	Annual Cycle: July 24–February 19 T_i (°C)	I_i (m)	$\|I_i\|/I_G$ (%)
(1)	(2)	(3)	(4) (2) × (3)	(5)	(6)	(7) (2) × (6)	(8)	(9)	(10) (2) × (9)	(11)
T_C	0.0005	8.5	0.004	1.7	4.5	0.002	0.9	33.9	0.017	2.3
T_G	0.0166	15.3	0.254	100.0	14.8	0.246	100.0	44.6	0.740	100.0
T_{GD}	−0.0015	18.6	−0.028	11.1	20.2	−0.031	12.4	23.0	−0.035	4.7
T_T	−0.0002	6.3	−0.001	0.4	2.4	0	0.2	29.8	−0.005	0.6
ΣI_i (Sum)			$\Sigma_1 = 0.229$			$\Sigma_2 = 0.217$			$\Sigma_3 = 0.718$	

Taking the mid-span deflection as an example, Table 5.9 presents the products of the sensitivity coefficient of D_T and temperature variation, namely, $I_i = k_i^{D_T} \times T_i$. The values in the last row are the algebraic sum of four effects. They are comparable to the total mid-span deflection obtained by the FE calculations that consider all four variables together, indicating that the combined effects of the temperature variables on D_T can be approximated by the linear superposition of the individual effects.

Columns 5, 8, and 11 of Table 5.9 present the magnitudes of the different effects divided by I_G. It can be seen that the effects of T_G and T_C on the mid-span deflection outweigh those of T_T and T_{GD} for both the annual and diurnal cycles. Moreover, the two dominant temperature variables, T_G and T_C, have opposing effects on the mid-span deflection and thus cancel out each other, resulting in Σ_3 being smaller than Σ_2 in Table 5.9. This indicates for cable-stayed bridges, the deflection variation

D_T caused by annual temperature changes may be smaller than those caused by daily temperature changes.

Table 5.10 compares the contributions of each temperature variable to the change in the girder length on the annual and daily cycles. The average girder temperature T_G is the only dominant factor for L_T on both cycles, and the annual change in the girder length is much larger than the daily one ($\Sigma_3 > \Sigma_1$ and $\Sigma_3 > \Sigma_2$ in Table 5.10) because the annual variation of T_G is greater than the daily variation.

5.4.3.2 Explanation of Monitoring Phenomena

According to the sensitivity coefficients in Table 5.8, the following multiple linear superposition formula is developed to estimate the temperature-induced variation in the girder length and mid-span deflection from the measured temperatures in February and July of 2012:

$$X_{\_a} = \sum k_i^X \cdot T_i + X_{\_0} \tag{5.16}$$

Here, X denotes L_T or D_T, $X_{\_0}$ is a constant associated with the initial state, subscript $_{\_a}$ (or $_{\_m}$ below) of X indicates the calculated (or measured) value, k_i^X is the sensitivity coefficients of X to the temperature variable T_i ($i = \{C, G, GD, T\}$). For the mid-span deflection, D_{T_a} is given by

$$D_{T\_a} = -0.0226 \cdot T_C + 0.0169 \cdot T_G + 0.0012 \cdot T_{GD} + 0.0036 \cdot T_T + D_{T\_0} \tag{5.17}$$

To ensure the same baseline as that of the scatter diagram in Figure 5.48b, D_{T_0} is estimated using the following equation based on measurements for both February and July:

$$D_{T_0} = \text{Avg}(D_{T_m}) - \text{Avg}(-0.0226T_C + 0.0169T_G + 0.0012T_{GD} + 0.0036T_T) \tag{5.18}$$

In Equation (5.18), Avg(·) denotes averaging, and D_{T_m} is the measured mid-span deflection. Figure 5.57b shows the scatter plot of D_{T_a} versus T_G, with Figure 5.48b overlaid for comparison.

Figure 5.57a compares the calculated and measured variations in the girder length. It can be seen from the figure that the linear superposition model generally reproduces the variation patterns of the measurement data, verifying the mechanism analysis for L_T. Figure 5.57b shows that the calculated mid-span deflection approximates the overall distribution pattern of the measured values, which verifies the mechanism analysis of the temperature effects on D_T and the corresponding sensitivity coefficients in Table 5.8. Hence, Equation (5.16) can simultaneously explain the annual and daily variations in the girder length and mid-span deflection of a cable-stayed bridge. The discrepancy between the calculated and measured deformations may be due to errors in the FE model, oversimplification of the temperature field, and measurement errors. Further studies are required to improve the accuracy of the physics-based Equation (5.16) in predicting thermal deformations.

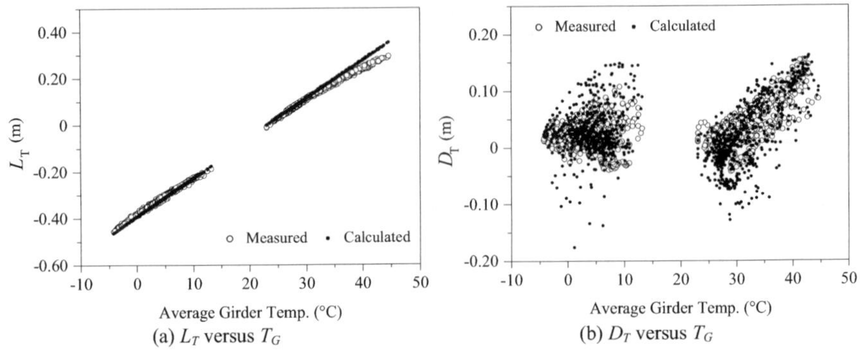

FIGURE 5.57 Calculated and measured changes in L_T and D_T: (a) L_T versus T_G and (b) D_T versus T_G.

Now, statistical correlations between the temperature and structural responses (L_T and D_T) are explained from the perspective of the mechanisms of the temperature effects:

1. The temperature-induced variation in the girder length depends on only the average girder temperature T_G. L_T is, therefore, linearly correlated with T_G, and the slope of the fitted linear relationship over a daily cycle is almost equal to that over an annual cycle, as shown in Figure 5.48a. As the annual variation of T_G is significantly larger than the diurnal variation, L_T takes on the same variation pattern.
2. The temperature-induced variation in the mid-span deflection is simultaneously determined by the average girder temperature T_G and cable temperature T_C. However, the proportional relationship between T_G and T_C is not constant over the daily and annual cycles (Figure 5.58), so the temperature-deformation correlation of D_T and T_G varies with the time scale. Further, the effects of T_C and T_G on the mid-span deflection are opposite; especially during the annual cycle, these two dominant temperature effects cancel each other, resulting in an unclear yearly variation pattern, as shown in Figure 5.48b.

5.4.3.3 Parametric Analysis

As derived from the plane geometric model in Section 5.4.2, the simplified formulas of the sensitivity coefficients of L_T and D_T can be extended to other two-tower cable-stayed bridges with different geometric arrangements, providing valuable insights into the temperature-induced deformation of such bridges. The variation of the mid-span deflection with temperature is illustrated as an example.

The above study reveals that the cable temperature T_C and the average girder temperature T_G dominate the temperature-dependent mid-span deflection with opposite

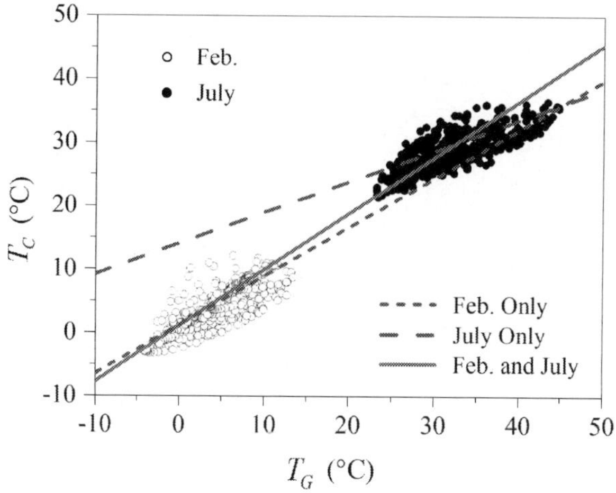

FIGURE 5.58 Fitted lines of the cable temperature as a function of the average girder temperature.

effects. As temperature increases, whether the girder mid-span goes upwards or downwards is determined by the relative effects of T_C and T_G, or the ratio $|I_C|/I_G$:

$$|I_C|/I_G = |\Delta T_C/\Delta T_G| \cdot (|k_C|/k_G) \tag{5.19}$$

Here, the sensitivity coefficients $k_C^{D_T}$ and $k_G^{D_T}$ are denoted as k_C and k_G for brevity. The temperature variations in Equation (5.19) are independent of the geometric dimensions, so we focus on the ratio of sensitivity coefficients $|k_C|/k_G$. Based on Equations (5.8) and (5.10), we can derive the following expression:

$$\frac{|k_C|}{k_G} = \frac{\xi - \sqrt{(1+\alpha^C)^2(\xi^2+1/4) - \left(\gamma_0 + 1/2 - \sqrt{(1+\alpha^C)^2(\xi^2+\gamma_0^2)-\xi^2}\right)^2}}{\sqrt{\xi^2+1/4-\left(1/2+\alpha^G(\gamma_0+1/2)\right)^2}-\xi} \approx \frac{\alpha^C}{\alpha^G}\left(\frac{2\xi^2}{\gamma_0}+1\right) \tag{5.20}$$

where $\xi = H_0/L_0$ is the ratio of the tower height above the deck and the center span length of the girder, γ_0 is the span ratio of the side span to the center span, and α^C and α^G are the thermal expansion coefficients of the cable and girder, respectively. A parametric analysis is performed to discuss the variation in $|k_C|/k_G$ with respect to ξ, γ_0, and α^G. The results are shown in Figure 5.59 (α^C is considered constant, as cables are always made of steel).

According to the China code "Guidelines for Design of Highway Cable-stayed Bridge" (Ministry of Transport of the People's Republic of China, 2007), symmetrical

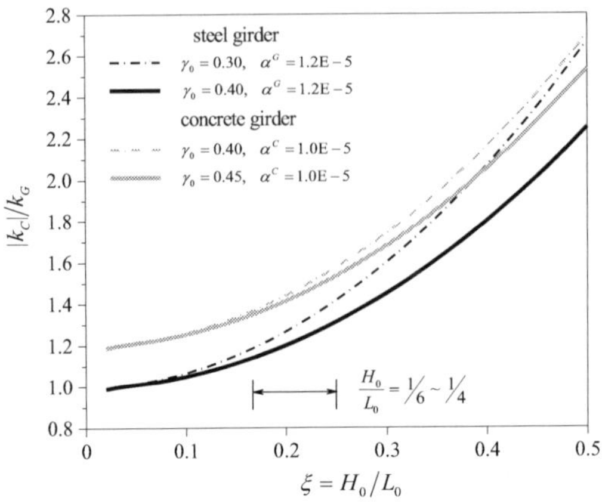

FIGURE 5.59 Variation of $|k_C|/k_G$ with H_0/L_0.

two-tower three-span cable-stayed bridges usually have γ_0 of 0.30–0.40 (steel girder) or 0.40–0.45 (concrete girder), and ξ of 1/6–1/4. In Figure 5.59, $|k_C|/k_G$ monotonously increases with increasing $\xi = H_0/L_0$ but decreases as γ_0 and α^G increase. For the ordinary two-tower cable-stayed bridges with ξ and γ_0 following the design code, $|k_C|$ is greater than k_G, and the ratio $|k_C|/k_G$ is between 1.14 and 1.42 for a steel girder bridge, and between 1.35 and 1.57 for a concrete girder bridge.

5.4.3.4 Summary

This section reveals the underlying mechanisms of the temperature effects on the deformation of cable-stayed bridges through a theoretical analysis via a plane geometric model. The Shanghai Yangtze River Bridge is used as the testbed. The mechanisms simultaneously explain the temperature-induced annual and daily variations of the girder length and mid-span deflection. The primary conclusions can be drawn as follows:

1. Within the normal temperature ranges, the temperature-induced responses of a cable-stayed bridge result from a combination of the thermal expansions/contractions of the individual components, and it can be well approximated by linear superpositions of a set of temperature variables. The temperature effects of a cable-stayed bridge can be investigated in terms of the following temperature variables: cable temperature T_C, average girder temperature T_G, girder vertical differential temperature T_{GD}, average tower temperature T_T, and temperature difference between opposite external surfaces of a tower T_{TD}. These temperature variables are more effective and intuitive than those recommended in structural design codes, namely, the uniform temperature variation of the bridge, the temperature difference between the cables and the deck (or tower), the temperature gradient of the girder, and the temperature gradient of the towers.

2. The temperature-induced longitudinal displacement of the girder is predominantly controlled by the average girder temperature T_G. The mid-span deflection is influenced by both the cable temperature T_C and average girder temperature T_G, but with opposite effects. An increase in the cable temperature causes the girder mid-span to move downward, while an increase in the average girder temperature causes the girder mid-span to move upward. Their combined effect depends on the temperature variation ranges of the two factors. The vertical differential temperature of the girder has little influence on the mid-span deflection.
3. The simplified plane geometric model can estimate the sensitivity coefficients of the dominant temperature variables, comparable to the FE solutions.

5.5 CHAPTER SUMMARY

This chapter comprehensively analyses the temperature behaviors of cable-stayed bridges through field monitoring, numerical investigation, and theoretical derivations. The temperature field characteristics and effects on two real cable-stayed bridges are studied using SHM data. A unified global 3D FE model of the Qingzhou Bridge is developed, which successfully replicates the temperature field and temperature-induced structural responses. Finally, the temperature field of cable-stayed bridges is then simplified into five representative variables, and the mechanisms by which each variable affects structural deformation are studied. These mechanisms provide a good explanation for the longitudinal girder displacement and mid-span deflection observed in the Shanghai Yangtze River Bridge.

REFERENCES

ANSYS Inc., 2009, *ANSYS 12.0*. Canonsburg, PA, ANSYS Inc.

Beckmann, R., Mella, R. and Wenman, M., 2013, Mesh and timestep sensitivity of fracture from thermal strains using peridynamics implemented in Abaqus. *Computer Methods in Applied Mechanics and Engineering* **263**: 71–80.

Cachot, E., Vayssade, T., Virlogeux, M., Lancon, H., Hajar, Z. and Servant, C., 2015, The Millau viaduct: ten years of structural monitoring. *Structural Engineering International* **25(4)**: 375–380.

Cao, Y. H., Zhang, Y., Zhao, Y. and Wang, M. L., 2011, Distributed health monitoring system for Zhanjiang Bay Bridge. *Advances in Structural Engineering* **14(1)**: 1–12.

Cho, S., Yim, J., Shin, S. W., Jung, H.-J., Yun, C.-B. and Wang, M. L., 2013, Comparative field study of cable tension measurement for a cable-stayed bridge. *Journal of Bridge Engineering* **18(8)**: 748–757.

European Committee for Standardization CEN, 2003, *Eurocode 1: Actions on Structures, Part 1-5: General Actions — Thermal Actions*. EN1991-1-5: 2003E. Brussels, European Committee for Standardization CEN.

Friswell, M. and Mottershead, J. E., 1995, *Finite Element Model Updating in Structural Dynamics*. New York, Springer.

Ministry of Transport of the People's Republic of China, 2007, *Guidelines for Design of Highway Cable-stayed Bridge*. JTG/T D65-01-2007. Beijing, China Communications Press.

Ni, Y. Q., Hua, X. G., Wong, K. Y. and Ko, J. M., 2007, Assessment of bridge expansion joints using long-term displacement and temperature measurement. *Journal of Performance of Constructed Facilities* **21(2)**: 143–151.

Wong, K. Y., 2004, Instrumentation and health monitoring of cable-supported bridges. *Structural Control and Health Monitoring* **11(2)**: 91–124.

Wong, K. Y., 2007, Design of a structural health monitoring system for long-span bridges. *Structure and Infrastructure Engineering* **3(2)**: 169–185.

Zhang, J., Xu, Y. L., Xia, Y. and Li, J., 2008, A new statistical moment-based structural damage detection method. *Structural Engineering and Mechanics* **30(4)**: 445–466.

Zhou, Y., Sun, L. and Peng, Z., 2015, Mechanisms of thermally induced deflection of a long-span cable-stayed bridge. *Smart Structures and Systems* **15(3)**: 505–522.

Zhou, Y. and Sun, L. M., 2019, A comprehensive study of the thermal response of a long-span cable-stayed bridge: From monitoring phenomena to underlying mechanisms. *Mechanical Systems and Signal Processing* **124**: 330–348.

Zhu, J. S. and Meng, Q. L., 2017, Effective and fine analysis for temperature effect of bridges in natural environments. *Journal of Bridge Engineering* **22(6)**: 04017107.

6 Temperature Behaviors of Arch Bridges

NOTATIONS

A_i, A_j	Two surfaces
$F_{i,j}$	View factor between two surfaces A_i and A_j
N	Number of pixels
ΔF_n	Delta-view factor for each pixel
θ_i, θ_j	Angles between the direction of two elementary areas and the normal direction of each elementary areas
dA_i, dA_j	Elementary areas

6.1 PREFACE

An arch bridge is an efficient bridge type utilizing the material's compressive capacity and the arch's geometry stiffness. The longest arch bridge in the world is the Tian'e Longtan Bridge, completed in 2024 in China with a main span of 600 m. Arch bridges are generally built in the valleys or have strong abutments due to the relatively large horizontal forces at the foot supports. These bridges can also be self-anchored. The temperature may considerably influence the deformation and stresses of arch bridges (Duan et al., 2011; Yarnold et al., 2012; Zhou et al., 2018).

Due to the structural complexity, the temperature behaviors are generally investigated through field monitoring and numerical simulation approaches (Li et al., 2023). This chapter uses the footbridge at the Hong Kong Polytechnic University as a testbed for field monitoring and numerical investigations.

6.2 FIELD MONITORING STUDIES OF POLYU FOOTBRIDGE

6.2.1 SHM System of the Footbridge

The footbridge linking Blocks Z and X in the Hong Kong Polytechnic University main campus was open to the public on October 15, 2019. The bridge has a total length of 84.24 m, consisting of a 64.26 m long main span that straddles the highway underneath and a 19.98 m long side span (Xia et al., 2021). As shown in Figures 6.1 and 6.2, the bridge deck is supported by two butterfly-shaped steel tube arches. One end of the arches meets the deck beams, which are supported on a tower at Block X. The other end of the deck beams rests on the podium floor of Block Z. The arches, beams, and columns are made of steel, whereas the bridge deck is composed of reinforced concrete.

DOI: 10.1201/9781003480136-6

FIGURE 6.1 Bird's view of the footbridge.

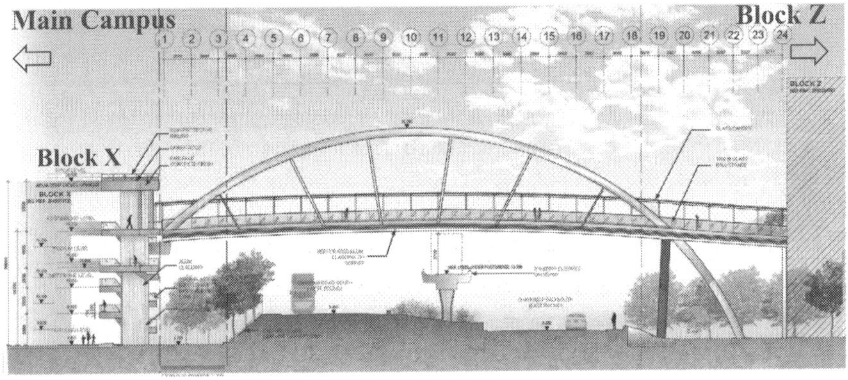

FIGURE 6.2 Elevation of the footbridge.

An SHM system was installed on the bridge, along with the construction progress of the structure. The system started data collection on September 28, 2019, and worked normally from October 13, 2019, after testing and debugging were completed. The sensor subsystem consists of 88 permanently installed sensors of 13 types (Xia et al., 2021). The monitoring items are generally classified into three categories: environmental parameters (air temperature, air pressure, humidity, rain, and solar radiation), loads (winds, pedestrians, and structural temperature), and bridge responses (stress/strain, displacement, and acceleration). The detailed layout of the sensors is shown in Figure 6.3. In particular, one pyranometer was installed on the

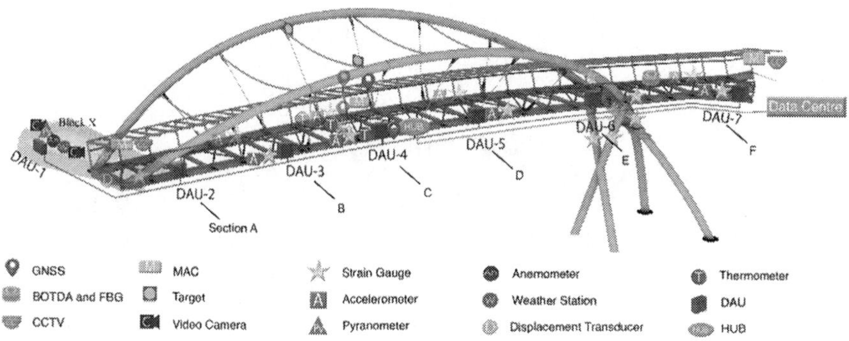

FIGURE 6.3 Layout of sensors and DUAs on the arch footbridge.

roof of Block X to measure the real-time solar radiation intensity. Seven independent data acquisition units (DAUs) were installed on the roof of Block X and Sections A to F of the bridge girder to collect data from the surrounding sensors. Among them, 20 vibrating wire strain gauges were installed on the arch bridge and distributed on steel girders, arches, and the concrete slab. The strain gauges measure the strain of the arch bridge members, which are used to calculate the internal force and then evaluate the safety and fatigue condition of the bridge.

6.2.2 Monitored Temperature Distributions

6.2.2.1 Air Temperature Variation

The measured temperature data in 2020 are first examined. The daily maximum and minimum temperature of all air temperature sensors in the year are shown in Figures 6.4 and 6.5. The highest temperature of the year appeared in July and August, while the lowest was in January. The maximum temperature difference around the year was 35°C. The two hottest and coldest days of the year are then selected to analyze the temperature and temperature-induced responses of the arch bridge.

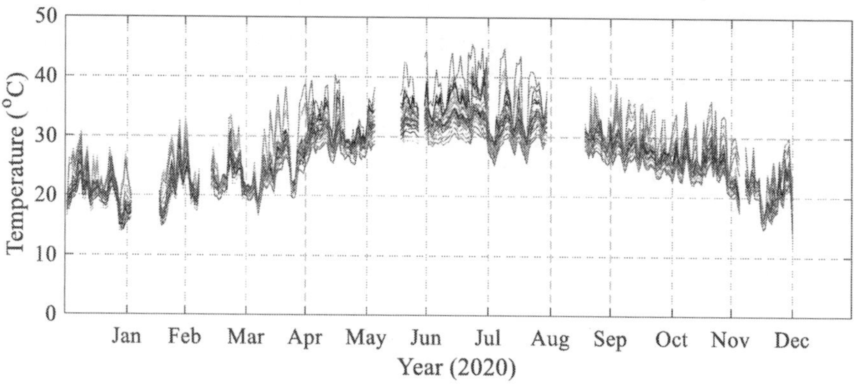

FIGURE 6.4 Daily maximum temperature in 2020.

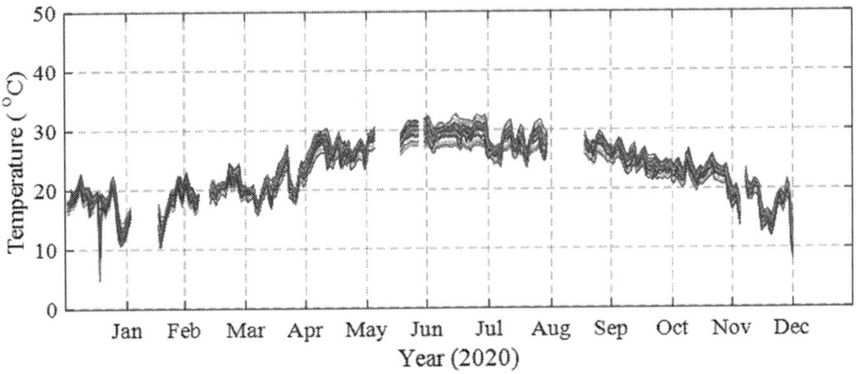

FIGURE 6.5 Daily minimum temperature in 2020.

6.2.2.2 Daily Temperature Variation

August 25 and February 17 represent the hottest and coldest days in 2020, respectively. The air temperature, solar radiation intensity, and average wind speed in the two days are shown in Figure 6.6. The air temperature varied between 28.5°C and 33.1°C on August 25, and between 9.7°C and 17°C on February 17. The latter had a larger daily temperature variation. These environmental data will be used to determine the thermal boundary conditions for the transient heat-transfer analysis of the arch bridge.

6.2.2.3 Measured Structural Temperature

The temperature data at the eastern beam, western beam, and the concrete slab of Section C are compared as an example. The locations of the strain gauges and thermometers are shown in Figure 6.7. For strain measurement, in Section C, SG–C2, SG–C4, and SG–C6 are under the top beam flange, and SG–C7, SG–C8, and SG–C9 are inside the concrete deck. At each position, a thermometer was installed to monitor the structural temperature and account for the temperature effects in the measured strains.

Figure 6.8a shows the temperature of the eastern beam of the footbridge measured on August 25, 2020. The calculated results will be discussed later. The lowest temperature was 30.1°C, and the highest temperature was 35.6°C. Similarly, the temperature of the western beam is shown in Figure 6.8b. The lowest temperature was 29.8°C, and the highest was 36.2°C. The measured temperature variation of the beams on the day was about 6°C.

The temperature at the middle height of the concrete slab is shown in Figure 6.9. The measured temperature variation on the day was less than 3°C, smaller than the steel beams. The reason is that the concrete slab is paved by timber panels. Consequently, the temperature variation was smaller.

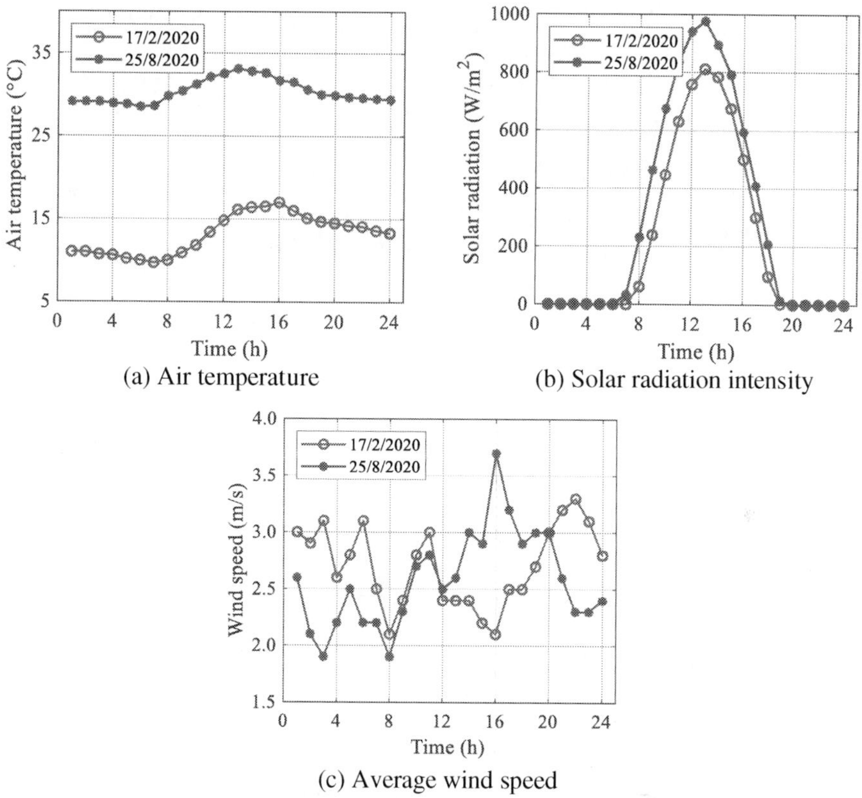

FIGURE 6.6 Meteorological data on February 17 and August 25, 2020: (a) air temperature, (b) solar radiation intensity, and (c) average wind speed.

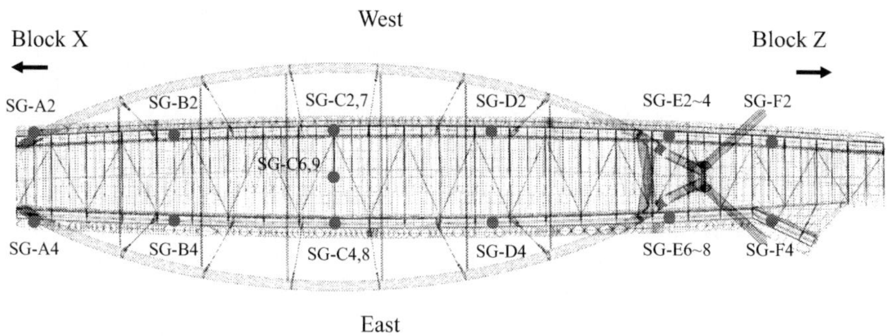

FIGURE 6.7 Layout of the vibrating wire strain gauges and thermometers.

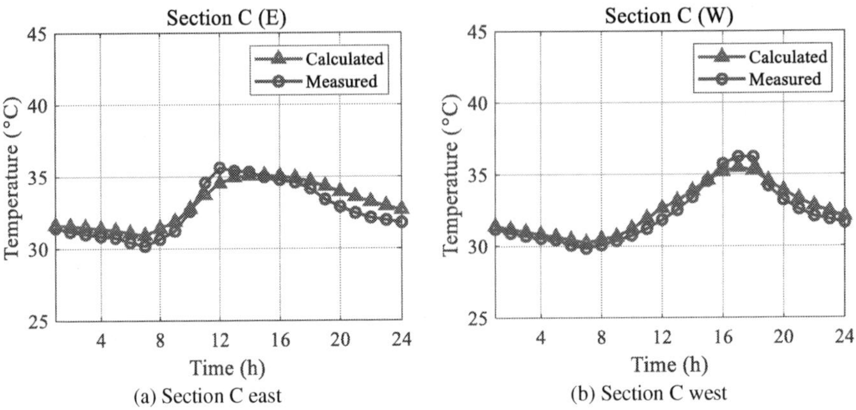

FIGURE 6.8 Temperature of the eastern and western beams on August 25, 2020: (a) Section C east and (b) Section C west.

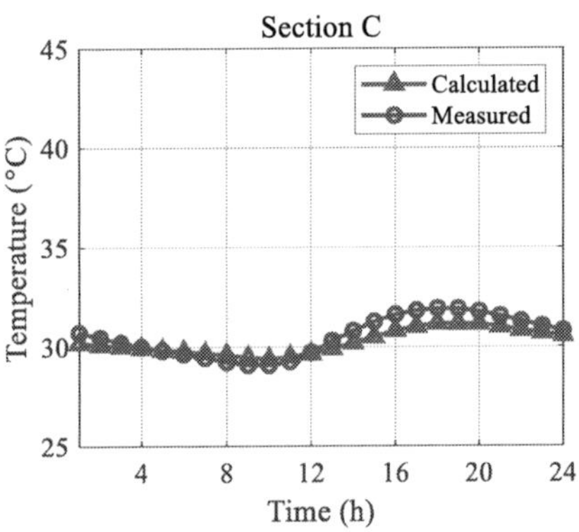

FIGURE 6.9 Temperature of the concrete slab on August 25, 2020: (a) Section C east and (b) Section C west.

The measured temperatures of the eastern and western beams on February 17 are shown in Figure 6.10. The lowest temperature at the east beam was 12°C, and the highest was 16.7°C. Similarly, the measured lowest temperature at the west beam was 12.2°C, and the maximum was 22.6°C. The west beam had a higher temperature than the east beam because several buildings in the southeast of the bridge blocked the sunshine in the morning of winter. The west beam also had a more significant temperature variation (about 10°C) than the latter. In addition, the temperature variation on that day was larger than that on August 25 because the air temperature variation on the former was larger (Figure 6.6a).

The temperature of the middle of the concrete slab of the footbridge is shown in Figure 6.11. The temperature variation of the slab was about 3.0°C, which was smaller than that of the beams because of the timber pavement on the concrete slab.

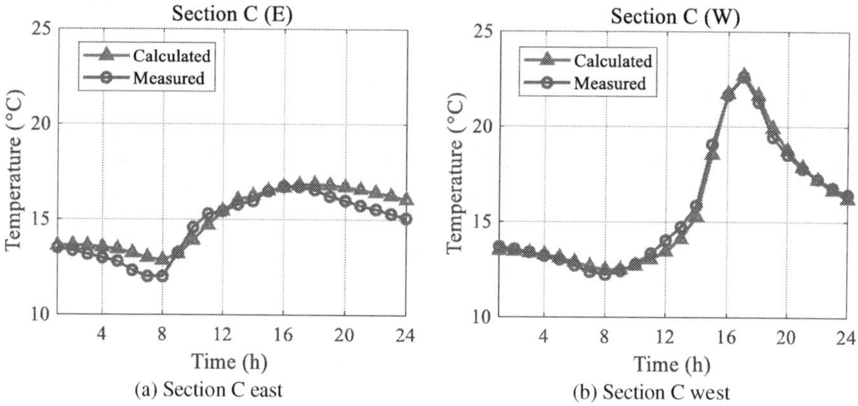

FIGURE 6.10 Temperature of the eastern and western beams on February 17, 2020.

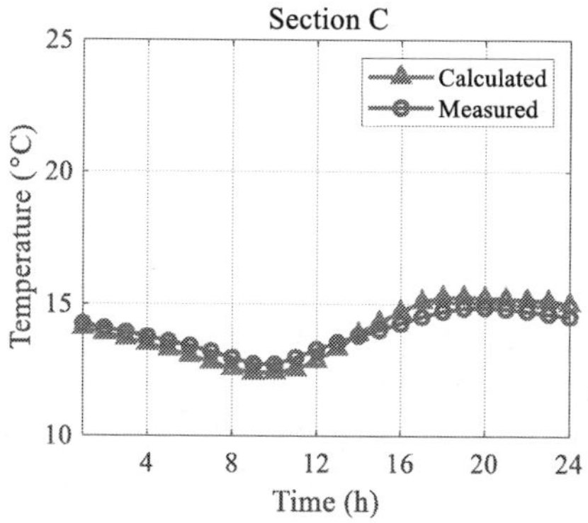

FIGURE 6.11 Temperature of the concrete slab on February 17, 2020.

The measured temperature data at all cross sections of the bridge are summarized in Table 6.1. Only maximum and minimum values are given for simplicity.

6.2.3 Temperature-Induced Structural Responses

6.2.3.1 Stress

The hourly averaged strain is used because temperature tends to cause static strain in a bridge, whereas pedestrians and winds cause dynamic strain. The stresses at sensors C2 (western beam) and C4 (eastern beam) in Section C on the two days are shown in Figure 6.12. Here, the measured stress is calculated from the strain after the free thermal expansion of the component is removed. In addition, the stress change is plotted by using the stress at 1:00 as the reference. The stresses of the beams decreased (in compression) when the temperature rose in the morning. The stresses of the beams increased (in tension) when the temperature decreased in the afternoon.

TABLE 6.1
Measured and Simulated Temperature of Each Section (Unit:°C)

Cross Section (Direction)	August 25					February 17				
	Maximum		Minimum		RMSE	Maximum		Minimum		RMSE
	Mea.	Sim.	Mea.	Sim.		Mea.	Sim.	Mea.	Sim.	
A (E)	35.5	36.5	28.6	30.7	1.894	–	17.5	–	12.4	–
A (W)	38.9	36.7	30.8	28.3	2.433	16.3	18.3	12.4	10.7	1.616
B (E)	35.9	35	30.2	30.9	0.852	15.8	16.3	11.7	12.8	0.675
B (W)	36.4	37.2	29.9	30.8	1.043	19.2	20.3	11.9	12.4	1.155
C (E)	35.6	35.1	30.2	30.9	0.698	16.7	16.8	12.0	12.8	0.604
C (W)	36.2	35.5	29.8	30.2	0.524	22.6	22.6	12.2	12.4	0.299
D (E)	35.6	35.2	30.3	30.9	0.703	16.7	16.8	12.2	12.8	0.440
D (W)	39.1	39.2	32.1	31.3	0.857	21.9	21.4	12.2	12.6	0.658
E (E)	37.3	37.1	30.2	30.9	0.747	17.0	16.7	12.4	12.9	0.421
E (W)	34.1	34.5	29.9	30.8	0.987	22.0	22.3	12.4	12.4	0.705
F (E)	35.8	36.6	30.7	30.2	0.908	17.5	16.8	12.8	12.8	0.402
F (W)	35.4	34.1	30.8	30.7	0.369	18.0	17.4	13.9	12.5	0.634

Note: E is for the east side, and W is for the west; Mea and Sim denote the measured and simulated data, respectively; RMSE refers to the root mean square error.

The stress variation of the west beam occurred 4–6 hours later than that of the east beam. The reason is that the temperature variation of the west beam lagged the east beam. Moreover, the stress variations of the east and west beams were 4 and 6 MPa in the summer and 7 and 10 MPa in the winter. This result is because the west beam had a higher temperature change than the east beam on both days, and winter had a higher daily temperature variation than summer, as shown in Table 6.1. Therefore, the effects of temperature in winter were more critical than those in summer for the footbridge.

6.2.3.2 Displacement

Only the displacement of the expansion joint is presented here. The measured expansion joint displacement and simulation results on the two days are shown in Figure 6.13. The displacements of the expansion joint were very small, about 1.2 and 1.5 mm on August 25 and February 17, respectively. Figures 6.9 and 6.11 show that the temperature change of the concrete slab within one day (summer and winter) was very small, about 2°C. The resulting free thermal expansion of the deck can be estimated as $84 \times 2 \times 1.0 \times 10^{-5} = 1.7 \times 10^{-3}$ m, which is close to the measured value of 1.5 mm.

6.2.4 Summary

In this section, the measured temperature and temperature-induced structural displacement and strain on both the hottest and coldest days of 2020 are presented. The differences between the measured highest and lowest temperatures at the beams on

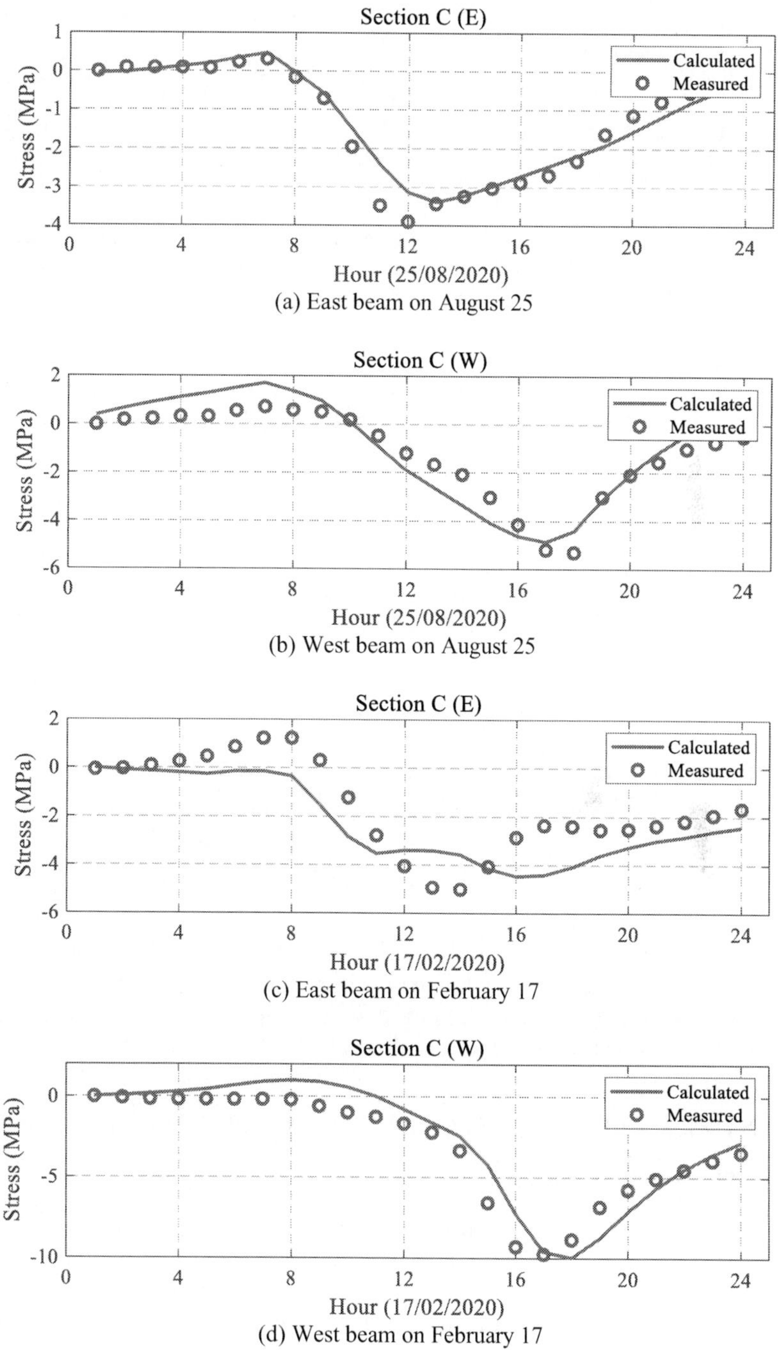

FIGURE 6.12 Stresses at Section C on August 25 and February 17, 2020: (a) east beam on August 25, (b) west beam on August 25, (c) east beam on February 17, and (d) west beam on February 17.

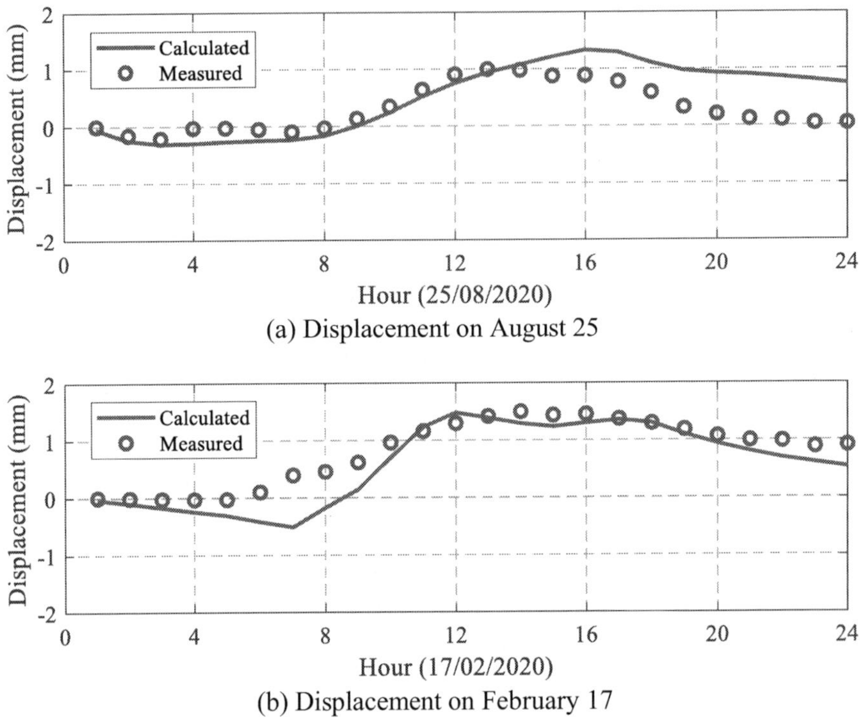

FIGURE 6.13 Displacement of the expansion joint on August 25 and February 17, 2020: (a) displacement on August 25 and (b) displacement on February 17.

the hottest and coldest days were about 6°C and 5°C, respectively. On both days, the east and west beams had similar temperatures, whereas the maximum temperature occurred at different times due to the influence of the bridge orientation. In different seasons, the temperature variation trends were similar. The displacements of the expansion joint on the hottest and coldest days were the same. It is mainly governed by the temperature variation of the concrete slab.

6.3 NUMERICAL INVESTIGATIONS OF THE FOOTBRIDGE

6.3.1 FE Model of the Footbridge

A 3D FE model was developed using the commercial FE analysis software ANSYS. The arches, main beams, deck, and columns are modeled using solid elements for refined analyses, such as buffeting analysis, global 3D heat-transfer analysis, etc. The full 3D FE model consists of 8,785 elements and 22,585 nodes, as shown in Figure 6.14.

The FE model will be used for the unified heat-transfer and structural analysis, as introduced in Chapter 3. The workflow is shown in Figure 6.15. In the heat-transfer analysis, the FE model consists of 3D thermal elements with appropriate thermal boundary conditions. The temperature distribution of the bridge is then calculated

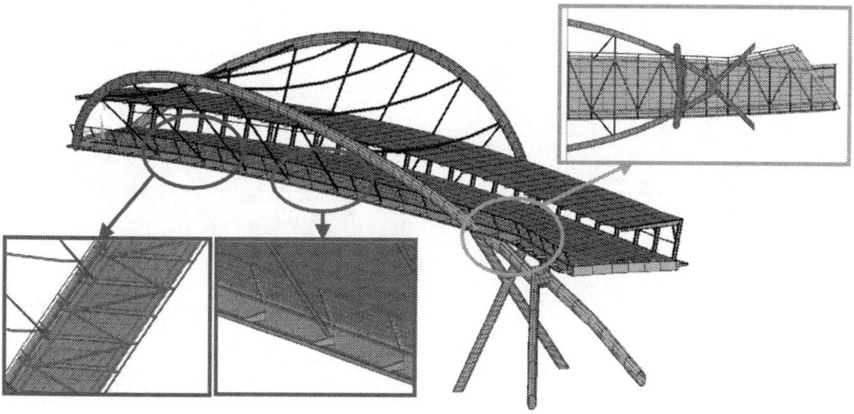

FIGURE 6.14 3D view of the FE model (Xia et al., 2022).

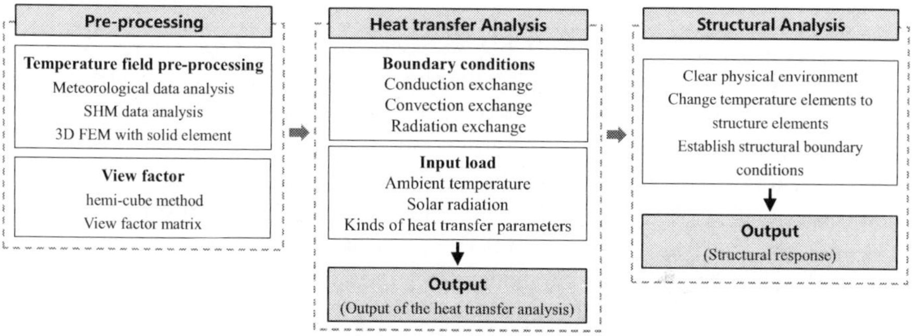

FIGURE 6.15 Workflow of the unified FE simulation.

and verified using the measured temperature data. In the structural analysis, the configuration and mesh of the FE model remain unchanged, and the 3D thermal elements change to 3D mechanics elements. The calculated temperature in the previous step is automatically assigned to the FE model, and the temperature-induced responses (such as displacement and stresses) are thus obtained. The two analyses will be presented later.

6.3.2 Thermal Boundary Conditions and Initial Temperature Condition

The thermal boundary conditions are generally determined using the measured meteorological conditions, following the methods introduced in Chapter 3. A special case worth considering is the shading effects of the shelter. A shelter made of steel frames and glass roof was designed, covering half of the pathways. The shelter affects the solar radiation received by the footbridge deck, and this effect varies with the solar conditions. The view factor method is adopted to consider the shading effects fully.

The radiation view factor describes how one object can "see" another object, which can be used to compute the radiation heat load (Gupta et al., 2017). Consider two

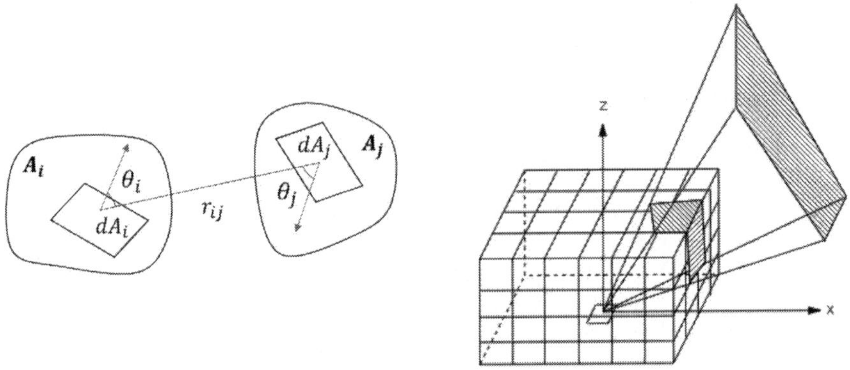

(a) Heat exchange between elementary areas (b) Hemi-cube method

FIGURE 6.16 View factor calculation: (a) heat exchange between elementary areas and (b) hemi-cube method.

surfaces A_i and A_j, the heat exchange between elementary areas dA_i and dA_j is shown in Figure 6.16a. The view factor $F_{i,j}$ between two surfaces A_i and A_j is the fraction of the energy emitted diffusely by surface A_i that reaches surface A_j. The view factor is a result of the orientation of surfaces in space. It depends mostly on the geometry of the body and the incident and reflected radiations on the body. The view factor is completely geometric and independent of surface qualities and temperature.

A structure consists of a large number of elements, and the irradiation and shade faces of the elements change continuously with the sun's orientation. Thus, applying solar radiation intensity to the FE model is the most difficult and challenging task in thermal analysis. Common methods are as follows: (1) direct integration method, (2) unit sphere method, (3) view factor by using the ray casting method, (4) cross string method, (5) view factor by using the Monte Carlo method, and (6) view factor by using the algebraic rule and matrix formulation. This study adopts the hemi-cube method (Gupta et al., 2017), as shown in Figure 6.16b.

The view factor is calculated using the hemi-cube method as

$$F_{i,j} = \sum_{n=1}^{N} \Delta F_n = \cos\theta_i \cos\theta_j / \left(\pi r_{ij}^2\right) \cdot \Delta A_j \tag{6.1}$$

where N is the number of pixels, and ΔF is the delta-view factor for each pixel.

The speed and accuracy of the hemi-cube method for the view factor computation can be affected by changing the size and number of discrete areas on the faces of the hemi-cube. All view factors are required to calculate the radiation exchange in the enclosure of surfaces. The view factors on surfaces can be written in the following matrix form:

$$\begin{bmatrix} F_{11} & F_{12} & \cdots & F_{1N} \\ F_{21} & F_{22} & \cdots & F_{2N} \\ \vdots & \vdots & \ddots & \vdots \\ F_{N1} & F_{N2} & \cdots & F_{NN} \end{bmatrix} \tag{6.2}$$

In this way, solar radiation from all surfaces can be obtained. The initial temperatures of the model are determined by a pre-analysis lasting 24 hours of the prescribed days.

6.3.3 Temperature Distributions

6.3.3.1 Temperature on the Hottest Day

As aforementioned, the temperatures on August 25, the hottest day in 2020, are studied. The most important task in the transient heat-transfer analysis is to determine the trajectory of the sun. First, according to the design of the footbridge, its azimuth is about 21° west-north, as shown in Figure 6.17.

The geographical coordinates of the footbridge are about 114.17° east longitude and 22.3° north latitude. From the Hong Kong Observatory, sunrise and sunset on

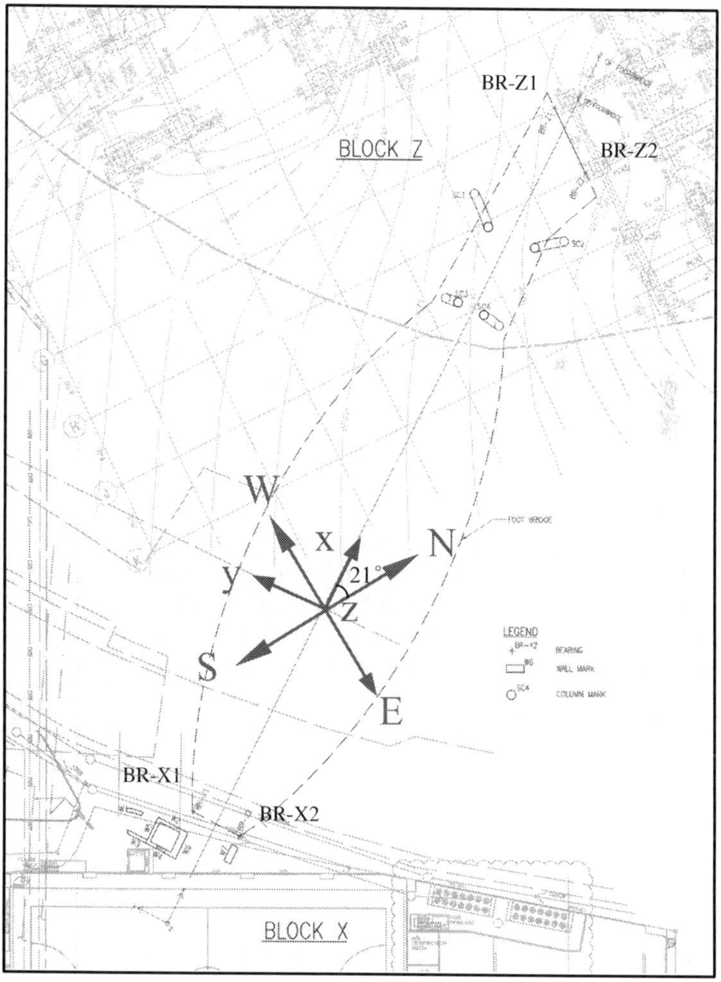

FIGURE 6.17 Orientation of the footbridge.

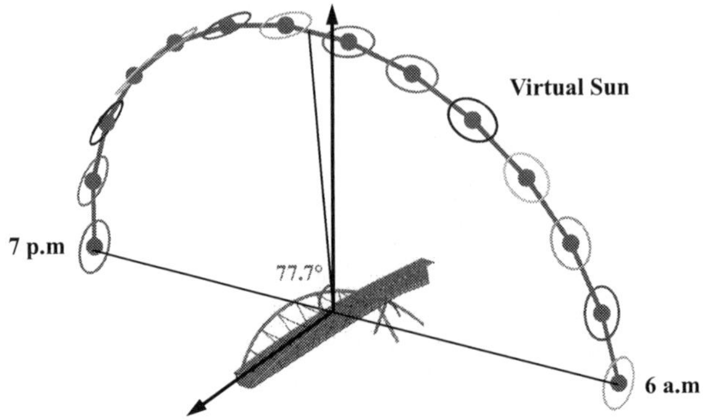

FIGURE 6.18 Solar orbit on August 25, 2020.

August 25, 2020, were at 6:03 and 18:46, respectively. The maximum solar altitude angle was 77.7° at about 12:26.

The virtual solar orbit in different hours based on the abovementioned information is shown in Figure 6.18. In each hour, the view factors of all surface faces of the FE model are calculated according to Equation (6.2). The heat exchange is then formulated using Equation (6.1). The solar radiation and ambient temperature are measured by the monitoring system.

The heat-transfer analysis is conducted on the FE model using the boundary condition on the day. The temperature distribution of the whole bridge in each hour is obtained and illustrated in Figure 6.19. The arches had higher temperatures than other components because they received the sunlight directly. As the solar orbit changed, the east side had a higher temperature than the west side in the morning and lower in the afternoon.

The simulated temperatures at the eastern beam, western beam, and the concrete slab of Section C on August 25 are compared with the measured results in Figures 6.8–6.9. The simulated lowest temperature of the eastern beam of the footbridge is 30.9°C, and the highest temperature is 35.1°C. Their discrepancies with the measurement counterparts are less than 1.0°C, and the variation trends are very close. At the western beam, the simulated lowest temperature is 30.2°C, and the highest is 35.5°C. The variation trends are also consistent with the measurement. The simulated temperature at the middle height of the concrete slab agrees well with the measured ones, with a difference of less than 0.5°C.

6.3.3.2 Temperature on the Coldest Day

From the Hong Kong Observatory, sunrise and sunset on February 17, 2020, occurred at 6:53 and 18:21, respectively. The maximum solar altitude angle was 55.1° at about 12:37. Figure 6.20 shows the solar orbit at different hours.

The temperature distribution of the whole bridge on the day is then calculated and shown in Figure 6.21. The main arches again had higher temperatures than other components because they received the sunlight directly.

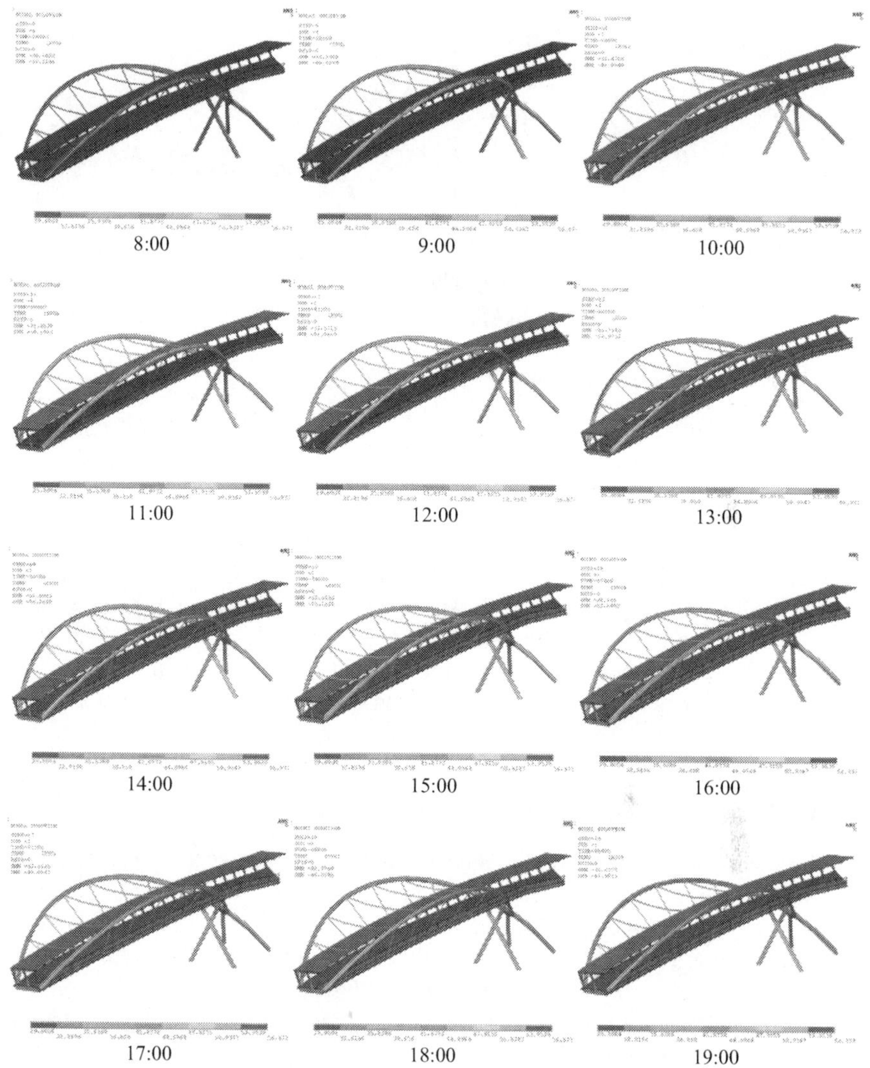

FIGURE 6.19 Temperature distributions of the footbridge on August 25, 2020.

The measured and simulated temperatures on the day are compared and shown in Figures 6.10 and 6.11. The simulated lowest temperature at the east beam was 12.8°C, and the highest was 16.8°C. The corresponding values on the western beam were 12.5°C and 22.7°C, respectively. The variation trend of simulated temperature was consistent with that of measured temperature. The simulated temperatures of the middle of the concrete slab of the footbridge also agreed well with the measured results, with a difference of less than 0.5°C.

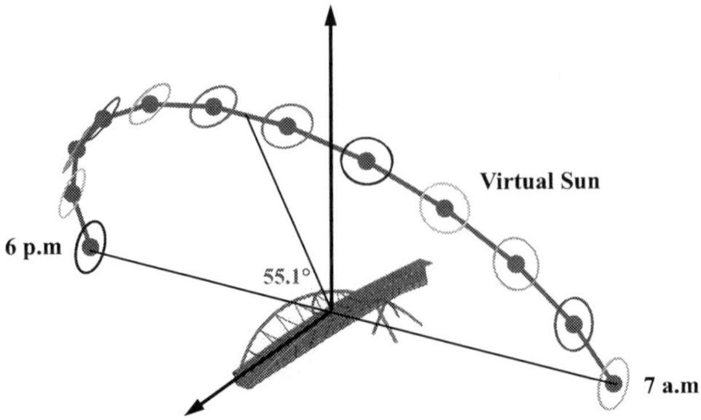

FIGURE 6.20 Solar orbit on February 17, 2020.

The measured and simulated temperature data at all cross sections of the bridge are compared and shown in Table 6.1. A significant discrepancy between the measured and simulated temperature is observed in Section A because a different type of sensor with relatively low accuracy was installed in the section due to the small space there. These sensors will be replaced in the future. Except that in Section A, the maximum discrepancy was 1.4°C on the west side of Section F. The RMSE of the whole day is also calculated for different sections. Most RMSE values were less than 1.0°C, with the largest being less than 1.2°C.

6.3.4 Temperature-Induced Structural Responses

As aforementioned, the structural analysis shares the same element mesh and configuration as the heat-transfer analysis. The FE model for the heat-transfer analysis can be used for the structural analysis directly by revising the thermal elements as the mechanical elements, and the results of the heat-transfer analysis are transferred to the FE model as the input of the structural analysis.

The procedure of transferring the heat-transfer analysis to the structure analysis in ANSYS is described as follows:

1. Save the result file after the temperature field calculation.
2. Clear the physical environment variables and change the temperature unit to the structural unit.
3. Apply the structural boundary conditions. Bearing-X1 (BR-X1) can move in the x-direction only. BR-X2, BR-Z1 and BR-Z2 can move in x- and y-directions. All bearings can rotate freely. Four feet are fixed to the ground. The bearing location is shown in Figure 6.17.
4. The calculated temperature results are input to the FE model as the node temperature load. The time step and the time interval should be consistent with the temperature field calculation.

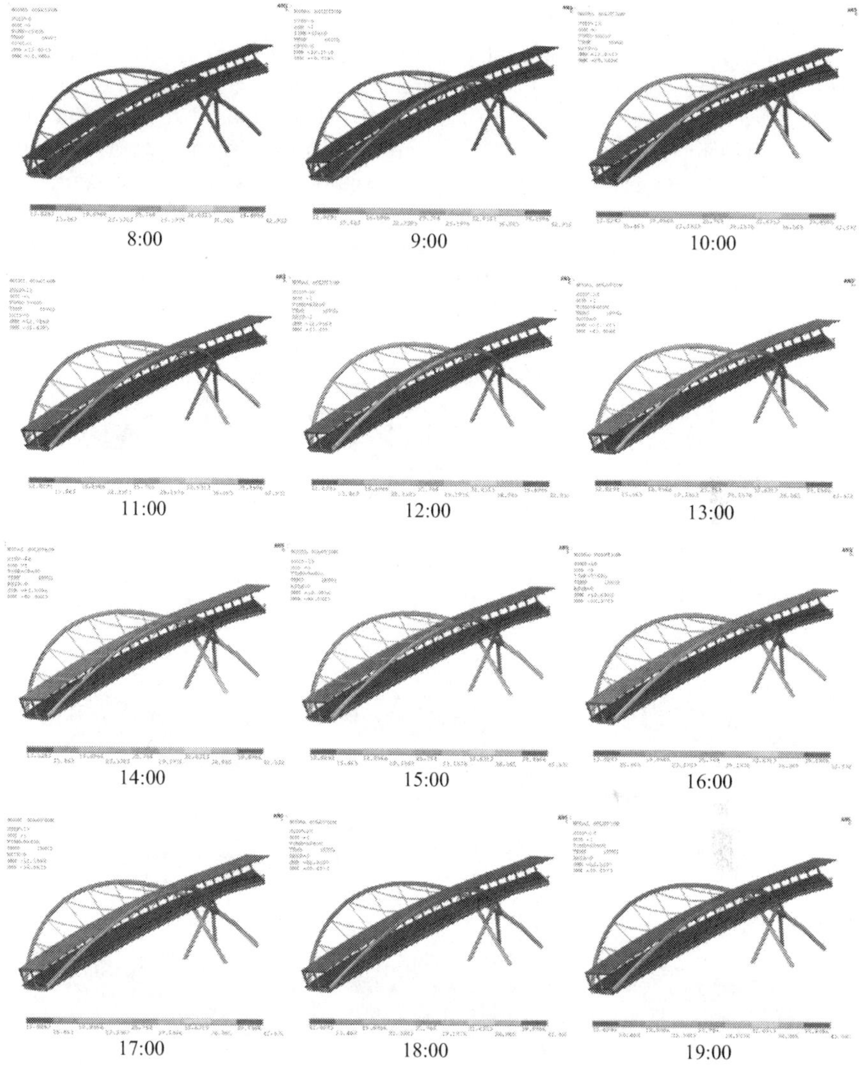

FIGURE 6.21 Temperature distributions of the footbridge on February 17, 2020.

5. The element types are revised from SOLID70 (3D thermal solid element) to SOLID45 (3D structural solid element with eight nodes, each having three lateral displacement degrees of freedom).

When the temperature load is applied to the FE model, the structural analysis is conducted to obtain the structural responses (stresses and displacement). The results are shown and discussed in the following subsections. The heat-transfer and structural analyses over a 24-hour period (one point per hour) take 86 and 76 seconds, respectively, using a laptop with a CPU of Intel® Core i7–1260P and 16GB RAM.

6.3.4.1 Stresses on the Hottest Day

The calculated structural stress of the whole bridge from 8:00 to 19:00 on August 25, 2020, is shown in Figure 6.22. The main arches exhibited higher structural stresses than other components because they had a higher temperature. As the solar orbit changed, the stress of the main arch changed accordingly. The two legs also had a relatively high stress, probably because the deformation of the feet was constrained.

The maximum tensile stress of the whole footbridge was 31.7 MPa at 13:00 on the upper flange of the eastern beam. At the same time, the maximum compressive stress of the whole footbridge was 90.2 MPa at the bottom of the east column.

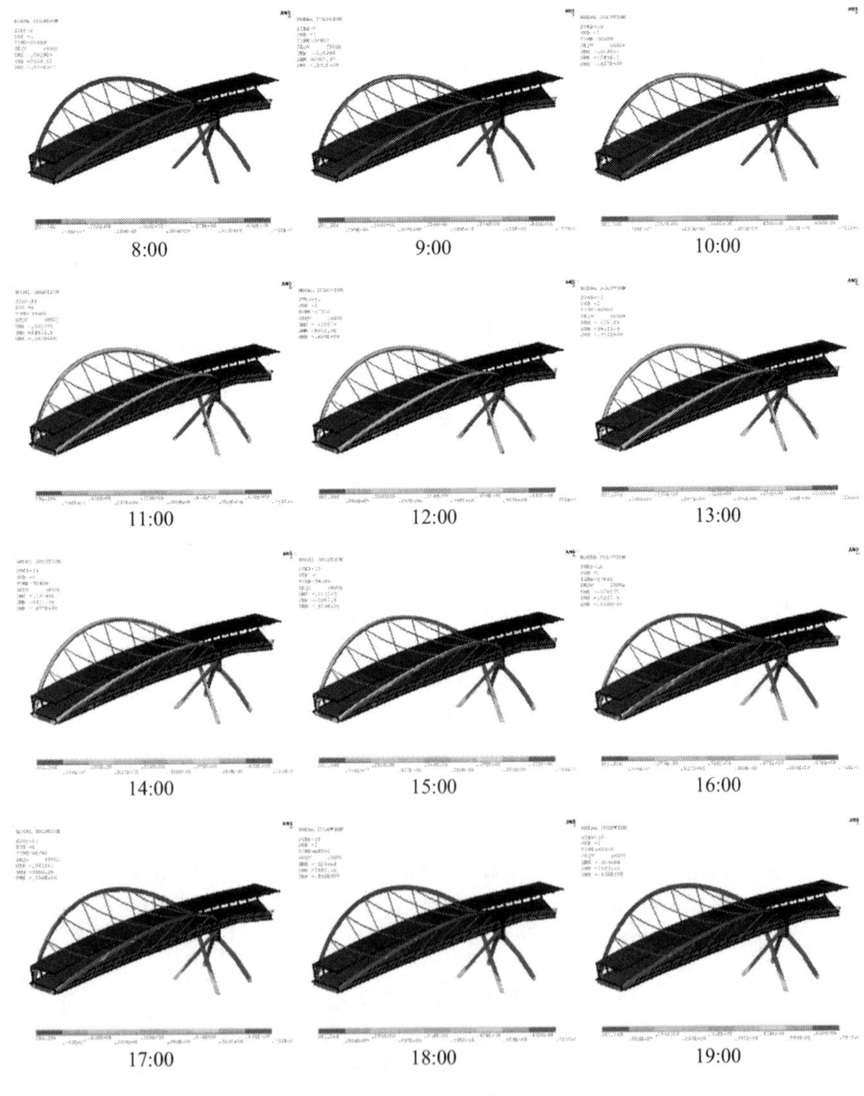

FIGURE 6.22 Structural stresses of the footbridge on August 25, 2020.

6.3.4.2 Stresses on the Coldest Day

The calculated structural stress distribution of the bridge from 8:00 to 19:00 on February 17 is shown in Figure 6.23. Again, the main arches exhibited a higher structural stress than other components. As the solar orbit changed, the structural temperature and the stress distribution of the main arches changed accordingly. The two legs also had a relatively high stress, as explained above.

The maximum tensile stress of the whole footbridge was 44.6 MPa at 13:00 on the upper flange of the eastern beam. At the same time, the maximum compressive stress of the whole footbridge was 82.8 MPa at the bottom of the east column.

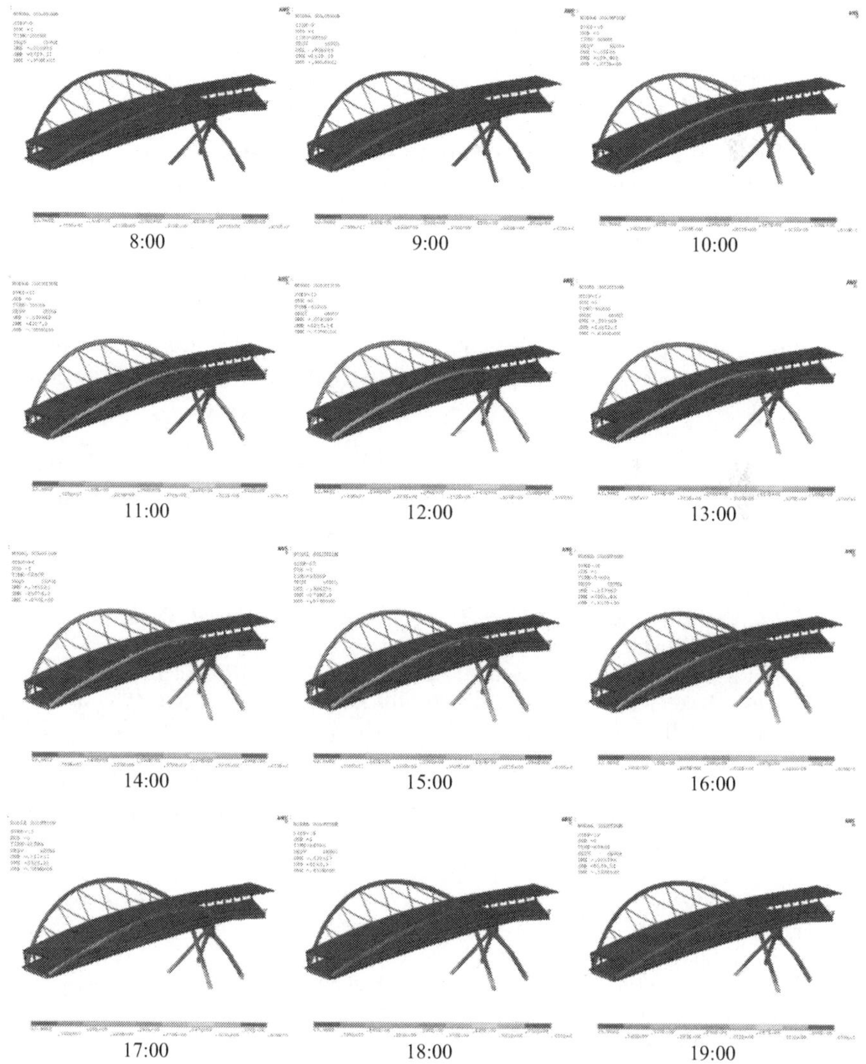

FIGURE 6.23 Structural stresses of the footbridge on February 17, 2020.

The maximum tensile stress on the coldest day was higher than on the hottest day. Conversely, the maximum compressive stress on the hottest day was higher than on the coldest day. Nevertheless, the maximum tensile stress on the coldest and hottest days occurs in the same locations. The same is observed on the maximum compressive stress.

The numerical and measured stress results are compared in Figure 6.12. The results are close on both days, and the variation trend is consistent.

6.3.4.3 Temperature-Induced Displacement

The measured expansion joint displacement and simulation results on the two days are shown in Figure 6.13. The calculated results match well with the measurement counterparts, verifying the measurement and numerical results.

6.4 CHAPTER SUMMARY

In this chapter, the temperature behaviors of the PolyU Footbridge are numerically investigated following the integrated heat-transfer and structural analysis approach. The calculated results of the temperature analysis and structural response agree well with the monitoring results. In different seasons, the discrepancy of the measured and simulated temperatures is nearly less than 1.0°C, and their variation trends are similar. The RMSE of the temperature in each section is less than 1.2°C. The maximum discrepancy of the stress does not exceed 1.5 MPa. Overall, the comparison results show the accuracy and practicability of the numerical method without human intervention of the methods.

REFERENCES

Duan, Y. F., Li, Y., Qiang, X. Y., 2011, Strain-temperature correlation analysis of a tied arch bridge using monitoring data. In *2011 International Conference on Multimedia Technology*, Hangzhou, China, pp. 6025–6028.

Gupta, M. K., Bumtariya, K. J., Shukla, H., Patel, P., Khan, Z., 2017, Methods for evaluation of radiation view factor: A review. *Materials Today: Proceedings* **4(2)**: 1236–1243.

Li, L., Chen, B., Zhou, L., Xia, Q., Zhou, Y., Zhou, X., Xia, Y., 2023, Thermal behaviors of bridges—A literature review. *Advances in Structural Engineering* **26(6)**: 985–1010.

Xia, Q., Wu, W. L., Li, F. N., Xia, Y., Ding, X. L., Lam, W. H., Chung, W. H., Xu, Y. L., 2021, System design and demonstration of performance monitoring of a butterfly-shaped arch footbridge. *Structural Control and Health Monitoring* **28(7)**: e2738.

Xia, Q., Wu, W. L., Li, F. N., Zhou, X. Q., Xu, Y. L., Xia, Y., 2022, Temperature behaviors of an arch bridge through integration of field monitoring and unified numerical simulation. *Advances in Structural Engineering* **25(16)**: 3492–3509.

Yarnold, M. T., Moon, F. L., Dubbs, N. C., Aktan, A. E., 2012, *Evaluation of a Long-Span Steel Tied Arch Bridge Using Temperature-Based Structural Identification*. Stresa, International Association for Bridge Management and Safety.

Zhou, G. D., Yi, T. H., Chen, B., Chen, X., 2018, Modeling deformation induced by thermal loading using long-term bridge monitoring data. Journal of Performance of Constructed Facilities **32(3)**: 04018011.

7 Temperature Behaviors of Beam Bridges

NOTATIONS

D_k	Mid-span deflection of the kth span
$D_1^{\lim}$	Limit of mid-span deflection
D_1^{odd}, D_1^{even}	Deflection of the first span in odd- and even-numbered span beams
D_u^{odd}, D_u^{even}	Mid-span deflection of odd- and even-numbered span beams
E	Elastic modulus
I	The second moment of inertia
L_{org}	Original free length
M_T	Fixed-end moment caused by the temperature difference ΔT
M_k, M_u	Moment at joints k and u
T	Temperature of the beam
h	Depth
i_1, i_0	Relative flexural stiffness of main and side spans
n	Number of spans
x_1, x_2	Roots to equations $f_i(x)=0$, $i=1, 2, 3$ and 4
u	A positive integer related to the span number
z_k	Rotation at joint k
ΔT	Differential temperature change between top and bottom surfaces
α	Linear expansion coefficient
α_k	A sequence with a recurrent form
ε	Strain
φ_1, φ_2	Constants
λ_1, λ_2	Constants
ξ	Side-to-main span ratio

7.1 PREFACE

Beam bridges have straightforward structural components: abutments (piers) and girders (decks). The girders, which can be in the form of solid plates, box girders, or trusses, are either simply supported or continuously span over the piers or abutments. Beam bridges are widely used in small- and medium-span bridges because of their advantages of simple configuration, convenient construction, and low maintenance cost. A simply supported beam bridge generally crosses no more than 50 m long, whereas a continuous type can stride a span of up to 300 m. In this chapter, temperature behaviors of beam bridges will be presented from the perspective of field monitoring, numerical simulation, and theoretical analysis.

DOI: 10.1201/9781003480136-7

7.2 FIELD MONITORING STUDIES OF A VIADUCT

In addition to three cable-stayed bridges, continuous beam bridges are also the key components of the 55-km long HZMB. In this section, the field monitoring results of a 370 m long three-span viaduct adjacent to the Qingzhou Cable-stayed Bridge will be presented.

7.2.1 Sensors on the Viaduct

The span arrangement and typical cross sections of the viaduct are shown in Figure 7.1a. Dual box sections, with cantilever arms on both sides of the girder and varying section heights (ranging from 4.5 to 6.5 m) as shown in Figure 7.1b, are adopted. The steel top plates are covered by a layer of 70 mm thick asphalt concrete and reinforced with U-ribs inside the box cell. The bottom plates are reinforced with plate ribs. Moreover, along the longitudinal direction, the diaphragms shown in the left half of Figure 7.1c are arranged at every 10 m, and the annular diaphragms shown in the right half of Figure 7.1c are further arranged at every 2.5 m between them. The entire girder rests on four concrete piers, including one shared with the Qingzhou Bridge on the west end.

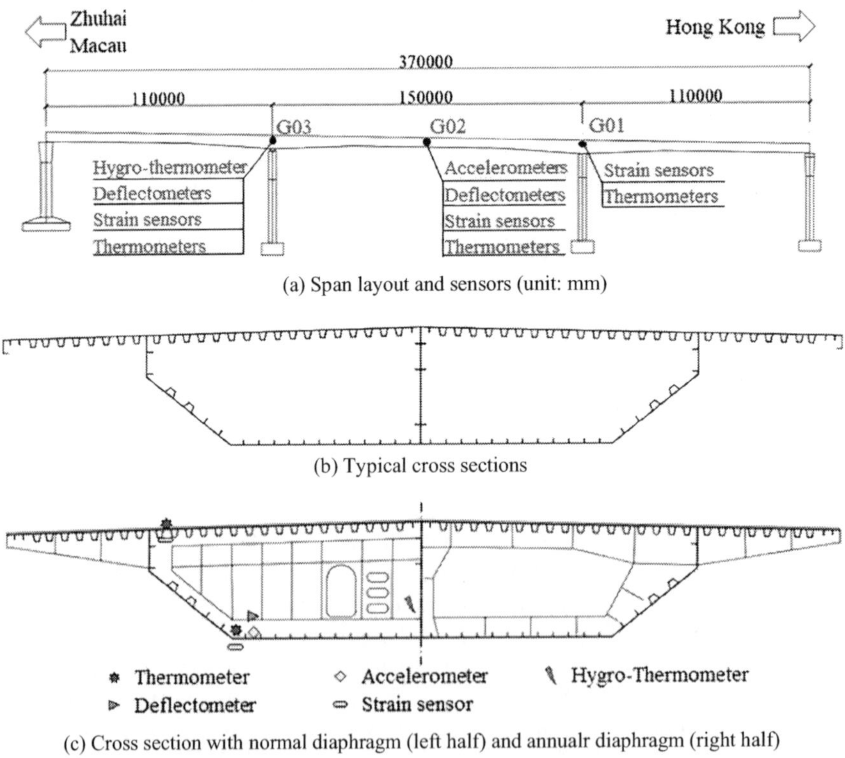

FIGURE 7.1 Span arrangement and typical cross sections of the viaduct: (a) span layout and sensors (unit: mm), (b) typical cross sections, and (c) cross section with normal diaphragm (left half) and annular diaphragm (right half).

The viaduct contains 4 deflectometers, 2 uni-axial accelerometers, 12 strain sensors, 12 structural temperature sensors, and 1 hygro-thermometer arranged in three sections as marked in Figure 7.1a. The specific locations of these sensors within the three sections are shown in Figure 7.1c. Only a half is shown here as these sensors are symmetrically deployed on the cross section except the hygro-thermometer. The vibration of the viaduct is measured at a sampling frequency of 50 Hz while the others at 10 Hz. As the structural temperature, deflection, and strain responses vary slowly, they are averaged hourly.

7.2.2 Monitored Temperature Distribution

The field monitored structural temperatures from December 2 to 8, 2020 are presented in this section.

7.2.2.1 Measured Environmental Conditions

As the viaduct is close to the Qingzhou Bridge, the external ambient temperature, humidity, wind direction, and speed are similar to the latter, which are shown in Figure 7.2. During the prescribed period, the ambient temperature ranged from 13.7°C to 22.1°C, and the humidity ranged from 48.3% to 67.7%. The wind generally blew from the north direction, and the maximum speed was less than 15 m/s.

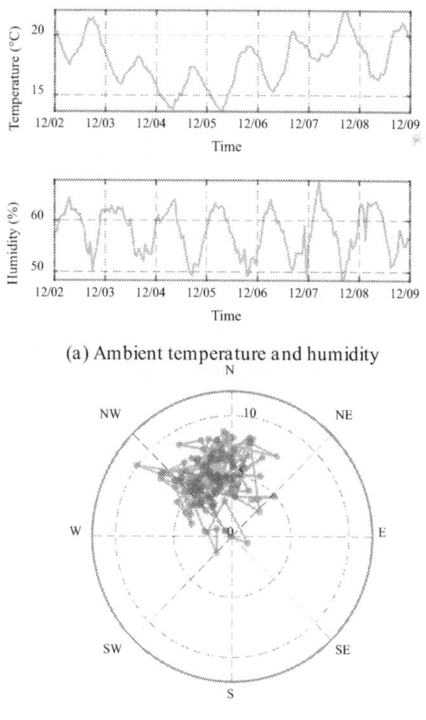

(a) Ambient temperature and humidity

(b) Wind direction and speed

FIGURE 7.2 Environmental conditions of the viaduct from December 2 to 8, 2020: (a) ambient temperature and humidity and (b) wind direction and speed.

The direct and diffuse solar radiation, extracted from the total solar radiation collected by the Hong Kong Observatory, is shown in Figure 7.3. The direct radiations were similar during the first six days, and they were greater than the last day, indicating the former days were sunny days while the last day was cloudy.

The ambient temperature and humidity inside the girder box of the viaduct during the prescribed period is shown in Figure 7.4. The temperature and humidity are more stable than the external conditions, ranging from 16.1°C to 22.3°C and 39.2% to 48.6%, respectively.

7.2.2.2 Temperature Distribution of the Main Girder

The structural temperature was measured at four points at three girder sections as marked in Figure 7.1. The four points in each section measured the structural temperatures of the top and bottom plate near the side web at both the north and south sides. Figure 7.5 shows the measured results on December 2 and 8, 2020. The top plate temperatures were much higher than the bottom plate during the daytime, showing a remarkable vertical temperature gradient. Moreover, the temperature of the south side was higher than the north side, especially during the daytime, indicating a non-negligible transverse temperature gradient. This is because the Sun moved to the south semi-sphere in the winter and thus the south side received more solar radiation than the north.

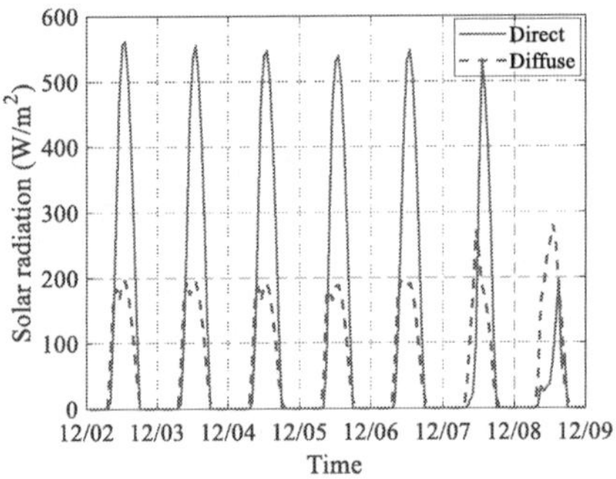

FIGURE 7.3 Solar radiation during December 2 to 8, 2020.

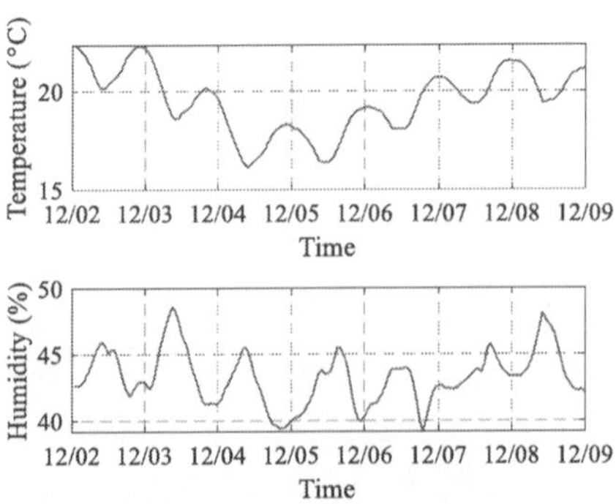

FIGURE 7.4 Ambient temperature and humidity inside the girder box.

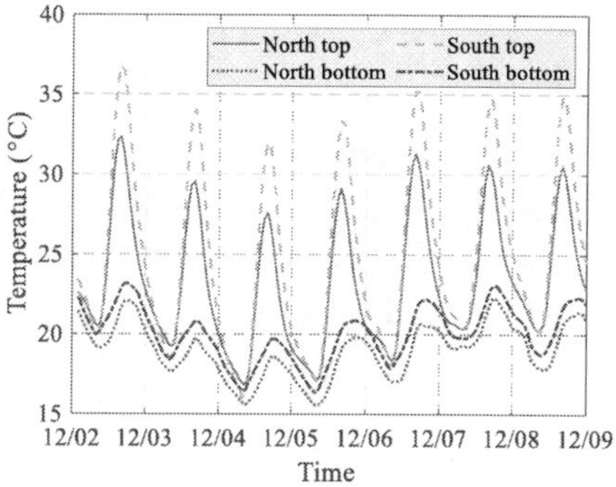

FIGURE 7.5 Measured temperature at section G02.

7.2.3 Temperature-Induced Structural Responses

The temperature-induced structural deflections and strain/stresses are presented in this section.

7.2.3.1 Temperature-Induced Displacement

The displacement transducers were installed at sections G03 and G02, as marked in Figure 7.1, measuring the vertical displacement above the pier and the mid-span of the girder, respectively.

The pair of deflectometers in G03 measured the thermal expansion of the concrete piers. The downward direction is defined as positive. The displacements are shown in Figure 7.6. The ambient temperature decreased from midnight to the morning. Correspondingly, the pier cooled down and contracted, showing an ascending part in Figure 7.6a. The temperature then started to rise till around 15:00 and the piers expanded during this period, showing a descending part in Figure 7.6a.

The deflectometers in G02 measured the mid-span deflection of the girder, which was mainly due to the temperature difference of the section. The deflection, as shown in Figure 7.6b, decreased in the early morning and then increased to the maximum in the afternoon, in a good agreement with the change in the temperature difference

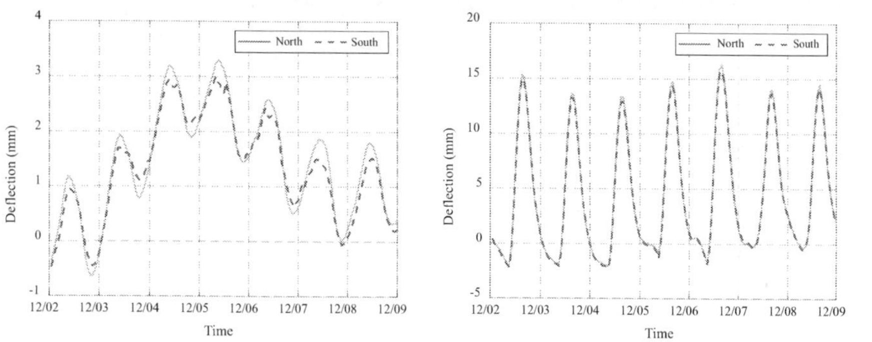

(a) Vertical displacement at G03 section (b) Vertical displacement at G02 section (mid-span)

FIGURE 7.6 Measured temperature-induced displacement of the beam: (a) vertical displacement at G03 section and (b) vertical displacement at G02 section (mid-span).

(Figure 7.5). The maximum deflection was 17 mm within one day. The underlying mechanism is discussed in Section 7.3.

7.2.3.2 Temperature-Induced Strain/Stress

The measured longitudinal strains at section G02 are shown in Figure 7.7. The traffic load during this period is negligible due to the COVID-19 pandemic, and thus the strain variations in the figure were purely induced by changing temperature. The measured strains at 00:00 on December 3, 2020 are taken as baselines and removed from the measurement. The baselines are specially chosen for the future consistency with Section 7.4, in which the first day is taken as a pre-analysis to obtain the initial conditions. The strain sensor at the south bottom plate malfunctioned and its result is not shown. Within a day, the temperature variation caused a strain variation around 100~180 $\mu\varepsilon$ in the top plate and 50 $\mu\varepsilon$ in the bottom plate. The top plate strain variations were more significant than the bottom counterpart due to the more remarkable temperature variation in the top plate, as shown in Figure 7.5.

The temperature-induced elastic stresses are calculated according to Equations (5.1) to (5.4) and shown in Figure 7.8. Structural temperature rises in the daytime induced tensile stresses in the bottom plate while compressive stress in the top plate. The maximum compressive stress could reach 18 MPa.

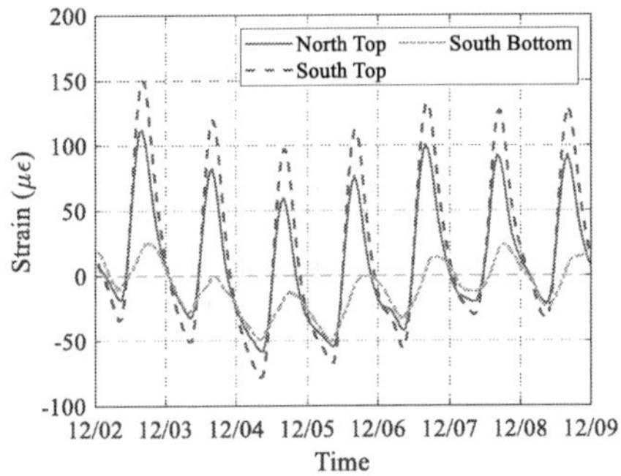

FIGURE 7.7 Measured strain at G02 section (baseline removed).

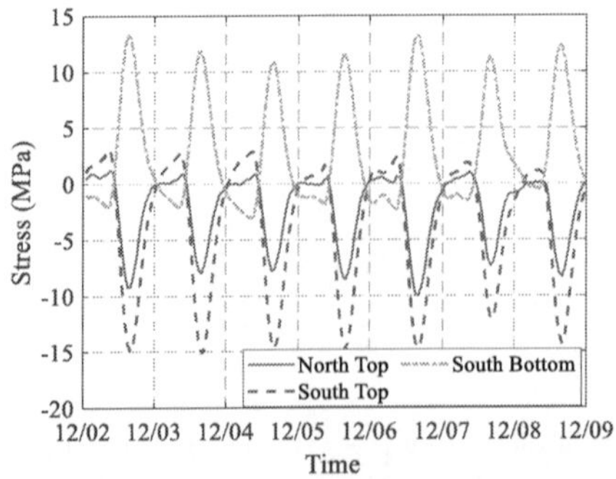

FIGURE 7.8 Temperature-induced elastic stress (baseline removed).

7.2.4 Summary

In this section, the field monitoring results of a viaduct of the HZMB are presented. Obvious vertical and transverse temperature gradients were observed. The temperature rise in the daytime induced downward deflection of the main-span girder. The temperature also induced elastic stresses up to 18 MPa.

7.3 ANALYTICAL FORMULATIONS

In this section, the temperature-induced bridge displacements of beam bridges will be analytically studied.

According to Eurocode 1991-1-5 (European Committee for Standardization CEN, 2003), the temperature action on a structural element can be decomposed into four constituent components: (1) uniform temperature change, (2) linearly varying temperature difference along the height, (3) linearly varying temperature difference along the width, and (4) nonlinear temperature difference component. Component 1 leads to an axial deformation. This effect is straightforward for beams, namely, the axial thermal expansion or contraction at beam ends $\delta L_{org} = L_{org} \cdot \alpha \cdot \delta T$, where L_{org}, α, and T respectively represent the original free length, linear expansion coefficient, and temperature of the beam, and $\delta(\cdot)$ denotes the quantity change (Roeder, 2003). Components 2 and 3 induce bending curvature, but in most cases, the latter is negligible compared with the former. Component 4 corresponds to self-equilibrated stress on the cross section, which does not cause global responses of the structure. This section focuses on component 2 of simply supported and continuous beams.

Although the differential temperature-induced deflection has been discussed on various beams, e.g., the scaled bridge model by Kulprapha and Warnitchai (2012) and several short-span bridges by Burdet (2010), these case studies are for specific beams but not generalized. For example, the variation in the beam deflection with the span number is unavailable.

This section is intended to develop analytical solutions to the beam deflection induced by the temperature difference across the depth. It will have the following aims: (1) to gain insights into the thermal deformation of continuous beams with any number of spans and (2) to provide a simple and precise formula to calculate the differential temperature-induced beam deflection.

All derivations in this section are based on the classical Bernoulli–Euler beam theory. Consider a prismatic n-span beam subjected to a differential temperature change ΔT between the top and bottom surfaces, as shown in Figure 7.9. A positive ΔT means the top surface is warmer than the bottom by ΔT(°C). Introduce a positive integer u, the smallest integer not less than $n/2$. Then the odd- and even-numbered n can be expressed as $n = 2u - 1$ and $n = 2u$, respectively. An n-span beam has $n + 1$ joints, each corresponding to a support of the beam. When the beam consists of odd-numbered spans (Figure 7.9a), the uth span is designated as the main span with a length of l_1, while the remaining spans are identical side spans with a length of l_0. When the beam has even-numbered spans (Figure 7.9b), both the uth and $(u + 1)$th spans are main spans, while the other spans have identical lengths of l_0.

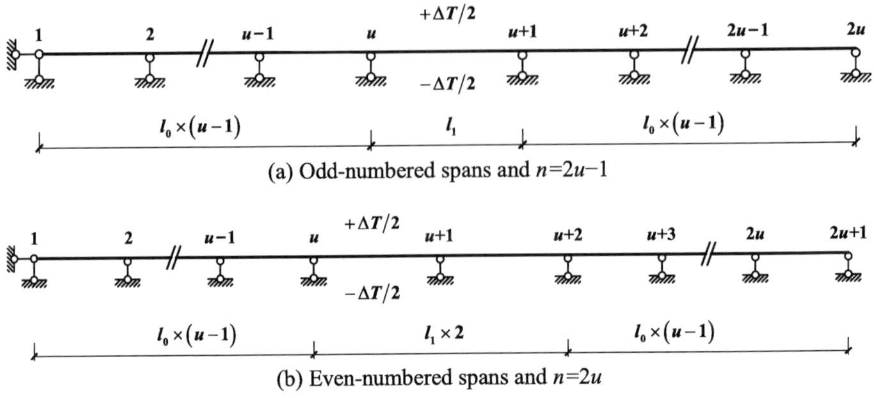

FIGURE 7.9 Analytical model of an n-span beam: (a) odd-numbered spans and $n=2u-1$ and (b) even-numbered spans and $n=2u$.

The beam has a constant cross section with the depth h and the second moment of inertia I. It is made up of the same material with elastic modulus E and linear expansion coefficient α. The relative flexural stiffnesses of the main and side spans are denoted as $i_1=EI/l_1$ and $i_0=EI/l_0$, respectively. The side-to-main span ratio is defined as $\xi=l_0/l_1$. Then, one has $i_1=i_0\xi$.

The mid-span deflection of the kth span beam, denoted by D_k, can be separately solved as follows according to whether the span number n is odd or even.

7.3.1 Odd-Numbered Beam Spans

As shown in Figure 7.9a, the joints on the beam are sequentially numbered as 1, 2,…, $2u$ from left to right. The rotation at joint k ($1\le k\le 2u$) is denoted as z_k, which is positive for a clockwise rotation. z_k can be solved by substituting the slope-deflection equations (Leet et al., 2021) into the moment equilibrium equations at each joint.

At joint 1, the moment equilibrium leads to:

$$4i_0z_1+2i_0z_2+M_T=0 \tag{7.1}$$

where M_T is the fixed-end moment caused by the temperature difference ΔT in a single span beam. M_T is positive if it makes the bottom of the beam stretched. According to Appendix A, M_T is:

$$M_T=\alpha\Delta T/h\cdot EI \tag{7.2}$$

At joint $k=2, 3,\ldots, u-1$:

$$2i_0z_{k-1}+8i_0z_k+2i_0z_{k+1}=0 \tag{7.3}$$

At joint u:

$$2i_0 z_{u-1} + (4i_0 + 4i_1) z_u + 2i_1 z_{u+1} = 0 \tag{7.4}$$

Due to the symmetry of the structural configuration and temperature action, the beam deforms in a symmetric pattern, which leads to $z_k = -z_{2u+1-k}$ ($k = 1, 2,\ldots, u$). Therefore, $z_{u+1} = -z_u$ and Equation (7.4) becomes:

$$2i_0 z_{u-1} + (4i_0 + 2i_1) z_u = 0 \tag{7.5}$$

The rotation at joints $k = u+2, u+3,\ldots, 2u$ can then be obtained from those at joints $k = u-1, u-2,\ldots, 1$, respectively. The number of independent unknowns and the number of simultaneous equations are both u.

To solve for z_k ($k = 1, 2,\ldots, u$), a sequence α_k with the recurrent form is introduced:

$$a_k = 1/(4 - a_{k-1}) \tag{7.6}$$

and $a_0 = 2 - \xi$. Then Equation (7.5) can be rewritten as:

$$z_u = -2i_0/(4i_0 + 2i_1) \cdot z_{u-1} = -1/(2+\xi) \cdot z_{u-1} = -a_1 \cdot z_{u-1} \tag{7.7}$$

By substituting $z_u = -a_1 \cdot z_{u-1}$ into Equation (7.3) with k being $u-1, u-2,\ldots,$ and 2 in sequence, one has the following recurrent formula:

$$z_k = -2i_0/(8i_0 - 2i_0 \cdot a_{u-k}) \cdot z_{k-1} = -1/(4 - a_{u-k}) \cdot z_{k-1} = -a_{u-k+1} z_{k-1} \tag{7.8}$$

According to Appendix B, the general expression of a_k ($k = 1, 2,\ldots, u-1$) can be solved as:

$$a_k = \left(\lambda_1 \varphi_1^{k-1} + \varphi_2^{k-1}\right) / \left(\lambda_1 \varphi_1^{k} + \varphi_2^{k}\right) \tag{7.9}$$

where the constants $\lambda_1 = (\sqrt{3}+\xi)/(\sqrt{3}-\xi)$, $\varphi_1 = 2+\sqrt{3}$, and $\varphi_2 = 2-\sqrt{3}$.

By substituting $z_2 = -a_{u-1} \cdot z_1$ into Equation (7.1), z_1 can be solved as:

$$z_1 = -(\alpha l_0 \Delta T)/(2\sqrt{3}h) \cdot \left(\lambda_1 \varphi_1^{u-1} + \varphi_2^{u-1}\right) / \left(\lambda_1 \varphi_1^{u-1} - \varphi_2^{u-1}\right) \tag{7.10}$$

Applying the recurrence relation in Equation (7.1) yields

$$z_k = (-1)^k \alpha l_0 \Delta T / (2\sqrt{3}h) \cdot \left(\lambda_1 \varphi_1^{u-k} + \varphi_2^{u-k}\right) / \left(\lambda_1 \varphi_1^{u-1} - \varphi_2^{u-1}\right) \tag{7.11}$$

The above equation applies to $k = 1, 2,\ldots, u$.

According to the slope-deflection equations (Leet et al., 2021), the end moments of each span beam can be expressed analytically. As no concentrated moment is acting at the joints, the left end moment of the kth ($k = 2, 3,\ldots, u$) span beam equals the right end moment of the $(k-1)$th span beam. Therefore, only the joint moment is presented, which is assumed to be positive if it stretches the beam bottom side. The moment at joint k ($k=1, 2,\ldots, u-1$) is

$$M_k = 4i_0 z_k + 2i_0 z_{k+1} + M_T = \left(1+(-1)^k \frac{\lambda_1 \varphi_1^{u-k} - \varphi_2^{u-k}}{\lambda_1 \varphi_1^{u-1} - \varphi_2^{u-1}}\right) \cdot \frac{\alpha EI \Delta T}{h} \tag{7.12}$$

The moment at joint u is

$$M_u = 4i_1 z_u - 2i_1 z_u + M_T = \left(1+(-1)^u \frac{\xi}{\sqrt{3}} \cdot \frac{\lambda_1 + 1}{\lambda_1 \varphi_1^{u-1} - \varphi_2^{u-1}}\right) \cdot \frac{\alpha EI \Delta T}{h} \tag{7.13}$$

Equation (7.13) is the same as Equation (7.12) when $k=u$. Therefore, Equation (7.12) is the unified formula to calculate the joint moment for $k=1, 2,\ldots, u$. With the symmetric condition, one has $M_{2u+1-k}=M_k$ ($k=1, 2,\ldots, u$).

The downward displacement of the beam is defined to be positive. The mid-span deflection of the kth ($k=1, 2,\ldots, u-1$) span beam can be calculated by the method of virtual work combined with the graphical "Product Integrals" method (Leet et al., 2021):

$$\begin{aligned} D_k &= \frac{1}{EI} \frac{l_0}{4} \frac{l_0}{2} \frac{M_k + M_{k+1}}{2} - \frac{\alpha \Delta T}{h} \frac{l_0}{2} \frac{l_0}{4} \\ &= (-1)^k \left(\alpha l_0^2 \Delta T\right) / (16h) \cdot \left(1+\sqrt{3}\right)\left(\lambda_1 \varphi_1^{u-k-1} + \varphi_2^{u-k}\right) / \left(\lambda_1 \varphi_1^{u-1} - \varphi_2^{u-1}\right) \end{aligned} \tag{7.14}$$

For the uth span beam, the mid-span deflection is

$$D_u = \frac{1}{EI} \frac{l_1}{4} \frac{l_1}{2} \frac{M_u + M_u}{2} - \frac{\alpha \Delta T}{h} \frac{l_1}{2} \frac{l_1}{4} = (-1)^u \frac{\alpha l_0^2 \Delta T}{8\sqrt{3}\xi h} \frac{\lambda_1 + 1}{\lambda_1 \varphi_1^{u-1} - \varphi_2^{u-1}} \tag{7.15}$$

The symmetric condition gives the mid-span deflection of the $(2u-k)$th span beam as $D_{2u-k}=D_k$ ($k=1, 2,\ldots, u-1$).

Furthermore, when the main and side spans have identical span lengths, i.e., $\xi=1$ and $\lambda_1 = (\sqrt{3}+1)/(\sqrt{3}-1) = \varphi_1$, Equations (7.1) and (7.2) can be unified into one formula as ($k=1, 2,\ldots, u$):

$$D_k = (-1)^k \alpha l_0^2 \Delta T / (16h) \cdot \left(1+\sqrt{3}\right)\left(\varphi_1^{u-k} + \varphi_2^{u-k}\right) / \left(\varphi_1^u - \varphi_2^{u-1}\right) \tag{7.16}$$

The formulas to calculate the joint rotation z_k (Equation 7.11), joint moment M_k (Equation 7.12), and mid-span deflection D_k (Equations 7.14–7.16) are valid for any positive integer u. For a simply supported beam, i.e., $n=1$ and $u=1$, these equations generate correct results with $\xi=1$.

7.3.2 Even-Numbered Beam Spans

The preceding procedure applies to a beam with even-numbered spans, as shown in Figure 7.9b, with slight modification. The moment equilibrium equations at joint 1 and joint k ($k = 2, 3, \ldots, u-1$) are identical to Equations (7.1) and (7.3), respectively. Due to symmetry, one has $z_k = -z_{2u+2-k}$ ($k = 1, 2, \ldots, u$) and $z_{u+1} = 0$. The equilibrium equation at joint u thus becomes

$$2i_0 z_{u-1} + (4i_0 + 4i_1) z_u = 0 \tag{7.17}$$

The number of independent unknowns and the number of simultaneous equations are also u.

Similar to the case of odd-numbered n, the relation between z_k and z_{k-1} can also be expressed by $z_k = -a_{u-k+1} \cdot z_{k-1}$ ($k = 2, 3, \ldots, u$), where the sequence a_k satisfies the condition of Equation (7.6) but with $a_0 = 2 - 2\xi$. From Appendix B, one has the general expression of a_k ($k = 1, 2, \ldots, u-1$):

$$a_k = \left(\lambda_2 \varphi_1^{k-1} + \varphi_2^{k-1}\right) / \left(\lambda_2 \varphi_1^{k} + \varphi_2^{k}\right) \tag{7.18}$$

where the constant $\lambda_2 = (\sqrt{3} + 2\xi)/(\sqrt{3} - 2\xi)$.

By replacing λ_1 with λ_2 in Equation (7.11), the joint rotation for the even-numbered n case can be obtained:

$$z_k = (-1)^k \alpha l_0 \Delta T / \left(2\sqrt{3}h\right) \cdot \left(\lambda_2 \varphi_1^{u-k} + \varphi_2^{u-k}\right) / \left(\lambda_2 \varphi_1^{u-1} - \varphi_2^{u-1}\right) \tag{7.19}$$

where $k = 1, 2, \ldots, u$.

According to the slope-deflection equations, the joint moment at joint k ($k = 1, 2, \ldots, u$) is

$$M_k = 4i_0 z_k + 2i_0 z_{k+1} + M_T = \left(1 + (-1)^k \frac{\lambda_2 \varphi_1^{u-k} - \varphi_2^{u-k}}{\lambda_2 \varphi_1^{u-1} - \varphi_2^{u-1}}\right) \cdot \frac{\alpha EI \Delta T}{h} \tag{7.20}$$

and at joint $u+1$:

$$M_{u+1} = -2i_1 z_u - 4i_1 z_{u+1} + M_T = \left(1 + (-1)^{u+1} \frac{\xi}{\sqrt{3}} \cdot \frac{\lambda_2 + 1}{\lambda_2 \cdot \varphi_1^{u-1} - \varphi_2^{u-1}}\right) \cdot \frac{\alpha EI \Delta T}{h} \tag{7.21}$$

Due to symmetry, one has $M_{2u+2-k} = M_k$ ($k = 1, 2, \ldots, u$).

The mid-span deflection of the kth ($k = 1, 2, \ldots, u-1$) span beam is similarly calculated as:

$$\begin{aligned} D_k &= \frac{1}{EI} \frac{l_0}{4} \frac{l_0}{2} \frac{M_k + M_{k+1}}{2} - \frac{\alpha \Delta T}{h} \frac{l_0}{2} \frac{l_0}{4} \\ &= (-1)^k \alpha l_0^2 \Delta T / (16h) \cdot \left(1 + \sqrt{3}\right) \left(\lambda_2 \varphi_1^{u-k-1} + \varphi_2^{u-k}\right) / \left(\lambda_2 \varphi_1^{u-1} - \varphi_2^{u-1}\right) \end{aligned} \tag{7.22}$$

and for the uth span beam:

$$D_u = \frac{1}{EI} \frac{l_1}{4} \frac{l_1}{2} \frac{M_u + M_{u+1}}{2} - \frac{\alpha \Delta T}{h} \frac{l_1}{2} \frac{l_1}{4} = (-1)^u \frac{\alpha l_0^2 \Delta T}{16\sqrt{3}\xi h} \cdot \frac{\lambda_2 + 1}{\lambda_2 \varphi_1^{u-1} - \varphi_2^{u-1}} \tag{7.23}$$

The symmetric condition leads to $D_{2u+1-k} = D_k$ ($k = 1, 2, \ldots, u$).

Furthermore, when the main and side spans are equal in length, i.e., $\xi = 1$ and $\lambda_2 = \left(\sqrt{3}+2\right)/\left(\sqrt{3}-2\right) = -\varphi_1^2$, the moment at joint k has a unified formula for $k = 1, 2, \ldots, u+1$:

$$M_k = \left(1 + (-1)^k \cdot \left(\varphi_1^{u-k+1} + \varphi_2^{u-k+1}\right) / \left(\varphi_1^u + \varphi_2^u\right)\right) \cdot (\alpha EI\Delta T)/h \tag{7.24}$$

So does the mid-span deflection of the kth span beam for $k = 1, 2, \ldots, u$:

$$D_k = (-1)^k \alpha l_0^2 \Delta T/(16h) \cdot \left(1 + \sqrt{3}\right)\left(\varphi_1^{u-k} - \varphi_2^{u-k+1}\right) / \left(\varphi_1^u + \varphi_2^u\right) \tag{7.25}$$

Also, Equation (7.19) applies to $k = u+1$ as it produces $z_{k+1} = 0$ at $\xi = 1$.

For a two-span continuous beam, i.e., $n = 2$ and $u = 1$, the above formulas for calculating the joint rotation, joint moment, and mid-span deflection are still valid with $\xi = 1$.

7.3.3 Parametric Analysis and Discussions

This section discusses the physical meanings of the derived formulas in the previous two sections through parametric analysis. As the beam deforms symmetrically, only the left half of the structure is investigated.

7.3.3.1 Contributing Factors

Since the beams consisting of an odd or even number of spans correspond to different formulas, the parity of the span number n matters. Besides, the beam deformation and moment are also closely related to the material properties, structural geometry, and temperature difference.

From Equations (7.11), (7.12), (7.14), (7.15), and (7.19)–(7.23), the joint rotation z_k, joint moment M_k, and mid-span deflection D_k are directly proportional to the linear expansion coefficient α and temperature difference ΔT, but inversely proportional to the beam depth h. The temperature-dependent rotation z_k and deflection D_k are proportional to l_0 and l_0^2, respectively, and both are independent of the flexural stiffness EI. The joint moment M_k has no relation with the absolute value of the span length, but depends on the span length ratio ξ implicitly, which is incorporated in the parameters λ_1 and λ_2. Meanwhile, M_k is directly proportional to EI, which implies that the strain ε of the beam is independent of EI as a result of $\varepsilon \propto M_k/(EI)$.

These temperature difference-induced responses of the beam (the joint rotation, moment, strain, and deflection) differ from the external force-induced responses. For example, the force-induced global deformation or local strain in a beam is inversely proportional to the flexural stiffness, while the internal moment is not. Additionally, z_k, M_k, and D_k also vary with the ordinal number k, cardinal number u, and span ratio ξ. For brevity, the dependency of D_k on k, u, and ξ is investigated next.

7.3.3.2 Deflection Shape

The deflection direction of the kth span beam depends on the sign of D_k. Assuming a positive temperature difference ($\Delta T>0$), the sign of D_k is only determined by the leading term $(-1)^k$ or $(-1)^u$ in Equations (7.14), (7.15), (7.22), and (7.23). To prove this, a quadratic function $f_1(x)$ can be defined as

$$f_1(x)=(x+1)\left(x\cdot\varphi_1^{u-1}-\varphi_2^{u-1}\right) \tag{7.26}$$

where $u\geq 1$. The roots of $f_1(x)=0$ are $x_1=-1$ and $x_2=\varphi_2^{2(u-1)}$ (as $\varphi_1=\varphi_2^{-1}$). As $\varphi_2=2-\sqrt{3}\approx 0.268<1$, one has $0\leq x_2\leq 1$. As $\varphi_1=2+\sqrt{3}>0$, the graph of $f_1(x)$ is a parabola opening upward. Therefore, $f_1(x)>0$ outside the region $x\in[-1,1]$.

On the other hand, recalling $\lambda_1=(\sqrt{3}+\xi)/(\sqrt{3}-\xi)$ and $\lambda_2=(\sqrt{3}+2\xi)/(\sqrt{3}-2\xi)$, one has $|\lambda_1|>1$ and $|\lambda_2|>1$ with $\xi>0$. As a result, $f_1(\lambda_1)>0$ and $f_1(\lambda_2)>0$, implying $(\lambda_1+1)/\left(\lambda_1\cdot\varphi_1^{u-1}-\varphi_2^{u-1}\right)$ and $(\lambda_2+1)/\left(\lambda_2\cdot\varphi_1^{u-1}-\varphi_2^{u-1}\right)$ in Equations (7.15) and (7.23) are positive, and the sign of D_k is solely dependent on $(-1)^u$.

Likewise, the following quadratic function $f_2(x)$ can be introduced as:

$$f_2(x)=\left(x\cdot\varphi_1^{u-k-1}+\varphi_2^{u-k}\right)\left(x\cdot\varphi_1^{u-1}-\varphi_2^{u-1}\right) \tag{7.27}$$

The roots of $f_2(x)$ are $x_1=-\varphi_2^{2(u-k)-1}$ and $x_2=\varphi_2^{2(u-1)}$. As $u\geq 1$ and $0<\varphi_2<1$, one has $-1<-\varphi_2\leq x_1\leq 0<x_2\leq 1$ for $k=1, 2,\ldots, u-1$, and $f_2(x)$ corresponds to a similar parabola as that of $f_1(x)$. When $|x|>1$, $f_2(x)>0$, so $f_2(\lambda_1)>0$ and $f_2(\lambda_2)>0$. Thus, $(\lambda_1\varphi_1^{u-k-1}+\varphi_2^{u-k})/(\lambda_1\varphi_1^{u-1}-\varphi_2^{u-1})$ and $(\lambda_2\varphi_1^{u-k-1}+\varphi_2^{u-k})/(\lambda_2\varphi_1^{u-1}-\varphi_2^{u-1})$ in Equations (7.14) and (7.22) are always positive, and the sign of D_k is determined by $(-1)^k$.

The above discussions show that the positive temperature difference ΔT always induces an upward deflection in the outermost span beam ($k=1$ and n), while the deflection shape of the middlemost span beam ($k=u$) depends on whether u is an odd or even number. From the first to uth spans, the beams deflect upward and downward alternately, and the deflection shape of the $(u+1)$th to nth span beams can be determined by symmetry condition with respect to the vertical center line of the entire beam. Therefore, for a beam with odd-numbered spans, i.e., $n=2u-1$, the deflection of each span is always opposite to the adjacent spans. For a beam with $n=2u$, the uth and $(u+1)$th span beams deflect in the same direction, but the remaining spans' deflection shapes keep the span-by-span alternation pattern. Figure 7.10 illustrates the cases of $n=1$–6.

7.3.3.3 Deflection Magnitude of Each Span

This subsection investigates the mid-span deflection magnitude. First, all spans are assumed to have equal length by setting $\xi=1$. The influence of the side-to-main span ratio ξ will be investigated later.

When the beam consists of an odd number of spans with $\xi=1$, the mid-span deflection of the kth ($k=1, 2,\ldots, u$) span beam can be calculated by Equation (7.16). Given a certain u, the deflection D_k is proportional to the item $\varphi_1^{u-k}+\varphi_2^{u-k}$. As $\varphi_2=1/\varphi_1$, two functions $f_3(x)=x+1/x$ and $f_4(x)=\varphi_1^{u-x}$ are introduced to define a composite function, which produces $f_3(f_4(k))=\varphi_1^{u-k}+\varphi_2^{u-k}$. For $k=1, 2,\ldots, u$, $f_4(x)$ decreases

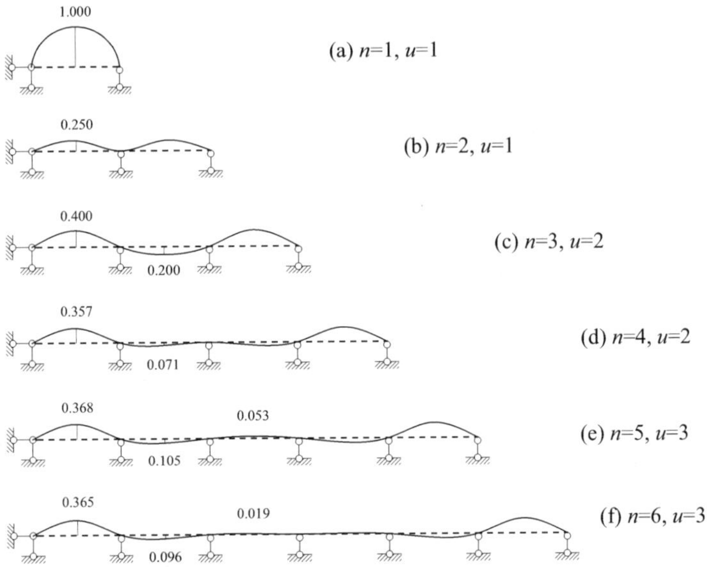

FIGURE 7.10 Thermal deflection of continuous beams with 1–6 spans ($\Delta T > 0$, $\xi = 1$, not to scale): (a) $n = 1$, $u = 1$, (b) $n = 2$, $u = 1$, (c) $n = 3$, $u = 2$, (d) $n = 4$, $u = 2$, (e) $n = 5$, $u = 3$, and (f) $n = 6$, $u = 3$.

with increasing k and $f_4(k) \geq 1$. Meanwhile, the derivative of $f_3(x)$ is $f_3'(x) = 1 - 1/x^2$, which is positive for $x \geq 1$. Consequently, $f_3(x)$ is a monotonically increasing function in the region. As a result, the composite function $f_3(f_4(k))$ decreases as k increases, which means the outermost span ($k = 1$) and the middlemost span ($k = u$) have the largest and smallest mid-span deflection, respectively.

When the beam contains an even number of spans of equal length, the mid-span deflection D_k is proportional to the item $\varphi_1^{u-k} - \varphi_2^{u-k+1}$, as shown in Equation (7.25). In this case, a new function $f_5(x) = x - \varphi_2/x$ is introduced, and the composite function $f_5\left(f_4(k)\right) = \varphi_1^{u-k} - \varphi_2^{u-k+1}$. As aforementioned, for $k = 1, 2, \ldots, u$, $f_4(x)$ is a monotonically decreasing function and $f_4(k) \geq 1$, and the derivative of $f_5(x)$ is $f_5'(x) = 1 + \varphi_2/x^2 > 0$. Therefore, the composite function $f_5(f_4(k))$ decreases with the increase of k, indicating that the outermost span beam experiences the largest thermal deflection at mid-span and the middlemost span has the least. This finding is consistent with that of the beams with odd-numbered spans.

The rotational restraints on the beam end significantly affect the beam deflection. The outermost joint receives the least rotational restraints, and thus, the outermost span has the largest deflection.

7.3.3.4 Influence of Number of Spans on Deflection Magnitude

This subsection examines the deflection of the first and uth span beams in terms of u. The discussion still focuses on beams with equal spans.

When the span number n is odd, substituting $k = 1$ into Equation (7.16) gives the largest deflection of all spans:

$$D_1^{\text{odd}} = -\frac{\alpha l_0^2 \Delta T}{16h} \frac{(\varphi_1^{u-1} + \varphi_2^{u-1})(1+\sqrt{3})}{\varphi_1^u - \varphi_2^{u-1}} = \frac{\sqrt{3}-1}{2}\left(1 + \frac{\varphi_1 + 1}{\varphi_1^{2u-1} - 1}\right) D_0 \tag{7.28}$$

where $D_0 = -\alpha l_0^2 \cdot \Delta T/(8h)$ represents the mid-span deflection of a simply supported beam with the span l_0, depth h, linear expansion coefficient α, and temperature difference ΔT. From Equation (7.28), the magnitude of D_1^{odd}, i.e. $|D_1^{\text{odd}}|$, becomes smaller with the increase of u. This is attributed to the rotational restraint on the right end of the first span beam due to the additional spans.

When n is even, the largest deflection can be found by substituting $k=1$ into Equation (7.25):

$$D_1^{\text{even}} = -\frac{\alpha l_0^2 \Delta T}{16h} \frac{(\varphi_1^{u-1} - \varphi_2^{u})(1+\sqrt{3})}{\varphi_1^u + \varphi_2^{u}} = \frac{\sqrt{3}-1}{2}\left(1 - \frac{\varphi_1 + 1}{\varphi_1^{2u} + 1}\right) D_0 \tag{7.29}$$

In contrast to the case of odd-numbered n, $|D_1^{\text{even}}|$ becomes larger as u increases. This is because the beam deflection is symmetric about the middlemost support $u+1$. Therefore, the support is equivalent to a fixed support without any rotation. With the increase of u, the outermost span is far from the fixed support and thus less restrained against rotation from the latter. The span deflects more.

As u approaches infinity, D_1^{odd} and D_1^{even} converge, though in opposite directions, to the same value:

$$D_1^{\text{lim}} = \left(\sqrt{3}-1\right)/2 \cdot D_0 \tag{7.30}$$

Therefore, D_0 and $D_1^{\text{lim}} \approx 0.366 D_0$ are, respectively, the upper and lower limits of the first span deflection of odd-numbered-span beams, while D_1^{lim} and $D_0/4$ are the corresponding limits of even-numbered-span beams. The conclusions are illustrated in Figure 7.10. An interesting observation is that both D_1^{odd} and D_1^{even} are rational number multiples of D_0 (see Appendix C), whereas their limit D_1^{lim} is an irrational number times D_0.

Similarly, the deflection of the middlemost span can be obtained by substituting $k=u$ into Equations (7.16) and (7.25), which represents the smallest mid-span deflection of all spans:

$$D_u^{\text{odd}} = (-1)^u \frac{\alpha l_0^2 \Delta T}{8h} \frac{1+\sqrt{3}}{\varphi_1^u - \varphi_2^{u-1}} = (-1)^{u-1} \frac{1+\sqrt{3}}{\varphi_1^u - \varphi_2^{u-1}} D_0 \tag{7.31}$$

$$D_u^{\text{even}} = (-1)^u \frac{\alpha l_0^2 \Delta T}{8h} \frac{1}{\varphi_1^u + \varphi_2^{u}} = (-1)^{u-1} \frac{1}{\varphi_1^u + \varphi_2^{u}} D_0 \tag{7.32}$$

Both $|D_u^{\text{odd}}|$ and $|D_u^{\text{even}}|$ decrease monotonically with the increase of u, and they approach zero as u goes infinity. This indicates that as $u \rightarrow +\infty$, the middlemost span is equivalent to a fixed-end beam that has no thermal deflection under the vertical temperature difference (see Figure A1a in Appendix A).

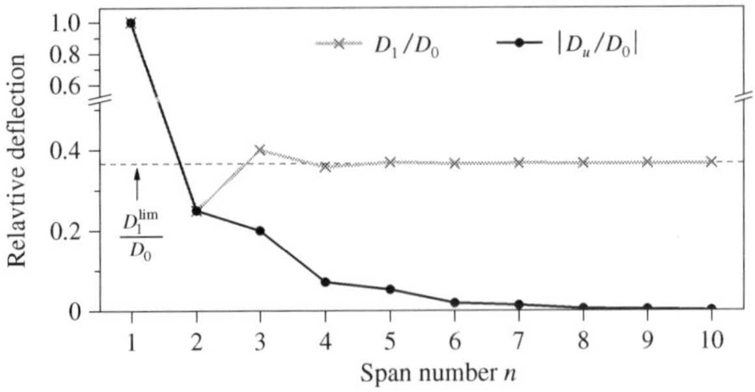

FIGURE 7.11 Relative deflection of the first and uth spans with span number n.

D_1 and D_u converge to their limits relatively fast. The deflection ratio of the outermost span, denoted as D_1/D_0, can be obtained from Equations (7.28) and (7.29). It is a function of the span number n, as shown in Figure 7.11. When $n \geq 5$ or $u \geq 3$, the oscillation in D_1/D_0 is almost invisible, demonstrating a fast convergence to $\left(\sqrt{3}-1\right)/2$.

For the middlemost span, the absolute value of the deflection ratio $|D_u/D_0|$ can be derived from Equations (7.31) and (7.32). Figure 7.11 shows that $|D_u/D_0|$ converges to zero quickly in a monotonically decreasing manner as n increases.

7.3.3.5 Influence of ξ on Deflection Magnitude

Due to the quick convergence of the thermal deflection with the increasing span number, the beams with $u=2$ are considered, i.e., the three-span and four-span beams, as examples to investigate the influence of the side-to-main span ratio ξ on the deflection magnitude. In this case, $k=1$ corresponds to the side span, while $k=2$ represents the main span.

By substituting $k=1$, $u=2$, and $\lambda_1 = (\sqrt{3}+\xi)/(\sqrt{3}-\xi)$ into Equation (7.14), the side-span deflection $D_1^{n=3}$ for the 3-span beam is obtained:

$$D_1^{n=3} = -\frac{\alpha l_0^2 \Delta T}{16h} \frac{(\lambda_1+\varphi_2)\left(1+\sqrt{3}\right)}{\lambda_1\varphi_1-\varphi_2} = -\frac{\alpha l_0^2 \Delta T}{16h} \frac{\xi+3}{2\xi+3} = \frac{1}{4}\left(1+\frac{3}{2\xi+3}\right) D_0 \tag{7.33}$$

Here, the side span length l_0 is assumed as a constant, while the main span length $l_1 = l_0/\xi$ varies with ξ. From Equation (7.33), the magnitude of $D_1^{n=3}$ decreases with the increase of ξ with the following limits:

$$\lim_{\xi \to 0^+} D_1^{n=3} = D_0/2 \tag{7.34}$$

$$\lim_{\xi \to +\infty} D_1^{n=3} = D_0/4 \tag{7.35}$$

$\xi \to +\infty$ means that the main span length is negligible. In this situation, the beam deforms like a 2-span beam, and the main span can be regarded as a fixed support to restrain the end rotation of the side span beam. This is equivalent to Figure 7.10b.

For the main span deflection, substituting $\lambda_1 = (\sqrt{3}+\xi)/(\sqrt{3}-\xi)$, $k=2$, and $u=2$ into Equation (7.15) yields:

$$D_2^{n=3} = \frac{\alpha l_0^2 \Delta T}{8\sqrt{3}\xi h}\frac{\lambda_1+1}{\lambda_1\varphi_1-\varphi_2} = \frac{\alpha l_1^2 \Delta T}{8h}\frac{\xi^2}{(2\xi+3)\xi} = \frac{1}{2}\left(1-\frac{3}{2\xi+3}\right)D_{0(l_1)} \tag{7.36}$$

where $D_{0(l_1)} = -\alpha l_1^2 \Delta T/(8h)$ represents the mid-span deflection of a simply supported beam of span l_1. Here, the main span length l_1 is assumed to be a constant, and the side span length $l_0 = \xi \cdot l_1$ varies with ξ. From Equation (7.36), $D_2^{n=3}$ increases with the increase of ξ, and approaches 0 and $-D_{0(l1)}/2$ as $\xi \to 0^+$ and $\xi \to +\infty$, respectively.

Based on Equations (7.33) and (7.36), the deflection ratio of the side-to-main spans is found to depend on the side-to-main span ratio ξ only, i.e.:

$$D_1^{n=3}/D_2^{n=3} = -(\xi+3)\xi/2 \tag{7.37}$$

As ξ is always positive, the magnitude of the deflection ratio $\left|D_1^{n=3}/D_2^{n=3}\right|$ exhibits a quadratic growth with ξ.

For the four-span beam, the relation between $D_1^{n=4}$ or $D_2^{n=4}$ and ξ has the same expressions as Equations (7.33) and (7.36) except replacing ξ by 2ξ.

$$D_1^{n=4} = -\frac{\alpha l_0^2 \Delta T}{16h}\frac{(\lambda_2+\varphi_2)(1+\sqrt{3})}{\lambda_2\varphi_1-\varphi_2} = \frac{1}{4}\left(1+\frac{3}{4\xi+3}\right)D_0 \tag{7.38}$$

$$D_2^{n=4} = \frac{\alpha l_0^2 \Delta T}{16\sqrt{3}\xi h}\frac{\lambda_2+1}{\lambda_2\varphi_1-\varphi_2} = -\frac{1}{2}\left(1-\frac{3}{4\xi+3}\right)D_{0(l_1)} \tag{7.39}$$

Therefore, $D_1^{n=4}$ and $D_2^{n=4}$ possess similar characteristics as their counterparts for three-span beams with the variation in ξ. The corresponding side-to-main-span deflection ratio can be calculated by $D_1^{n=4}/D_2^{n=4} = -(2\xi+3)\xi$. As ξ increases, the magnitude of $D_1^{n=4}/D_2^{n=4}$ increases approximately four times as fast as $\left|D_1^{n=3}/D_2^{n=3}\right|$ does.

7.3.3.6 Comparison with Beam Deflection Limits

This section compares the thermal deflection of beams with the design limits. According to Section 7.3.3.4, the maximum possible thermal deflection in equal-span beams is that of a simply supported beam, $D_0 = -\alpha l_0^2 \Delta T/(8h)$. D_0 can be normalized as a dimensionless parameter as:

$$D_0/l_0 = -1/8\,\alpha \Delta T\, l_0/h \tag{7.40}$$

where D_0/l_0 is the deflection-to-span ratio and l_0/h is the span-to-depth ratio. A typical value of l_0/h in bridges is around 20 (AASHTO, 2010). With $l_0/h=20$, Equation (7.40) leads to $D_0/l_0 = 2.5\alpha\Delta T$. As the thermal expansion coefficient α is in the order of 10^{-5}°C^{-1} and the temperature difference ΔT for bridges is generally about 10°C, the temperature difference-induced deflection of beams is in the order 10^{-4} of the span length.

The bridge serviceability usually defines a maximum acceptable deflection-to-span ratio of about 10^{-3} (AASHTO, 2010). Although it is much larger than the thermal effects, a maximum deflection is assumed to result from progressive accumulation of deterioration in bridges during a long operational period, e.g., over 10 years. In addition, the maximum allowable deflection during the serviceability limit state of beams is determined for extremely overloaded cases, rather than the actual operation. In this regard, the temperature-induced deflection may be comparable to, or even larger than, the live load effect and annual deterioration (Burdet, 2010). Therefore, the thermal deflection must be monitored during the operation. Its effect should be removed when assessing the live load effects.

7.3.4 A Case Study

The derived formulas are applied to the viaduct introduced in Section 7.2 for verification.

The viaduct has $l_1 = 150$ m, $l_0 = 110$ m, thus, $\xi = l_0/l_1 = 0.733$, thermal expansion coefficient $\alpha = 1.2 \times 10^{-5}$, and $u = 2$. As the viaduct has a varying girder section height, ranging from 4.5 m to 6.5 m, an "effective" height is adopted according to the covering span, i.e., $h_e = (5.5 \times 75 + 4.5 \times 75)/150 = 5$ m. The temperature difference between the average temperatures of the top and bottom plates (calculated from the respective averages of the north and south sides) is used as the temperature gradient. This temperature gradient, however, is smaller than the gradient temperature of the cross section as they are not the external surface temperatures.

Then, the theoretical mid-span deflections are calculated using Equation (7.15) and drawn in Figure 7.12. The theoretical calculation result matches well with the measurement. During the seven days, the RMSE between the theoretical calculation and measurement is merely 0.94 mm. The amplitude of the calculation is slightly smaller than the measurement, possibly due to the error in estimating the effective section height and the temperature gradients, as mentioned above. The effectiveness of the theoretical formulae is thus validated.

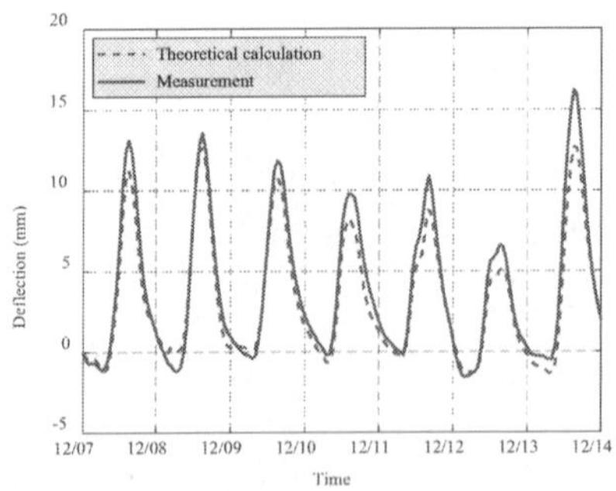

FIGURE 7.12 Calculated and measured mid-span deflection.

7.3.5 Summary

This section derives explicit theoretical solutions to the temperature difference-induced deflection of a prismatic beam with any number of spans. Based on the theoretical analysis, the following conclusions can be drawn.

1. The parity of the span number n significantly influences the thermal behavior of beam structures. The thermal responses of the odd- and even-numbered-span beams depend on the dimensionless parameters λ_1 and λ_2, respectively.
2. For a prismatic beam with n spans, the thermal mid-span deflection is symmetric about the centerline of the entire structure. For the first to uth spans, each span beam bends oppositely to its adjacent spans, and the first span always moves upward under the positive temperature difference. The deflection shape of the $(u+1)$th to nth spans can be determined by symmetry condition.
3. For an equal-span beam, the outermost (first or nth) span has the largest thermal deflection of all spans. For odd-numbered-span beams, the mid-span deflection D_1 of the outermost span decreases from D_0 to $D_0(\sqrt{3}-1)/2$ as u increases from 1 to $+\infty$. For even-numbered-span beams, D_1 increases from $D_0/4$ to $D_0(\sqrt{3}-1)/2$ with the increase in u.
4. For an n-span beam of equal spans, the middlemost span has the smallest deflection of all spans ($n \geq 3$), and its deflection at mid-span converges quickly to zero as n increases.

The analytical formulas proposed in this section are of both theoretical and practical value. They provide not only deep insights into the fundamental relation between thermal deflection and structural properties but also a convenient way to calculate the thermal deformation of beams. The influence of shear deformations in the Timoshenko beams deserves further investigation.

7.4 NUMERICAL SIMULATION OF THE VIADUCT

In this section, the temperature behaviors of the viaduct introduced in Section 7.2 are simulated numerically in ANSYS using the integrated method.

7.4.1 The FE Model

The refined FE model of the viaduct consists of 132,984 nodes and 142,272 elements, including 52,439 solid elements for the steel box girder, the asphalt concrete layer, and piers; and 89,833 shell elements for the diaphragms and ribs. The FE model of the bridge is shown in Figure 7.13. The webs, diaphragms, and U-ribs of the girder are simulated with three- or four-node shell elements (SHELL57 elements for the thermal analysis and SHELL181 for structural analysis). The top and bottom plates of the steel girder are simulated with six- or eight-node solid elements (SOLID70 for the thermal analysis and SOLID45 for structural analysis). The concrete piers are

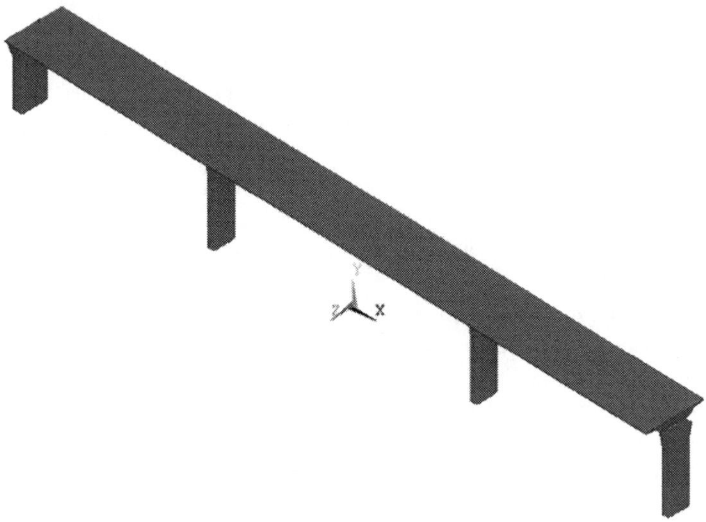

FIGURE 7.13 FE model of the viaduct.

TABLE 7.1
Material Parameters for the Heat-Transfer Analysis

Parameter	Unit	Asphalt	Steel	Concrete
Density	kg/m^3	2365	7850	2600
Specific heat capacity	J/kg/°C	1075	460	925
Thermal conductivity	W/m/°C	1.8	55	2.71
Emissivity coefficient	–	0.92	0.8	0.88
Absorptivity coefficient	–	0.90	0.685	0.65
Poisson's ratio	–	0.2	0.3	0.2
Young's modulus	GPa	12	206	31
Thermal expansion coefficient	–	1.0×10^{-5}	1.2×10^{-5}	1.0×10^{-5}

simulated with ten-node second-order tetrahedral elements (SOLID87 for the thermal analysis and SOLID92 for structural analysis). The material specifications for the thermal and structural analyses are listed in Table 7.1.

7.4.2 Model Verification

To verify the FE model, a simple structural analysis is conducted by replacing the thermal elements with their structural counterpart elements. The auxiliary facilities of the viaduct are not simulated in the FE model, but their masses can influence the modal properties of the viaduct. Their weight accounts for about 18% of the steel girder weights. Correspondingly, the density of the steel increases by 18%. The viaduct deflection due to the self-weight is shown in Figure 7.14. The maximum girder deflection is 0.233 m and occurs at the mid-span (G02) section, which is reasonable.

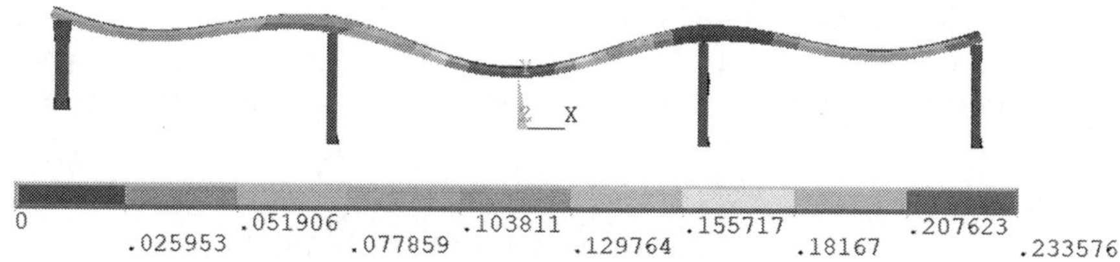

FIGURE 7.14 Deflection under self-weight of the viaduct.

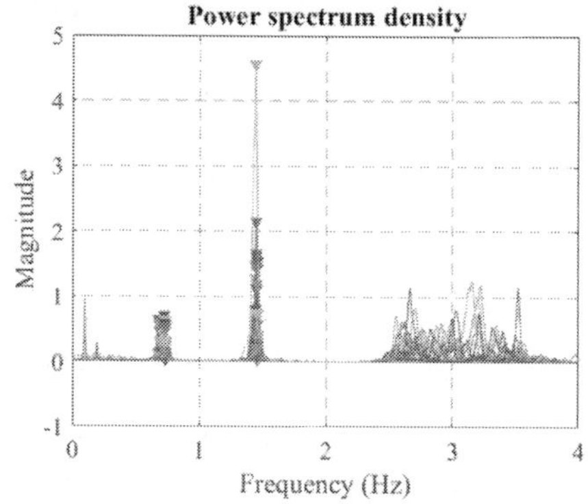

(a) Natural frequency extraction using peak-picking method

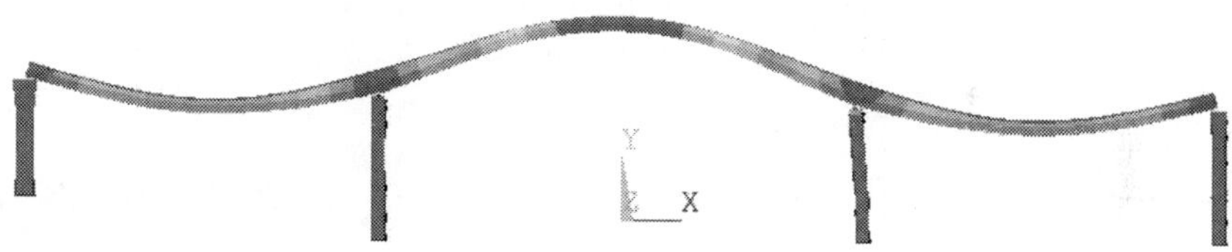

(b) First bending mode (frequency = 0.717 Hz, relative differentce = −0.1%)

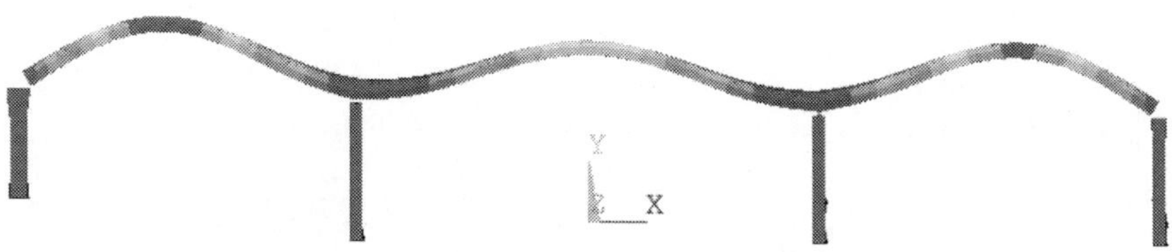

(c) Second bending mode (frequency = 1.476 Hz, relative difference= 1.7%)

FIGURE 7.15 The first two bending mode shapes of the viaduct: (a) natural frequency extraction using peak-picking method, (b) first bending mode (frequency = 0.717 Hz, relative difference = −0.1%), and (c) second bending mode (frequency = 1.476 Hz, relative difference= 1.7%).

The first two bending modes of the viaduct are extracted from the power spectrum density of the measured acceleration using the peak-picking method, as shown in Figure 7.15a. The two measured natural frequencies are 0.718 Hz and 1.451 Hz. The first two bending mode shapes calculated from the FE model are shown in Figures

7.15b and 7.15c, and their natural frequencies are 0.717 Hz and 1.476 Hz. The relative differences are −0.1% and 1.7% only. Therefore, the FE model is a good representation of the real viaduct, and the subsequent transient heat-transfer analysis is conducted.

7.4.3 Thermal Boundary Conditions and Initial Condition

Given the measured environmental conditions during December 2 and 8, 2020 in Section 7.2.2.1, the thermal boundary conditions for the heat-transfer analysis are determined. As introduced in previous chapters, a pre-analysis of 24 hours, that is, from 00:00 to 24:00 on December 2, is conducted. The thermal analysis on the subsequent three days, i.e., from 01:00 on December 3 to 24:00 of December 5, are conducted. The third kind thermal boundary conditions during the four days, including the equivalent air temperature and the overall heat-transfer coefficient, are applied to the FE model. Nine regions of a cross section shown in Figure 7.16 are defined as they are subjected to different inclinations and orientations. Different boundary conditions are applied to these regions. Regions 1 to 6, 8, and 9 have consistent surface inclination along the longitudinal direction. However, region 7 has three types of surface inclinations along the longitudinal direction due to the variations of section heights as shown in Figure 7.1a. Consequently, 11 groups of boundary conditions are applied to the external surfaces. Note that the direct solar radiations are mainly received by regions 1 and 2. The equivalent temperatures and overall heat-transfer coefficients of regions 1 and 2 are shown in Figure 7.17. The thermal boundary conditions of the piers external surfaces are applied in the same way.

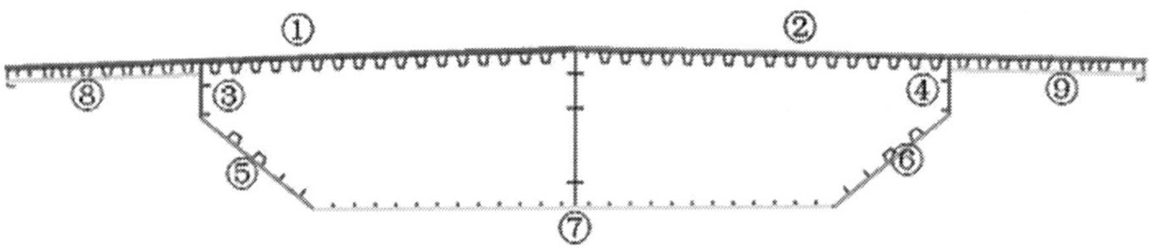

FIGURE 7.16 Regions for different boundary conditions within a cross section.

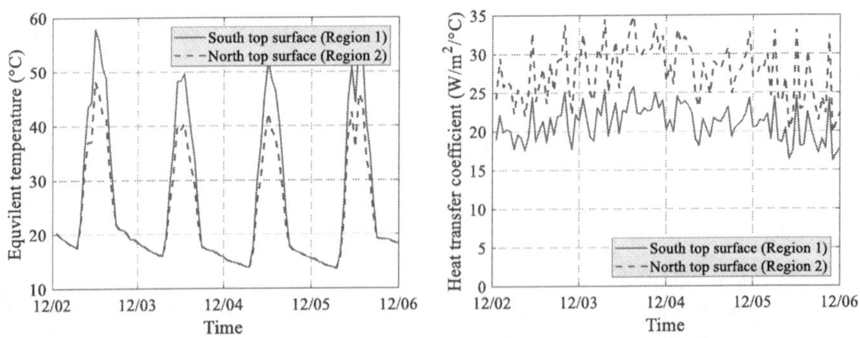

FIGURE 7.17 Thermal boundary conditions of top surfaces of the girder.

As for the internal surfaces, different from the method adopted in Chapter 5, the radiation matrix method is adopted in this chapter to account for the short-wave radiations among the internal surfaces. The method involves generating a matrix of form factors (view factors) between the radiating surfaces and using the matrix as a super-element in the thermal analysis (ANSYS, 2013; Zhou et al., 2015; Zhang et al., 2024). In this way, the irradiation heat transfer among the internal surfaces is automatically considered. As for the convection heat transfer, the air element method adopted by Zhou et al. (2015) is found to contribute little to the overall heat-transfer among the internal surfaces (Li et al., 2023) and thus omitted in this section. Altogether 96 enclosures are defined according to the normal diaphragms and the middle webs, and thus 96 MATRIX50 super-elements are involved in the thermal analysis.

A pre-analysis of 24 hours is conducted to obtain the initial conditions for all DOFs of the FE model. The transient heat-transfer analysis is conducted every 20 minutes, or three steps per hour. The analysis results are presented as follows.

7.4.4 Temperature Distribution

The computed temperature distribution of the viaduct at 12:00 on December 3, 2020 is shown in Figure 7.18a. The temperature distribution in section G02 is shown in Figure 7.18b. Both the vertical and transverse temperature gradients are clearly observed. In general, the south part of the girder has a higher temperature than the north counterpart. The top plate above the box cell has a higher temperature than the corresponding cantilever arm of the same side, possibly because the box can store the energy. At the instant, the highest temperature of the viaduct is 37.6°C (at the girder top), and the lowest temperature is 17.6°C at the pier. The computed temperatures corresponding to the sensor points at the mid-span section (G02) are shown in Figure 7.19. The calculated temperatures match well with the measurement counterparts. Specifically, the overall area-weighted RMSE of these points is merely 1.1°C, validating the accuracy of the numerical simulation and the effectiveness of the integrated analysis method. Then the temperature results are converted to the thermal loads to compute the temperature-induced structural responses.

7.4.5 Temperature-Induced Structural Responses

The computed temperature responses of all nodes are converted to the thermal loads in ANSYS, and then the structural analysis is conducted.

7.4.5.1 Temperature-Induced Deflection

The mid-span deflection of the girder is shown in Figure 7.20. The downward direction is defined as positive. With the temperature increase in the daytime, the central span deflected downward. The simulation results matched well with the field monitoring results and the overall RMSE between them is 2.57 mm, showing the accuracy of the simulation results. The RMSE accounts about 17.4% of the maximum measured deflection. The errors might be caused by several factors: The error propagated from the transient heat-transfer analysis, the difference between the FE model and the real viaduct, etc.

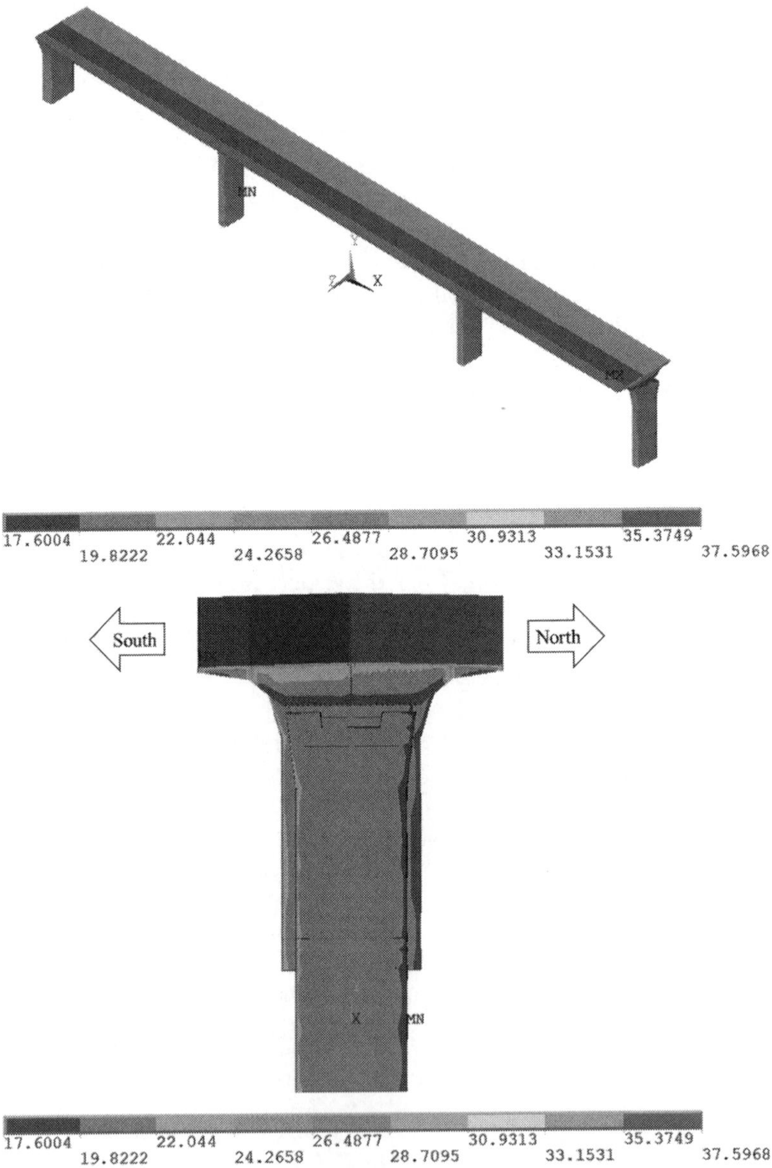

FIGURE 7.18 Calculated temperature distribution at 12:00 on December 3, 2020.

7.4.5.2 Temperature-Induced Strain/Stress

The thermal strain of the G02 section extracted from ANSYS (with the baseline removed) is compared with the measured thermal strain in Figure 7.20a. All simulated strains at the four representative points match well with the measured ones. The temperature increases of the girder in the daytime results in thermal expansion. The elastic thermal stresses are calculated and shown in Figure 7.20b. The strain sensor

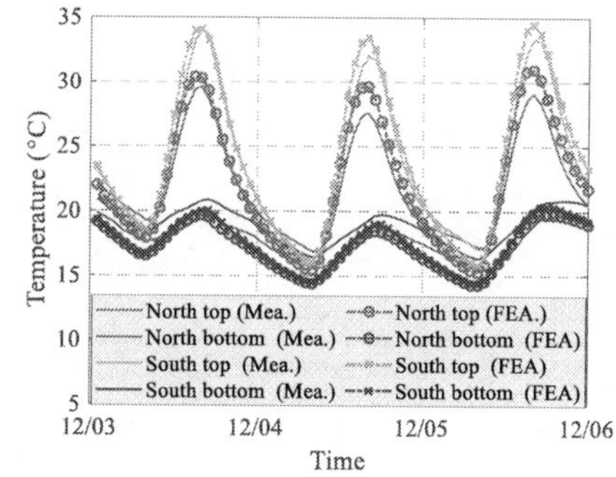

FIGURE 7.19 Measured and computed girder temperatures.

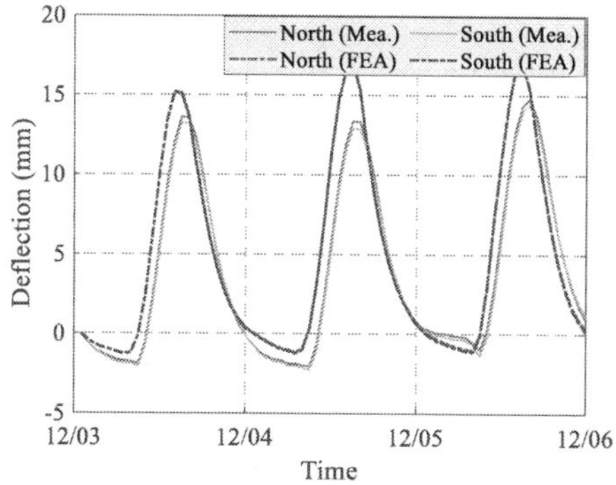

FIGURE 7.20 Measured and computed girder deflection.

at the south bottom plate malfunctions, and thus its result is not shown here. The temperature increase in the daytime results in the top plate in compression and the bottom plate in tension (Figure 7.21).

7.4.6 Summary

In this section, the thermal behavior of the viaduct in Section 7.2 is simulated in ANSYS through the integrated approach. Both the temperature and temperature-induced girder deflection and strain/stresses match well with the measurement counterparts.

7.5 CHAPTER SUMMARY

In this chapter, a three-span viaduct on the Hong Kong-Zhuhai-Macao Bridge is used as a testbed to comprehensively study its temperature behaviors through field monitoring, theoretical analysis, and numerical simulation. Both the theoretical and numerical results match well with the field monitoring results in terms of temperature distribution, structural displacement, and strain/stress. As the temperature rises in the daytime, the central span of the three-span viaduct deflects downward. Meanwhile, the temperature change causes compression in the top part of the girder and tension in the bottom part.

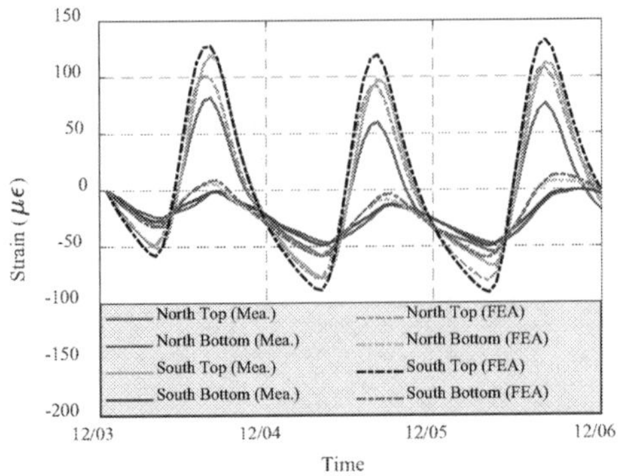

(a) Thermal strain on G02 section

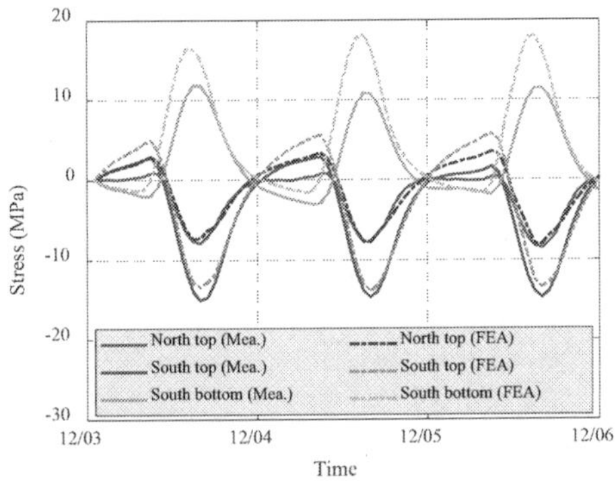

(b) Thermal stress on G02 section

FIGURE 7.21 Thermal strain and stresses on section G02: (a) thermal strain and (b) thermal stress.

REFERENCES

AASHTO, 2010, *AASHTO LRFD Bridge Design Specifications*. Washington, D.C, American Association of State Highway and Transportation Officials.

ANSYS, I., 2013, *ANSYS Mechanical APDL Thermal Analysis Guide Release 15.0*. Washington, D.C, American Association of State Highway and Transportation Officials

Burdet, O. L., 2010, Thermal effects in the long-term monitoring of bridges. In *IABSE Symposium: Large structures and Infrastructures for environmentally constrained and urbanised areas:* 7, Venice, Italy.

European Committee for Standardization CEN, 2003, *Eurocode 1: Actions on Structures, Part 1–5: General Actions — Thermal Actions*. EN1991-1-5: 2003E. Brussels, European Committee for Standardization CEN.

Kulprapha, N., Warnitchai, P., 2012, Structural health monitoring of continuous prestressed concrete bridges using ambient thermal responses. *Engineering Structures* **40**: 20–38.

Leet, K., Uang, C.-M., Lanning, J., 2021, *Fundamentals of Structural Analysis*. New York, McGraw-Hill Education.

Li, L., Shan, Y., Jing, Q., Yan, Y., Xia, Z., Xia, Y., 2023, Global temperature behavior monitoring and analysis of a three-tower cable-stayed bridge. *Engineering Structures* **295**: 116855.

Roeder, C. W., 2003, Proposed design method for thermal bridge movements. *Journal of Bridge Engineering* **8(1)**: 12–19.

Zhang, L., Shan, Y., Li, L., Wang, F., Xia, Y., 2024, Thermal boundary conditions for heat transfer analysis of bridges considering non-uniform distribution of internal air temperature by computational fluid dynamics. *Journal of Civil Structural Health Monitoring* **14(5)**: 1295–1310.

Zhou, L., Xia, Y., Brownjohn, J., Koo, K. Y., 2015, Temperature analysis of a long-span suspension bridge based on field monitoring and numerical simulation. *Journal of Bridge Engineering* **21**: 04015027.

APPENDICES

APPENDIX A

This appendix derives the fixed-end moments of a beam subjected to the temperature difference ΔT across the depth, as shown in Figure A1a. The beam has a constant cross section along the length. The span, depth, moment of inertia, elastic modulus, and linear expansion coefficient are L, h, I, E, and α, respectively.

By taking advantage of symmetry, the left half of the original beam is considered a cantilever beam of length $L/2$, as shown in Figure A1b. With the axial deformation neglected, the mid-span moment, X_1, is the only redundant force of the cantilever beam. Here, the moment stretching the beam bottom is assumed to be positive. Under the action of the temperature difference ΔT and the redundant moment X_1, the slope of the cantilever beam at the free end is zero.

By the flexibility method (also known as the force method), the compatibility equation of the cantilever beam is written as

$$\delta_{11}X_1 + \Delta_{1P} = 0 \tag{A1}$$

where the flexibility coefficient δ_{11} can be evaluated using the virtual work method by applying a unit virtual moment $X_1 = 1$, as shown in Figure A1c:

$$\delta_{11} = 1/(EI)\cdot 1\cdot L/2\cdot 1 = L/(2EI) \tag{A2}$$

and Δ_{1P} is the free-end slope produced by the temperature difference ΔT:

$$\Delta_{1P} = \alpha\cdot\Delta T/h\left(-L/2\cdot 1\right) = -\alpha L\cdot\Delta T/(2h) \tag{A3}$$

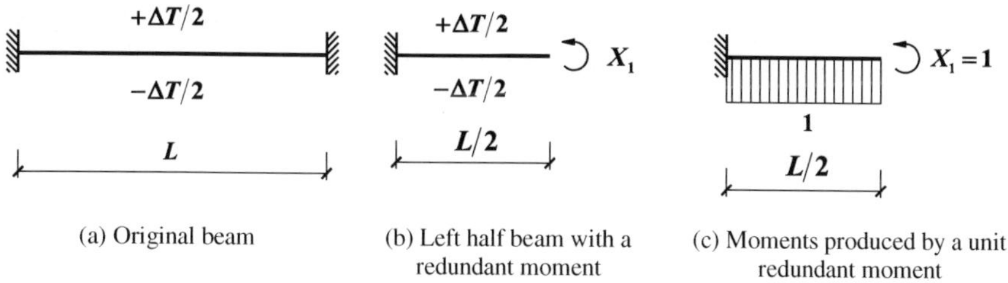

FIGURE A1 Analytical model of a fixed-end beam subjected to temperature difference ΔT: (a) original beam, (b) left half beam with a redundant moment, and (c) moments produced by a unit redundant moment.

The redundant moment X_1 is solved from Equation (A3) as

$$X_1 = -\Delta_{1P}/\delta_{11} = \alpha \cdot \Delta T/h \cdot EI \tag{A4}$$

X_1 is equal to the fixed-end moment of the beam in Figure A1a, which is independent of span L. The fixed-fixed beam in Figure A1a has no deformation but is subject to a constant moment X_1 along the span.

APPENDIX B

This appendix derives the general formula of a_k, which has the following recurrence relation:

$$a_k = 1/(4 - a_{k-1}) \tag{B1}$$

Subtracting a constant φ from both sides of Equation (B1) yields:

$$a_k - \varphi = 1/(4 - a_{k-1}) - \varphi \tag{B2}$$

The right-hand side of Equation (B2) is equivalent to

$$\frac{1}{4 - a_{k-1}} - \varphi = \frac{1 - \varphi(4 - a_{k-1})}{4 - a_{k-1}} = \frac{\varphi a_{k-1} + 1 - 4\varphi}{4 - a_{k-1}} = \frac{\varphi(a_{k-1} - \varphi) + 1 - 4\varphi + \varphi^2}{4 - a_{k-1}} \tag{B3}$$

Let $1 - 4\varphi + \varphi^2 = 0$, and the roots of this quadratic equation are $\varphi_1 = 2 + \sqrt{3}$ and $\varphi_2 = 2 - \sqrt{3}$. According to Equations (B2) and (B3), the following two equations are obtained:

$$a_k - \varphi_1 = \varphi_1 (a_{k-1} - \varphi_1)/(4 - a_{k-1}) \tag{B4}$$

$$a_k - \varphi_2 = \varphi_2 (a_{k-1} - \varphi_2)/(4 - a_{k-1}) \tag{B5}$$

Dividing Equation (B4) by Equation (B5) leads to

$$\frac{a_k - \varphi_1}{a_k - \varphi_2} = \frac{\varphi_1}{\varphi_2} \cdot \frac{a_{k-1} - \varphi_1}{a_{k-1} - \varphi_2} \tag{B6}$$

Therefore, the sequence $(a_k - \varphi_1)/(a_k - \varphi_2)$ is a geometric sequence with a common ratio φ_1/φ_2. As the initial value is $(a_0 - \varphi_1)/(a_0 - \varphi_2)$, the kth term of this geometric sequence is

$$\frac{a_k - \varphi_1}{a_k - \varphi_2} = \left(\frac{\varphi_1}{\varphi_2}\right)^k \cdot \frac{a_0 - \varphi_1}{a_0 - \varphi_2} \tag{B7}$$

For odd-numbered-span beams, $a_0 = 2 - \xi$ and a_k can be solved from Equation (B7) as

$$a_k = \left(\lambda_1 \varphi_1^{k-1} + \varphi_2^{k-1}\right) / \left(\lambda_1 \varphi_1^k + \varphi_2^k\right) \tag{B8}$$

where $\lambda_1 = (\sqrt{3}+\xi)/(\sqrt{3}-\xi)$. As $a_0 = 2-2\xi$ for even-numbered-span beams, a_k becomes

$$a_k = \left(\lambda_2 \varphi_1^{k-1} + \varphi_2^{k-1}\right) / \left(\lambda_2 \varphi_1^k + \varphi_2^k\right) \tag{B9}$$

where $\lambda_2 = (\sqrt{3}+2\xi)/(\sqrt{3}-2\xi)$.

APPENDIX C

This appendix demonstrates that Equations (7.16) and (7.25) produce the results in the form of cD_0, where c is a rational number.

Equation (7.16) can be rewritten as:

$$\begin{aligned} D_k &= (-1)^k \frac{\alpha l_0^2 \cdot \Delta T}{16h} \cdot \frac{\left(\varphi_1^{u-k} + \varphi_2^{u-k}\right)\left(1+\sqrt{3}\right)}{\varphi_1^u - \varphi_2^{u-1}} \\ &= (-1)^{k-1} \frac{D_0}{2} \cdot \frac{\left(\varphi_1^{u-k} + \varphi_2^{u-k}\right)\left(1+\sqrt{3}\right)\left(1-\varphi_2\right)}{\left(\varphi_1^u - \varphi_2^{u-1}\right) \cdot \left(1-\varphi_2\right)} \\ &= (-1)^{k-1} D_0 \cdot \frac{\varphi_1^{u-k} + \varphi_2^{u-k}}{\left(\varphi_1^u + \varphi_2^u\right) - \left(\varphi_1^{u-1} + \varphi_2^{u-1}\right)} \end{aligned} \tag{C1}$$

Note that $(1+\sqrt{3})(1-\varphi_2) = 2$ in Equation (C1). The binomial expansion of the item $\varphi_1^u + \varphi_2^u$ gives:

$$\begin{aligned} \varphi_1^u + \varphi_2^u &= \left(2+\sqrt{3}\right)^u + \left(2-\sqrt{3}\right)^u \\ &= \sum_{j=0}^{u} \frac{u!}{j!(u-j)!} \cdot 2^{u-j} \cdot \left(\sqrt{3}\right)^j + \sum_{j=0}^{u} \frac{u!}{j!(u-j)!} \cdot 2^{u-j} \cdot \left(-\sqrt{3}\right)^j \end{aligned} \tag{C2}$$

In Equation (C1), when j is odd, the $(\sqrt{3})^j$ and $(-\sqrt{3})^j$ related items cancel out; when j is even, both $(\sqrt{3})^j$ and $(-\sqrt{3})^j$ are integers. Consequently, $\varphi_1^u + \varphi_2^u$ produces an integer. Similarly, the items $\varphi_1^{u-k} + \varphi_2^{u-k}$ and $\varphi_1^{u-1} + \varphi_2^{u-1}$ in Equation (C1) are also integers, which means the coefficient of D_0 is a ratio of integers, i.e., a rational number.

Likewise, Equation (7.25) be rewritten as:

$$D_k = (-1)^k \frac{\alpha l_0^2 \cdot \Delta T}{16h} \cdot \frac{\left(\varphi_1^{u-k} - \varphi_2^{u-k+1}\right)\left(1+\sqrt{3}\right)}{\varphi_1^u + \varphi_2^u}$$

$$= (-1)^{k-1} \frac{D_0}{2} \cdot \frac{\left(\varphi_1^{u-k} - \varphi_2^{u-k+1}\right)\left(\varphi_1 - 1\right)}{\varphi_1^u + \varphi_2^u} \tag{C3}$$

$$= (-1)^{k-1} \frac{D_0}{2} \cdot \frac{\left(\varphi_1^{u-k+1} + \varphi_2^{u-k+1}\right) - \left(\varphi_1^{u-k} + \varphi_2^{u-k}\right)}{\varphi_1^u + \varphi_2^u}$$

Similarly, the coefficient of D_0 in Equation (C3) is also a rational number.

8 Long-Term Temperature Behaviors of the Tsing Ma Bridge

NOTATIONS

D	Unit temperature-induced longitudinal displacement
D_V	Unit temperature-induced deflection at the mid-span
L	Total length of the bridge
L_1	Distance between the sensor and fixed end
T, X	Bridge effective temperature and displacement
a, b	Fitting coefficient
h_{P1}, h_{P2}	Heights of the left and right towers
n	Sag ratio of the main cable
α	Thermal expansion coefficient
ε	Error

8.1 PREFACE

Onsite tests have revealed that the effects of temperature variations on bridge responses could be more significant than operational loads (Zhu et al., 2020). With the increasing global warming, the risk of bridges due to temperature is expected to become more significant as higher temperature may exert larger thermal loads to the bridges. More heat waves occurred worldwide in recent years. For example, from June to August 2022, persistent heatwaves affected Europe, the UK, China, and the USA. The highest temperature recorded was 47.0°C in Portugal on July 14, 2022. Temperature surpassed 40°C in the UK for the first time. High-temperature weather lasted over 60 days in China, and temperatures in some cities were over 44°C. In Hong Kong, July of 2022 was the hottest month since records began in 1884. The monthly mean maximum temperature, monthly mean temperature, and monthly mean minimum temperature were the highest of the corresponding monthly mean values on record. Moreover, this record was broken again in August 2023.

However, the current bridge design standards are based on previously recorded data and fail to consider the increasing forces. Therefore, bridges based on the old and current design standards may not be capable of resisting future extreme loads during their 100 or 120-year design lives. Particularly, some old bridges may exhibit deterioration after many years' service.

DOI: 10.1201/9781003480136-8

However, studies on the global warming-induced structural temperature of bridges and its direct consequence to bridge safety are few. How many structural temperature changed and how many temperature-induced responses occurred when the mean air temperature rose during the past few decades are unclear. One practical challenge is that the long-term actual data of bridges is lacking. Many studies only show the temperature data in one or several years, which cannot display the slowly varying climate change.

In this regard, the Tsing Ma Bridge with over 25 years of field monitoring data is taken as an example to demonstrate the long-term temperature behavior of bridges in this chapter. The long-term temperature data of Hong Kong city since 1886 is first described. The air and structural temperatures (bridge deck and main cable) will then be analyzed. Finally, the temperature effects on long-term bridge responses, including displacement and frequencies, and their correlations with temperature will be studied.

8.2 LONG-TERM METEOROLOGICAL DATA IN HONG KONG

Bridge temperature distribution and variation are highly related to meteorological conditions like temperature and solar radiation. The Hong Kong Observatory (HKO) has been recording the annual maximum, mean, and minimum air temperature since 1884 and the annual mean daily total solar radiation since 1979. The results are shown in Figures 8.1 and 8.2, respectively.

Linear regressions show that the annual maximum, mean, and minimum air temperatures have increased since 1884. Specifically, the maximum and mean air temperatures have increased more significantly at a rate of 0.74°C/decade and 0.30°C/decade, respectively, in the last 30 years. The annual mean daily total solar radiation also increased at the rate of 0.40 MJ/m^2/decade.

8.3 LONG-TERM MONITORED TEMPERATURES ON THE TSING MA BRIDGE

8.3.1 Air Temperature and Deck Effective Temperature

Figure 8.3 presents the monthly maximum, mean, and minimum air temperature and deck effective temperature for the bridge deck from 1997 to 2022. The air temperature is from sensor P3-TJS, and the deck effective temperature is the area-weighted average temperature at multiple measurement points in Section O in Figure 4.6.

The monthly deck effective temperature follows a similar seasonal variation pattern as the ambient air temperature. The maximum effective temperature exceeded the air temperature by about 9°C in summer and 4°C in winter, respectively. The mean effective temperature was about 2°C higher than the ambient air temperature. The minimum effective temperature was close to the ambient air temperature.

Figure 8.4 shows the annual maximum, mean, and minimum deck effective temperatures and the ambient air temperatures. A linear regression analysis shows that the maximum deck effective and ambient temperatures increased at rates of 0.51°C/decade and 0.72°C/decade, respectively. Similarly, the increase rates of mean deck

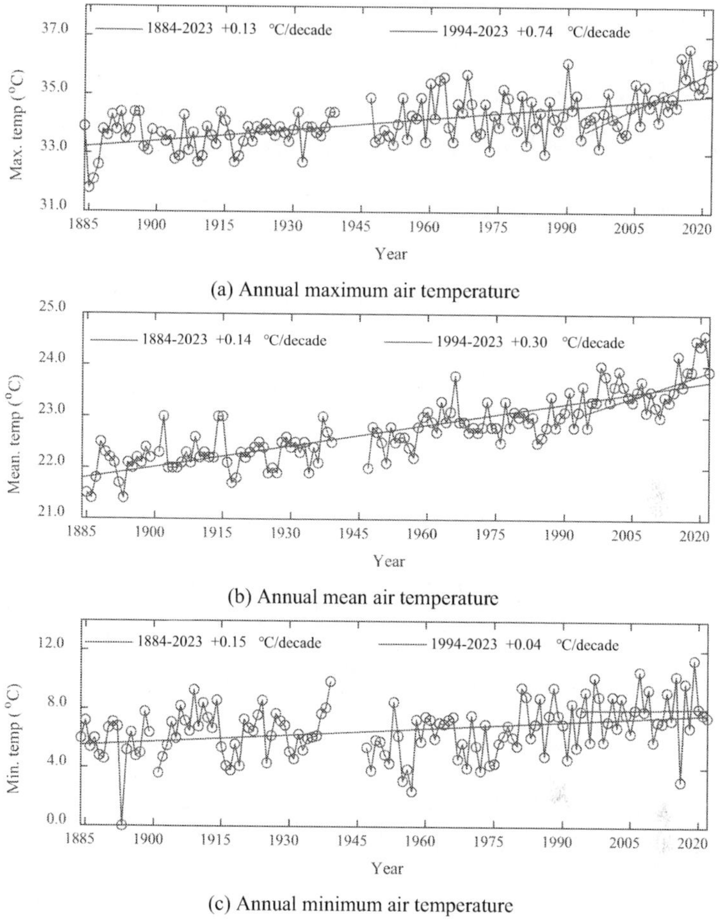

FIGURE 8.1 Annual air temperature since 1884: (a) annual maximum air temperature, (b) annual mean air temperature, and (c) annual minimum air temperature.

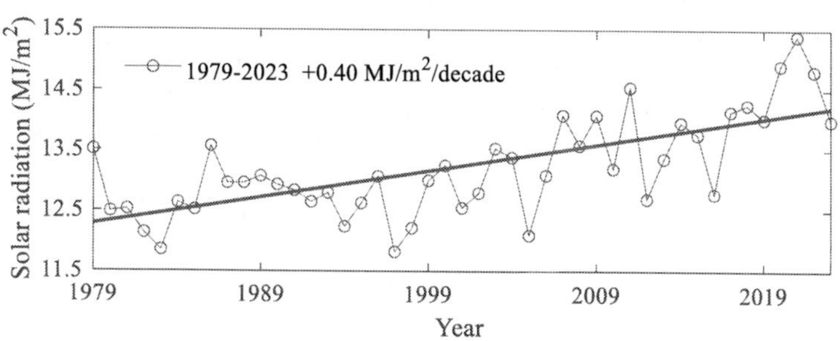

FIGURE 8.2 Annual mean daily total solar radiation since 1979.

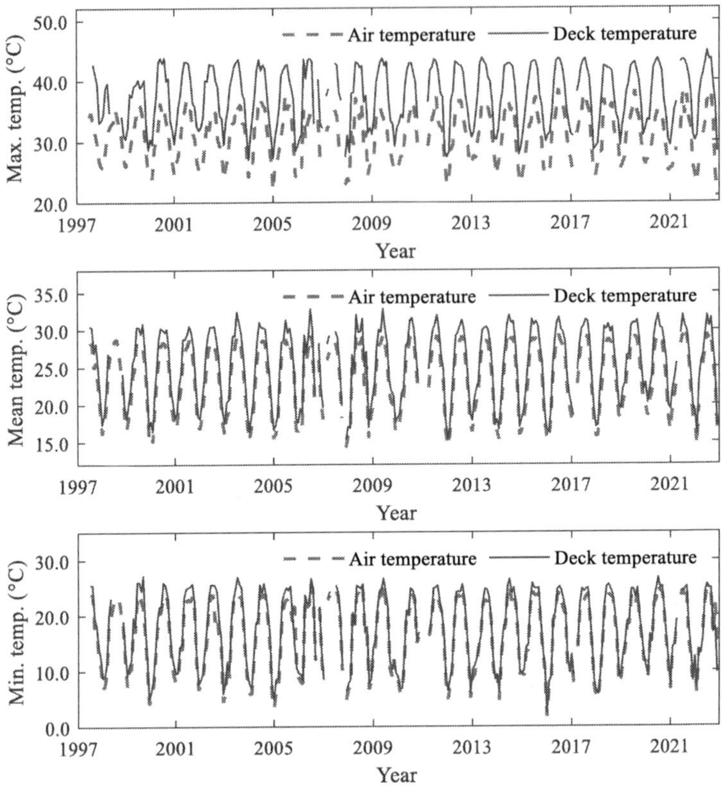

FIGURE 8.3 Monthly ambient and deck effective temperatures during 1997–2022.

effective and ambient temperatures were 0.16°C/decade and 0.20°C/decade, respectively; and the minimum deck effective and ambient temperatures increased by 0.20°C/decade and 0.50°C/decade. Therefore, the maximum temperature increased faster than the mean and minimum temperatures.

The design maximum and minimum effective temperatures of the deck are 46°C and –2°C, respectively (Xu and Xia, 2011). Although the measured effective temperatures remained within the design values, the annual maximum effective temperature was very close to the design limit. In particular, the maximum effective temperature in 2022 was 44.90°C, just 1.10°C lower than the design value. These temperature increases may be due to global warming during the past few decades, which was not considered in the original design code. Compared to the average values of 1995–2014, the IPCC AR6 report predicted the global surface temperature could increase by about 1.85°C and 3.55°C, respectively, by the end of this century under the intermediate (SSP2-4.5) and very high (SSP5-8.5) greenhouse gas emissions scenarios (Masson-Delmotte et al., 2021). Therefore, the maximum effective temperature of the bridge may likely exceed the design value in the future because the maximum temperature increases faster than the mean value. If the trend during

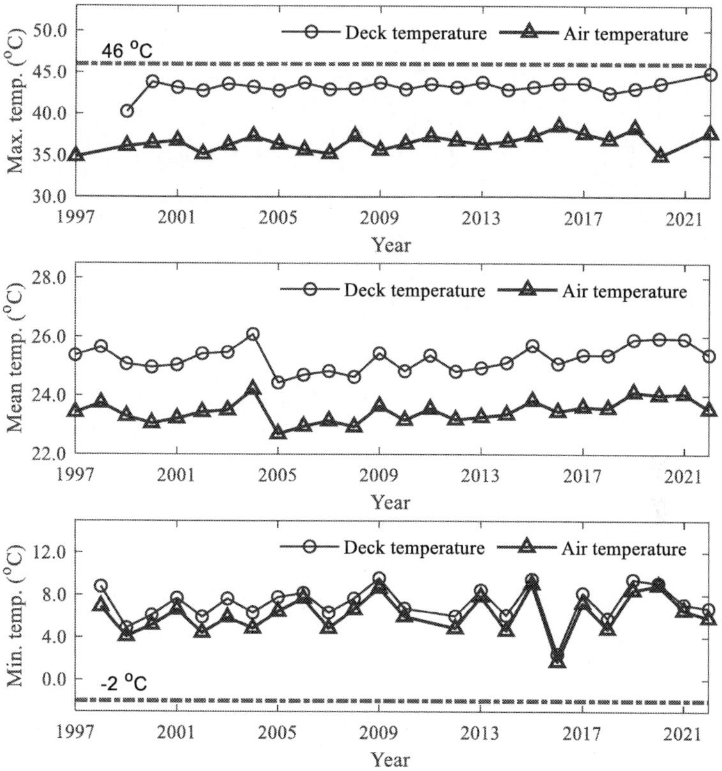

FIGURE 8.4 Annual ambient and deck effective temperatures during 1997–2022.

the past 25 years continues, the maximum effective temperature of the bridge deck may reach the design value (46°C) in 22 (=1.1/0.51 × 10) years.

8.3.2 Vertical Temperature Gradient

Figure 8.5 illustrates the monthly maximum vertical temperature gradients on the south and north sides of the Tsing Ma Bridge deck. The gradients exhibited seasonal variation, reaching its peak in summer at about 17°C and minimum in winter at about 8°C. The south and north side gradients were relatively similar in winter, and the north side gradient in summer was about 3°C higher than that on the south side. This is because the bridge is located south of the Tropic of Cancer, resulting in direct solar radiation on the north side of the bridge during summer, and thus a higher temperature gradient on the north side.

8.3.3 Main Cable Effective Temperature

Section E of the main cable (as shown in Figure 4.5) is equipped with nine temperature sensors. The effective temperature of the main cable is calculated from the measured data of these sensors. The monthly maximum, mean, and minimum air

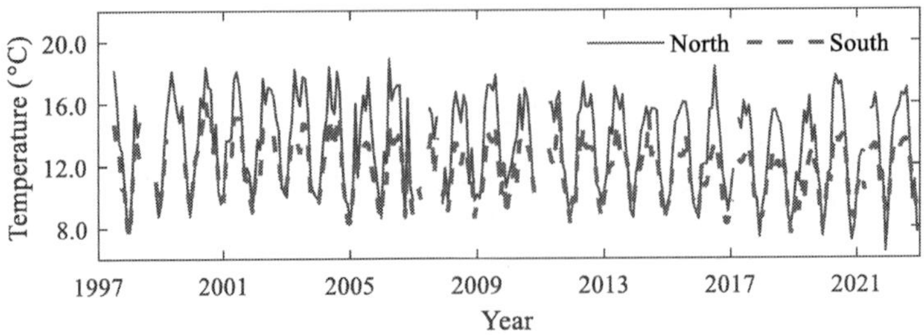

FIGURE 8.5 Monthly maximum deck vertical temperature gradient.

and cable effective temperatures are displayed in Figure 8.6. The maximum cable effective temperature is similar to the air temperature, both reaching around 36°C in summer and 24°C in winter. The mean cable temperature was about 2°C higher than the air temperature in summer and 1°C in winter, respectively. The minimum temperatures exceeded the air temperature about 2°C in summer, and they were similar in winter.

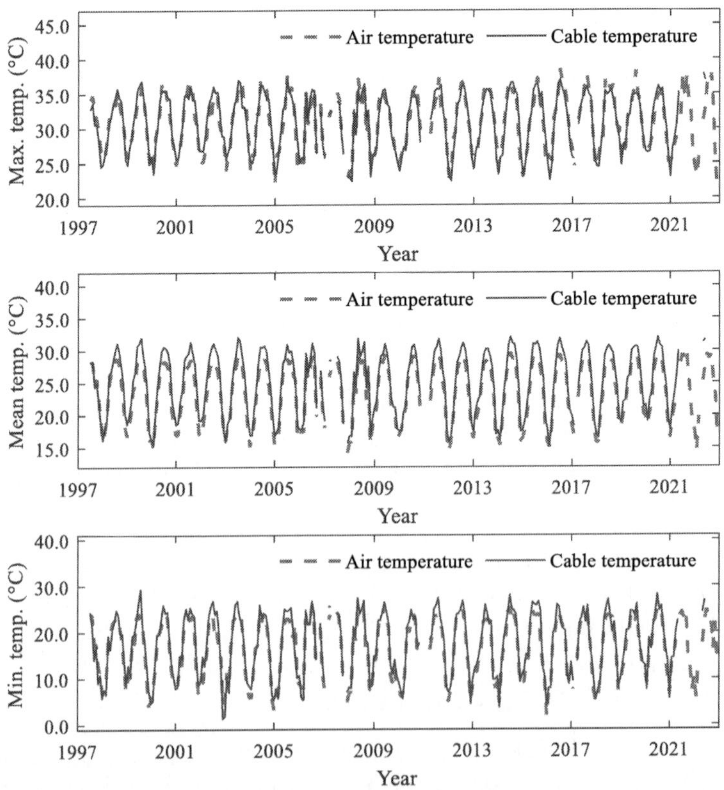

FIGURE 8.6 Monthly ambient and main cable effective temperatures during 1997–2022.

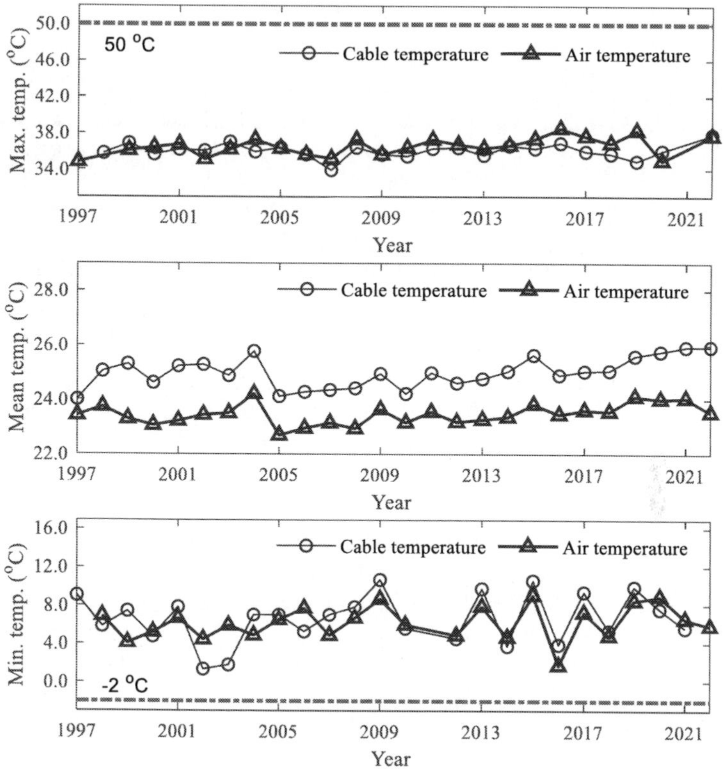

FIGURE 8.7 Annual air and main cable effective temperatures during 1997–2022.

Figure 8.7 illustrates the annual maximum, mean, and minimum air and cable effective temperatures. Linear regression analyses indicate that the maximum cable effective temperatures increased at a rate of 0.12°C/decade, much slower than the ambient temperature rise rate of 0.72°C/decade. The increase rates for the mean and minimum cable effective temperatures were 0.35°C/decade and 0.63°C/decade, respectively. Notably, the designed maximum and minimum effective temperatures for the cables are 50°C and –2°C, respectively. As of 2022, the maximum effective main cable temperature was 37.87°C, much lower than the design value.

8.4 TEMPERATURE-INDUCED BRIDGE RESPONSES

8.4.1 Displacement Responses

8.4.1.1 Displacement Sensor Layout

The locations of displacement sensors on the Tsing Ma Bridge are shown in Figures 4.8 for longitudinal transducers, 4.9 for level sensing stations, and 4.10 for GPS stations. For easy referencing, some selected displacement sensors are shown in Figure 8.8. In particular, level sensing stations (LV-TJN-01) measure the vertical

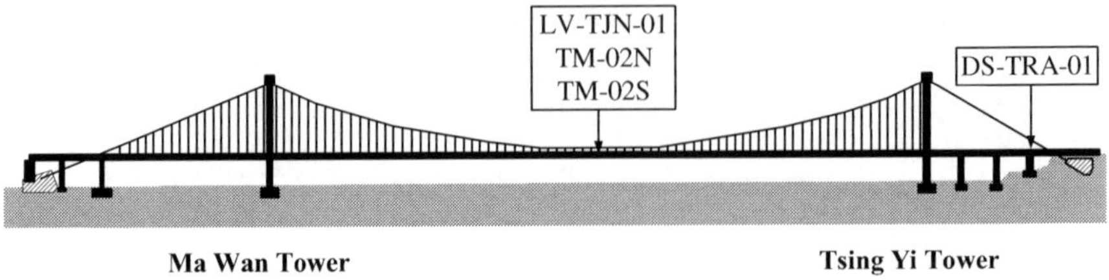

FIGURE 8.8 Locations of selected displacement sensors on the Tsing Ma Bridge.

deflection of the mid-span deck. The displacement transducer (DS-TRA-01) on the Tsing Yi side span measures the longitudinal displacement of the bridge deck. GPS stations (TM-02N/S) measure the lateral, vertical, and longitudinal displacements of the mid-span deck.

8.4.1.2 Long-Term Displacement

Figure 8.9 shows the monthly mean mid-span deflection and longitudinal displacement at the Tsing Yi abutment from 1997 to 2022, in which both GPS and level sensing station measurements are displayed. Note here in this chapter the downward deflection is defined as negative, and the expansion of the longitudinal displacement is defined as positive. The mid-span deflection of the GPS station and the level station match well, mutually verifying the long-term reliability of the two displacement sensors.

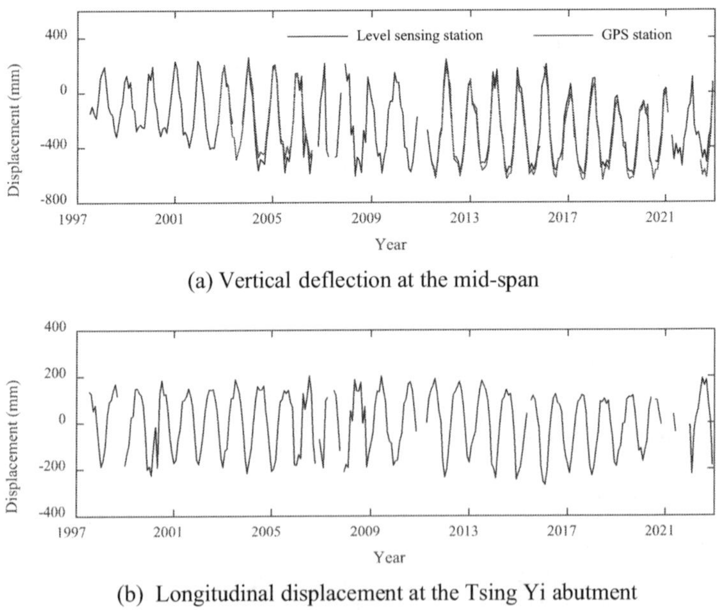

FIGURE 8.9 Monthly mean displacement during 1997–2022: (a) vertical deflection at the mid-span and (b) longitudinal displacement at the Tsing Yi abutment.

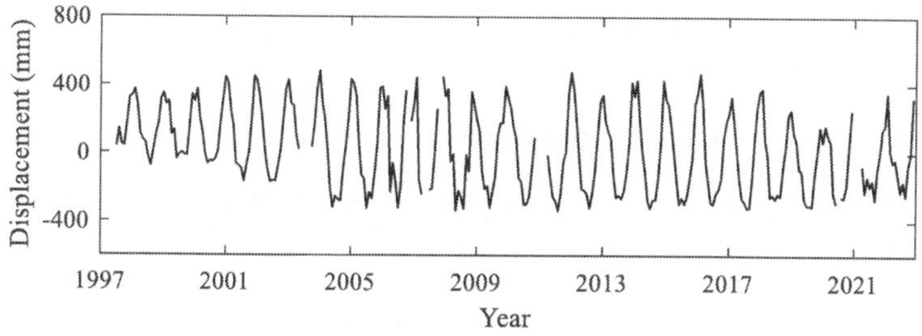

FIGURE 8.10 Monthly mean mid-span deflection in 1997–2022 (excluding traffic loads).

Both vertical and longitudinal displacements exhibit obvious seasonal variations due to the temperature effect. The maximum monthly mean downward deflection was approximately 600 mm in summer, and the maximum monthly mean upward displacement was about 300 mm in winter. The peak monthly mean longitudinal displacement reached approximately 200 mm toward the Tsing Yi tower in summer and 200 mm toward the Ma Wan tower in winter. Regarding the long-term bridge displacement, the mid-span deflection exhibits an obvious increasing trend. In contrast, the longitudinal displacement at the Tsing Yi abutment displays no identifiable long-term trends. The mid-span deflection without traffic load is shown in Figure 8.10, where data between 2:00 am and 4:00 am are selected as the traffic when the period is very low. The downward deflection shows an increasing trend over the first several years and stabilized afterward.

8.4.2 Frequency Variations

The varying traffic may affect the vibrational properties of a bridge as well and its effect should be removed. In this regard, the bridge deck acceleration responses of the Tsing Ma Bridge from 2004 to 2022, occurring between 2:00 and 4:00 when the deck effective temperature was around the annual mean (difference smaller than 0.3°C) are selected to calculate the natural frequencies of the bridge. This will minimize the effects of traffic on the frequencies. Figure 8.11 shows the frequency ratios of the first four modes versus the effective temperature from 2004 to 2022. It shows that the frequencies ratio decrease at the ratio of -1.74×10^{-5} /year.

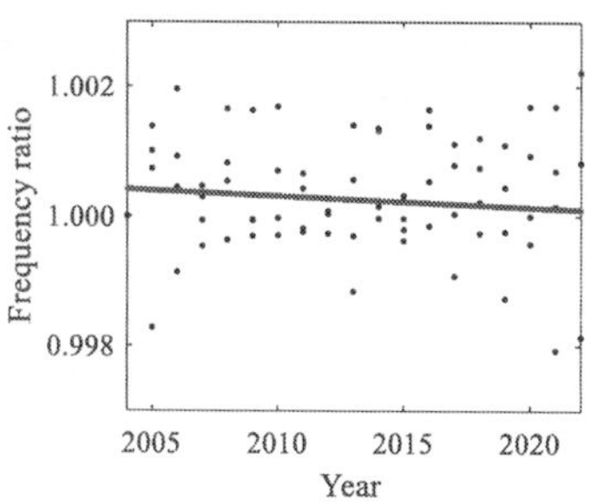

FIGURE 8.11 Variation in bridge frequencies from 2004 to 2022.

8.5 CORRELATION BETWEEN TEMPERATURES AND BRIDGE RESPONSES

8.5.1 Relationship between Temperature and Displacement

The relationship between bridge deck effective temperature and displacement is explored in this section. The corresponding linear regression can be expressed as (Xu and Xia, 2011):

$$X = aT + b + \varepsilon \tag{8.1}$$

where X and T are bridge displacement and effective temperature, respectively, a and b are the fitting coefficients, and ε is the error.

The relationship between daily mean effective temperature and deck mid-span deflection during 1997–2022 is shown in Figure 8.12a. Applying the data to Equation (8.1) leads to $a = -48.11$ mm/°C and $b = 1412.32$ mm. According to Zhou et al. (2020), the unit temperature-induced deflection at the mid-span of a suspension bridge can be simplified as

$$D_V = \alpha \times \left[-3L/(16n) + (h_{P1} + h_{P2})/2\right] \tag{8.2}$$

where α is the thermal expansion coefficient, L is the total length of the bridge, n is the sag ratio of the main cable, and h_{P1} and h_{P2} are the heights of the left and right towers, respectively. For the Tsing Ma Bridge, $\alpha = 1.2 \times 10^{-5}$/°C, $L = 2132$ m, $n = 0.0925$ and $h_{P1} = h_{P2} = 206.4$ m. The calculated deflection of the mid-span is -49.38 mm/°C, which is very close to the measured -48.11 mm/°C. These results indicate that the temperature is the primary contributor to bridge vertical displacement.

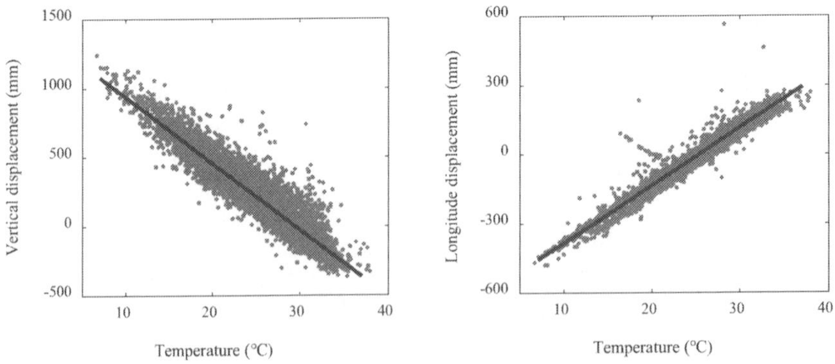

(a) Vertical (b) Longitudinal

FIGURE 8.12 Relationship between daily mean displacement and deck effective temperature: (a) vertical and (b) longitudinal.

The relationship between daily mean deck effective temperature and longitudinal displacement at the Tsing Yi abutment during 1997–2022 is shown in Figure 8.12b. According to Equation (8.1), the coefficients are estimated as $a = 25.52$ mm/°C and $b = -637.18$ mm. Since the bridge deck is fixed on the Ma Wan side and free to move on the Tsing Yi side, the unit temperature-induced longitudinal displacement at the Tsing Yi side can be calculated as

$$D_L = \alpha L_1 \tag{8.3}$$

where $L_1 = 2072$ m is the distance between the sensor DS-TRA-01 and the fixed end on the Ma Wan side. The calculated longitudinal displacement due to temperature variation is 24.86 mm/°C, which is also very close to the measured value of 25.52 mm/°C.

8.5.2 Relationship between Frequency and Temperature

The acceleration data is used to extract the vibrational properties of the bridge, including frequencies and mode shapes through modal analysis (Fu and He, 2001). Four days in different seasons in 2018 (namely, February 6, April 16, August 2, and November 21) are chosen to explore the relationship between the frequencies and effective temperature. Figure 8.13 shows the normalized frequencies of the first four vertical modes and the first two lateral modes. Both vertical and lateral frequencies decrease as the effective temperature increases. The vertical frequency ratio slope against temperature is -1.7×10^{-4}/°C, approaching half of the steel modulus temperature coefficient (-3.6×10^{-4}/°C), which is consistent with the existing research (Xia et al., 2012). The lateral frequency ratio slope is -6.7×10^{-4}/°C. This reflects the influence of the concrete towers, which have a temperature coefficient of modulus of -3.0×10^{-3}/°C.

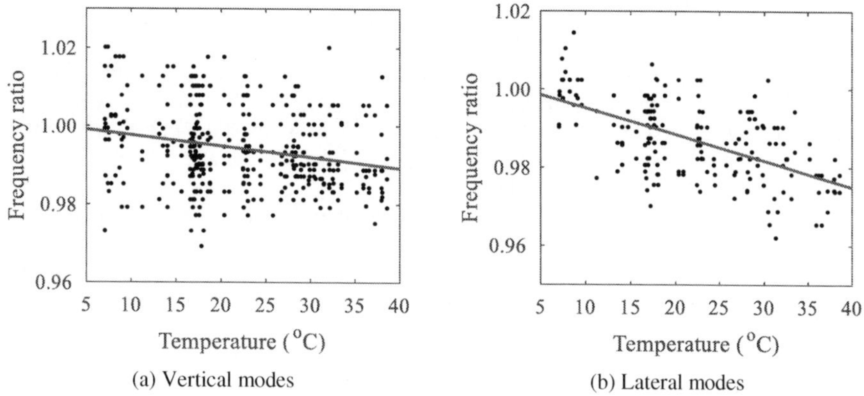

FIGURE 8.13 Relationship between the bridge frequency and deck effective temperature: (a) vertical modes and (b) lateral modes.

8.6 CHAPTER SUMMARY

This chapter investigated the long-term temperature behavior and its effects on the Tsing Ma Bridge using the monitoring data over two decades. The following conclusions can be drawn:

1. The maximum effective deck temperature has been increasing at a rate of 0.51°C/decade, reaching 44.9°C in 2022, which is close to the design limit of 46°C. In contrast, the maximum effective temperature of the main cable is 37.87°C in 2022, increasing at a much smaller rate of 0.12°C/decade and remaining well below the design limit of 50°C.
2. The downward deflection shows an increasing trend over the first several years and stabilized afterward, while the longitudinal displacement exhibits no obvious long-term trends. Bridge frequency ratios have decreased at the ratio of -1.74×10^{-5} per year from 2004 to 2022.
3. Bridge frequencies decreased with rising temperatures, with vertical frequency rations dropping at a rate of -1.7×10^{-4}/°C and lateral frequencies decreasing faster at -6.7×10^{-4}/°C. Measured temperature-induced deformations and displacements closely match theoretical calculations based on structural properties and thermal expansion coefficients.
4. According to the predicted the global surface temperature, the effective temperature of bridge deck may likely exceed the limits specified in current design codes by the end of this century, even under the intermediate greenhouse gas emissions scenarios. It is necessary to account for long-term temperature effects when designing new bridges to ensure their serviceability throughout their designed lifespan.

REFERENCES

Fu, Z.-F. and He, J., 2001, *Modal Analysis*. Amsterdam, Elsevier.

Masson-Delmotte, V., Zhai, P., Pirani, A., Connors, S. L., Péan, C., Berger, S., Caud, N., Chen, Y., Goldfarb, L. and Gomis, M., 2021, Contribution of working group I to the sixth assessment report of the intergovernmental panel on climate change. *Climate Change 2021: the Physical Science Basis* **2(1)**: 2391.

Xia, Y., Chen, B., Weng, S., Ni, Y. Q. and Xu, Y. L., 2012, Temperature effect on vibration properties of civil structures: a literature review and case studies. *Journal of Civil Structural Health Monitoring* **2(1)**: 29–46.

Xu, Y. L. and Xia, Y., 2011, *Structural Health Monitoring of Long-Span Suspension Bridges*. London, CRC Press.

Zhou, Y., Xia, Y., Chen, B. and Fujino, Y., 2020, Analytical solution to temperature-induced deformation of suspension bridges. *Mechanical Systems and Signal Processing* **139**: 106568.

Zhu, Q.-X., Wang, H., Mao, J.-X., Wan, H.-P., Zheng, W.-Z. and Zhang, Y.-M., 2020, Investigation of temperature effects on steel-truss bridge based on long-term monitoring data: case study. *Journal of Bridge Engineering* **25(9)**. 05020007.

9 Temperature Effects on Vibrational Properties of Bridges

NOTATIONS

E	Modulus of elasticity
I	Moment of inertia of the cross-sectional area
f, f_n	Frequency of order n
l	Length of the beam
$\beta_T, \beta_0, \varepsilon_f$	Intercept, slope, and error of the linear regression model
δ	An increment in the corresponding parameters
λ_n	Dimensionless parameter related to boundary conditions
μ	Mass per unit length
θ_T, θ_E	Thermal coefficients of linear expansion and modulus

9.1 PREFACE

The effects of varying temperatures on bridges' static responses (displacement and strain) have been studied in previous chapters. Changing temperature has also been observed to affect structural vibration properties, such as frequencies, mode shapes, and damping. Some studies have found that changes in vibration properties because of temperature variations could be more significant than those caused by a medium degree of structural damage (Salawu, 1997) or under normal operational loads (Xu et al., 2010). Therefore, the temperature effect on the structural vibration properties should be considered or removed if the vibration properties are used for condition assessment or damage identification (Hou and Xia, 2021). Failing to do this will cause false structural condition identification (Giraldo et al., 2006; Xia et al., 2006). Xia et al. (2012) reviewed temperature effects on vibration properties and statistical models considering varying temperatures.

9.2 VARIATIONS IN NATURAL FREQUENCY

Thermal effects on structural dynamic properties have been investigated since the 1970s. In a pioneer study, Adams et al. (1978) investigated the relation between temperature and axial resonant frequency of a bar. Since then, many researchers have studied temperature-induced variations in the natural frequency of structures because natural frequencies can be measured in practice more easily and accurately.

DOI: 10.1201/9781003480136-9

In field measurements, bridges have been paid more attention than other types of structures, possibly because bridges are exposed to the ambient environment directly.

Askegaard and Mossing (1988) studied a three-span reinforced concrete footbridge and observed a 10% seasonal change in frequency over 3 years. Cornwell et al. (1999) investigated the thermal variations in the dynamic properties of the Alamosa Canyon Bridge and found about 5% daily changes in the first three natural frequencies.

Peeters and De Roeck (2001) monitored the Z24 Bridge continuously for nearly a year and reported a bilinear relation between the first two frequencies and the structural temperature. They found that the two frequencies increased by about 10% when the temperature decreased from 0 to −7°C. For temperatures above 0°C, the first frequency decreased slightly when the wearing surface temperature went up, whereas the second frequency increased slightly when the deck soffit temperature went up.

Fu and DeWolf (2001) studied a two-span, slightly skewed composite bridge and found that the expansion bearings were approximately partially constrained below 15.6°C. The first three frequencies decreased by 12.3%, 16.8%, and 9.0%, respectively, as the temperature increased from −17.8°C to approximately 15.6°C, whereas they changed little as the temperature was above 15.6°C. The authors then simulated a thermal axial load and applied it to the girder eccentrically to calculate the frequencies under different temperatures. The change in frequencies agreed well with the measurement.

Ni et al. (2005), Ni et al. (2005), and Zhou et al. (2010) extracted 1-year modal properties of the Ting Kau cable-stayed bridge in Hong Kong. The relative variations in the measured modal frequencies under weak wind conditions ranged between 1.7% and 6.7% when bridge temperatures ranged between 3°C and 53°C. They concluded that the effective temperature was insufficient in formulating a good correlation between the modal frequencies and temperatures because of the existence of temperature gradient over the cross-section.

Macdonald and Daniell (2005) investigated variations in natural frequencies of the Second Severn Crossing cable-stayed bridge because of wind, temperature, and traffic loading. They reported no apparent trend between the natural frequencies and the mean bridge deck temperature because the temperature change was small.

Desjardins et al. (2006) studied the variations in frequencies of the Confederation Bridge (made of pre-stressed concrete) over 6 months. They reported a clear trend of reduction in the modal frequencies by about 4% when the average temperature of the concrete of the bridge varied from −20 to +25°C.

Liu and DeWolf (2007) reported that, during a 1-year measurement, the first three frequencies of a curved concrete box bridge decreased when concrete temperature increased. A linear regression analysis showed that frequencies decreased by 0.007, 0.008, and 0.007 Hz as temperature increased by one Fahrenheit degree, which is equivalent to 0.8%, 0.7%, and 0.3% per degree Celsius.

The Runyang Suspension Bridge with a 1490-m main span experienced about 2% variation in the first six modal frequencies during 10 months, as the ambient temperature of the steel bridge varied from −5 to +50°C (Deng et al., 2010). During 16 days of continuous monitoring of a cable-stayed bridge, Li et al. (2010) found that the first six frequencies varied by about 1.5%–3.2% as ambient temperature changed from −11.5 to +3.7°C.

9.3 VARIATIONS IN MODE SHAPE AND DAMPING

Previous studies demonstrated that natural frequencies of structures decrease with an increase in structural or ambient temperature, whereas the thermal effect on mode shapes is negligible. For example, the first two modal frequencies identified from the Bill Emerson Memorial Bridge decreased monotonically as the temperature went up linearly, whereas the mode shapes did not show any significant change (Song and Dyke, 2006).

In Ni et al.'s study (Ni et al., 2005) on the Ting Kau cable-stayed bridge, some mode shapes were stationary at different times, some mode shapes at some points were stationary, but other points were not, and some mode shapes had considerable variations. Quantitative analysis found that the mode shapes were out of proportion to temperature.

Xia et al. (2006) investigated the effects of temperature on a continuous concrete slab for nearly two years. They observed that damping ratios had a positive correlation with temperature, whereas mode shapes did not correlate with temperature, as ambient temperature affects the entire structure uniformly.

Xu and Wu's numerical analysis (Xu and Wu, 2007) showed that the mode shape curvature at some local points of a cable-stayed bridge changed slightly, by about 1%–8%, as seasonal temperature increased from −20 to +40°C. These changes in mode shapes were attributed to the changes in cable tensions and sags.

In a chamber experiment, Balmès et al. (2008) found that the variations in mode shapes were invisible when temperature decreased by about 17°C. As the effect of temperature on damping was concerned, no consistent conclusion has been drawn. One reason is that the damping of structures is more difficult to measure accurately than the frequencies and the mode shapes. A significant uncertainty in modal damping might mask the temperature effects (Balmès et al., 2008).

Li et al. (2010) found that some modal damping ratios of a cable-stayed bridge had a negative correlation with temperature, whereas others had a positive correlation. At the same time, the variation range of damping ratios could be quite large and could reach 280%.

9.4 QUANTITATIVE ANALYSIS

In the above literature, most studies show that an increase in temperature leads to a decrease in structural frequencies, while magnitude varies, depending on bridges, materials, and temperature range. Variations in natural frequencies of structures with temperature are caused by changes in material properties, particularly the modulus of elasticity. A single-span or multi-span prismatic beam made of an isotropic material is used as an example to quantify the effect of temperature on natural frequencies. The *n*-th undamped flexural vibration frequency of the beam is (Blevins, 1979):

$$f_n = \lambda_n^2/\left(2\pi l^2\right)\cdot\sqrt{EI/\mu} \tag{9.1}$$

where λ_n is a dimensionless parameter and is a function of the boundary conditions, l is the length of the beam, μ is the mass per unit length, E is the modulus of elasticity, and I is the second moment of inertia of the cross-sectional area. It is assumed that temperature variations will affect the geometry of the structure and the mechanical properties of the material but not affect mass and boundary conditions.

It can be shown that

$$\frac{\delta f_n}{f_n} = -2\frac{\delta l}{l} + \frac{1}{2}\frac{\delta E}{E} + \frac{1}{2}\frac{\delta I}{I} - \frac{1}{2}\frac{\delta \mu}{\mu} \tag{9.2}$$

where δ represents an increment in the corresponding parameters. Assuming that the thermal coefficient of linear expansion of the material is θ_T and the thermal coefficient of modulus is θ_E, one obtains

$$\frac{\delta l}{l} = \theta_T \delta T, \quad \frac{\delta E}{E} = \theta_E \delta T, \quad \frac{\delta I}{I} = 4\theta_T \delta T, \quad \frac{\delta \mu}{\mu} = -\theta_T \delta T \tag{9.3}$$

Here, the variations in Young's modulus with temperature are assumed linear for small temperature changes, variations in the moment of area are four times the variations in linear expansion, and the mass per unit length is inversely proportional to the length as the total mass is a constant. Consequently, Equation (9.2) yields

$$\frac{\delta f_n}{f_n} = \frac{1}{2}(\theta_T + \theta_E)\delta T \tag{9.4}$$

Equation (9.4) estimates the dimensionless rate of the frequency change with the temperature change. The linear thermal expansion coefficients (θ_T) of steel (Brockenbrough and Merritt, 2020), aluminum (Bird and Ross, 2015), and concrete (CEB-FIP, 1993) are 1.1×10^{-5}, 2.30×10^{-5}, and 1.0×10^{-5}°C^{-1}, respectively. The modulus thermal coefficients (θ_E) of the three materials are -3.6×10^{-4}, -5.6×10^{-4}, and -3.0×10^{-3}°C^{-1}, respectively. θ_E is much larger than θ_T for the three materials, which indicates that variations in natural frequency subjected to temperature change are controlled by θ_E. It also shows that the variation percentage of natural frequency is a function of the thermal coefficients of modulus only, regardless of modes, spans, and beam types (simply supported or cantilever beam). Consequently, theoretical variation percentages of the natural frequency of steel beams, aluminum beams, and RC beams are 0.018%, 0.028%, and 0.15% per degree Celsius, respectively. These are verified through laboratory experiments and field investigations in later subsections. The big difference in the modulus thermal coefficient contributes to significantly different observations in existing literature. For example, variations in natural frequencies of a steel beam and an RC beam are about 0.36% and 3.0%, respectively, under a temperature change of 20°C. The former is difficult to observe in practice and may be masked by measurement noise or other factors, especially for lower modes with small absolute frequencies.

In addition, structural temperature rather than air temperature is employed in the above analysis. As the temperature distribution of a large-scale structure is generally non-uniform and time-dependent, using ambient air temperature, structure surface temperature, structure interior temperature, or averaged temperature causes different quantitative results (Xia et al., 2011). Moreover, temperature variations inside a structure lag behind those on the surface because of temperature inertia, especially for concrete structures. Consequently, frequency changes and air temperature changes may not occur simultaneously. Xia et al. (2011) measured the non-uniform temperature field of an RC slab and calculated Young's modulus of each component according to the relation between Young's modulus and temperature. Natural frequencies of the structure were then calculated, which showed good agreement with the measurements. They concluded that consideration of the temperature distribution of a whole structure leads to more accurate results of the temperature effect on the vibration properties of a structure. When measurement of all components is difficult, a heat-transfer analysis can be carried out.

In the above quantitative analysis, the effect of the internal force on frequency is not considered. The internal force substantially impacts lower modal frequencies more than higher modes (Xia et al., 2006). When the thermal stress is negligible, its effect on frequency can be neglected. Temperature variations may affect boundary conditions, which are difficult to quantify. For bridges, if expansion joints are not prevented from moving, the boundary conditions can be considered unchanged.

9.5 CASE STUDIES

9.5.1 Laboratory-Tested Beams

Although temperature effects on dynamic properties have been extensively studied, they were conducted under different temperature conditions and thus might have different results. This section compares the effect of temperature on vibration frequencies of beam structures made of different materials, particularly, steel, aluminum, and RC, through laboratory experiments.

9.5.1.1 Description of the Models and Experiments

An RC slab, a steel beam, and an aluminum beam were constructed and exposed to solar irradiation. The configuration of the models is illustrated in Figure 9.1. The RC slab was simply supported, while the two metallic beams were cantilevered. Therefore, the slab and beams could move freely. It assumes that temperature variation did not affect the support conditions; only the geometric dimensions and Young's moduli of the structures were affected. Temperature variations along the horizontal direction of the slab were assumed to be insignificant. Consequently, a temperature gradient might occur only in the thickness direction. Seven thermocouples (T1–T7), with an even interval of 20 mm in the vertical direction, were embedded in the RC slab to collect the time-dependent temperature data.

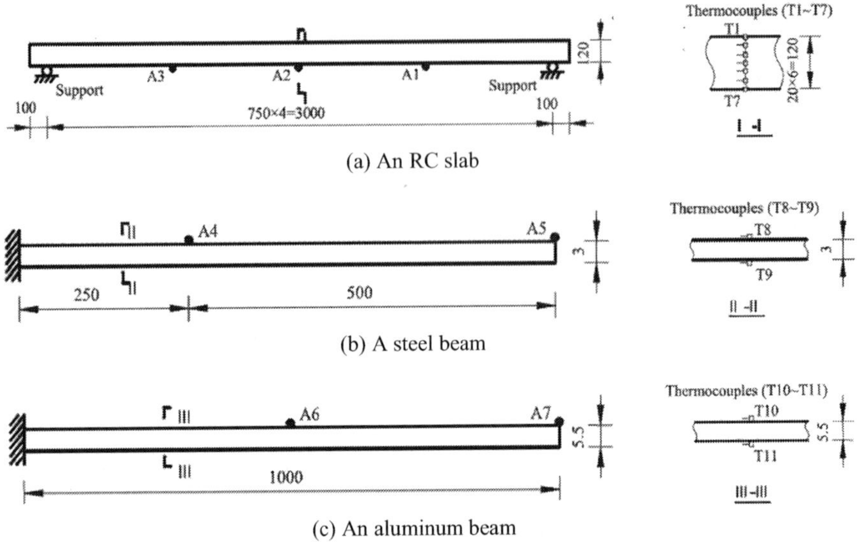

FIGURE 9.1 Configuration of the slab and the beams (unit: mm, A1–A7: accelerometers; T1–T11: thermocouples): (a) an RC slab, (b) a steel beam, and (c) an aluminum beam.

As the thermal conductivity coefficients of steel and aluminum are rather large and the thickness of the beams is small, the temperature difference between the top and the bottom surfaces was expected to be small. Therefore, only the top and the bottom surfaces of the two metallic beams were mounted with thermocouples (T8 and T9, and T10 and T11) to measure the temperature (Figure 9.1). A total of seven accelerometers (A1–A7) were firmly mounted on the models to measure their vibration and extract the modal properties.

Temperature variations were recorded from the embedded thermocouples. Modal testing was carried out using an instrumented hammer. Six impacts were exerted at each time to average the signal. Frequency response functions (FRFs) of the three models were derived in a frequency range of 0–500 Hz. Natural frequencies were then extracted from the FRFs using the global rational fraction polynomial method (Ewins, 2009).

9.5.1.2 Variations in Frequency with Respect to Temperature

The first four frequencies of the two metallic beams and the first two frequencies of the RC slab were extracted from the collected data at different times. Figure 9.2 shows the first four natural frequencies of the steel beam with respect to the surface average temperature. All natural frequencies decreased with an increase in surface temperature before 11:30 and increased slowly after that with a decrease in surface temperature. The variation trend matched the surface temperature well, although the frequency variations were small. For comparison of different modes, the variation percentage of the natural frequencies with respect to the surface temperatures is shown in Figure 9.3, in which the natural frequencies are divided by the first

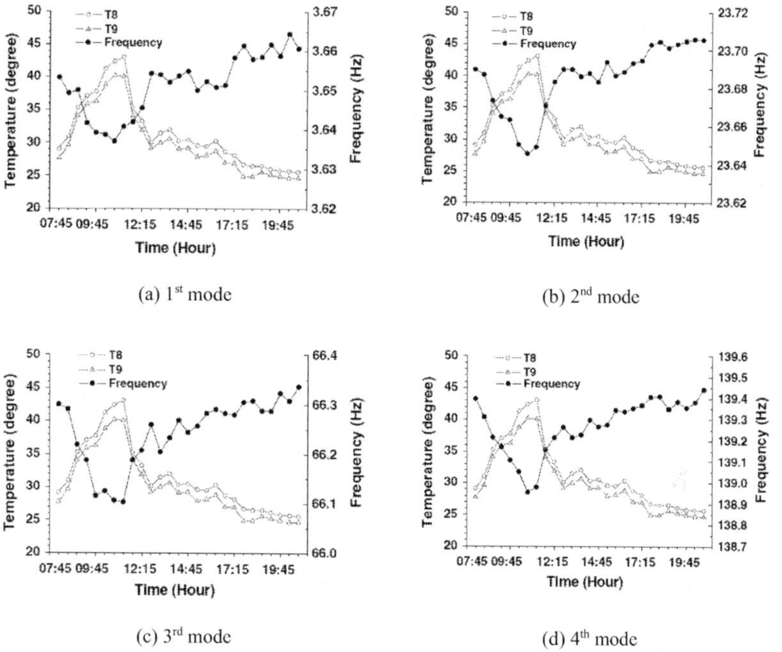

FIGURE 9.2 Variations in natural frequencies of the steel beam with respect to temperature: (a) first mode, (b) second mode, (c) third mode, and (d) fourth mode.

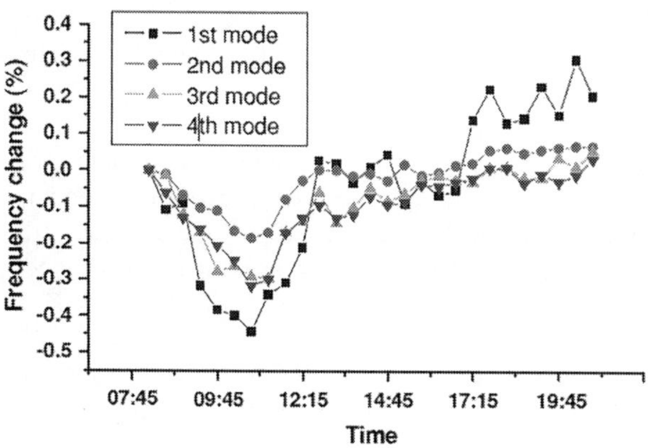

FIGURE 9.3 Variations in natural frequencies of the steel beam with respect to temperature.

measurement of the modes. Different modes have a similar relative frequency change (−0.45% to 0.3%).

Variations in the first four natural frequencies of the aluminum beam with respect to the temperature are plotted in Figure 9.4. Again, the variation trend of all four

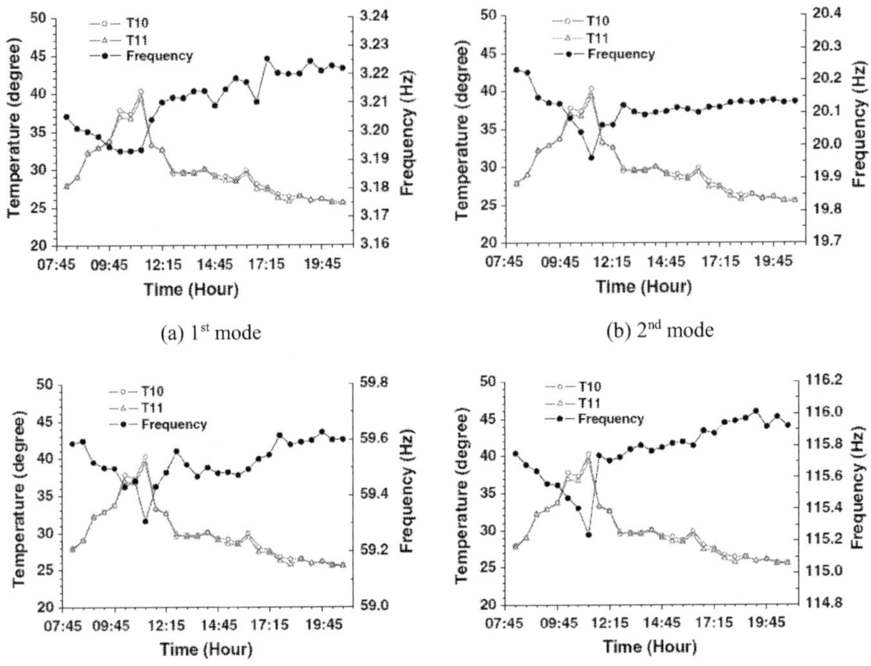

(a) 1st mode (b) 2nd mode

(c) 3rd mode (d) 4th mode

FIGURE 9.4 Variations in natural frequencies of the aluminum beam with respect to temperature: (a) first mode, (b) second mode, (c) third mode, and (d) fourth mode.

natural frequencies agrees well with the surface temperature. The higher modes match the temperature better than the lower modes. One reason is that the variation magnitude of the lower frequency was smaller than that of the higher frequency under the identical percentage change. For example, the absolute variation in the first frequency is 0.03 Hz only, whereas the counterparts for the second, third, and fourth frequencies are 0.25, 0.30, and 0.51 Hz, respectively. Consequently, the measurement noise has a larger effect on the lower modes than on the higher modes. Nevertheless, the experimental results still show a strong correlation between the frequencies and the temperature.

For the RC slab, variations in the first two natural frequencies and the averaged structural temperature are illustrated in Figure 9.5. Again, the natural frequencies generally decreased with an increase in temperature in the morning. They increased slowly in the afternoon with a decrease in temperature. In addition, frequency variations of the RC slab (2.5%) are much larger than those of the metallic beams.

9.5.1.3 Quantitative Relation between Frequency and Temperature

The linear regression technique is utilized to examine the relation between the frequencies and the structural temperature of the models. A linear regression model has the following form (Kottegoda and Rosso, 1997):

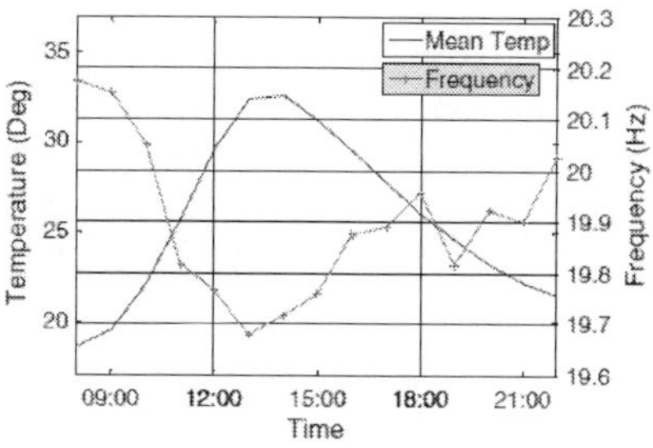

(a) 1st frequency

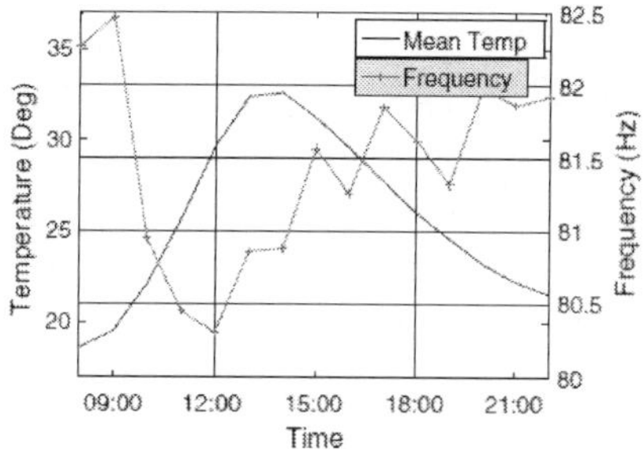

(b) 2nd frequency

FIGURE 9.5 Variations in natural frequencies of the RC slab with respect to temperature: (a) first frequency and (b) second frequency.

$$f = \beta_0 + \beta_T T + \varepsilon_f \tag{9.5}$$

where f is the frequency, T is the temperature explanatory variable, β_0 (intercept) and β_T (slope) are the regression coefficients, and ε_f is the regression error. In the regression analysis, the averaged temperature of the cross-section is adopted. The correlation coefficients of the first four natural frequencies of the steel beam are −0.93, −0.99, −0.96, and −0.97, respectively, implying an excellent linear correlation between the averaged temperature and the natural frequencies. The fitted regression lines of the steel beam with 95% confidence bounds are plotted in Figure 9.6.

The regression coefficients of the three structures are also obtained and listed in Table 9.1 for each mode. The slope β_T represents the frequency change rate with respect to temperature, and the intercept β_0 represents the frequency at 0°C. For comparison of different modes, β_T is normalized to β_0 and listed in Table 9.1. β_T/β_0 indicates the percentage of the frequency change when the structural temperature increases by a unit degree Celsius. The changes of the first frequency of the two metallic beams are larger than those of the higher modes. This may be because the measurement noise significantly affects the lower modes, as the lowest modal frequency has the smallest amplitude. Except for the first mode of the metallic beams, the modal frequencies of the steel beam, the aluminum beam, and the RC slab decrease by about 0.015%–0.020%, 0.027%–0.042%, and 0.11%–0.14%, respectively, when the structural temperature increases by unit degree Celsius. The averaged values are 0.018%, 0.03%, and 0.12%, respectively. These are very close to half of the modulus thermal coefficients of the material (θ_E = 0.036%, 0.056%, and 0.30% for steel, aluminum, and concrete, respectively) according to Equation (9.4). This verifies Equation (9.4) and indicates that the variations in the bending frequencies of a structure are mainly caused by the change in the modulus of the materials.

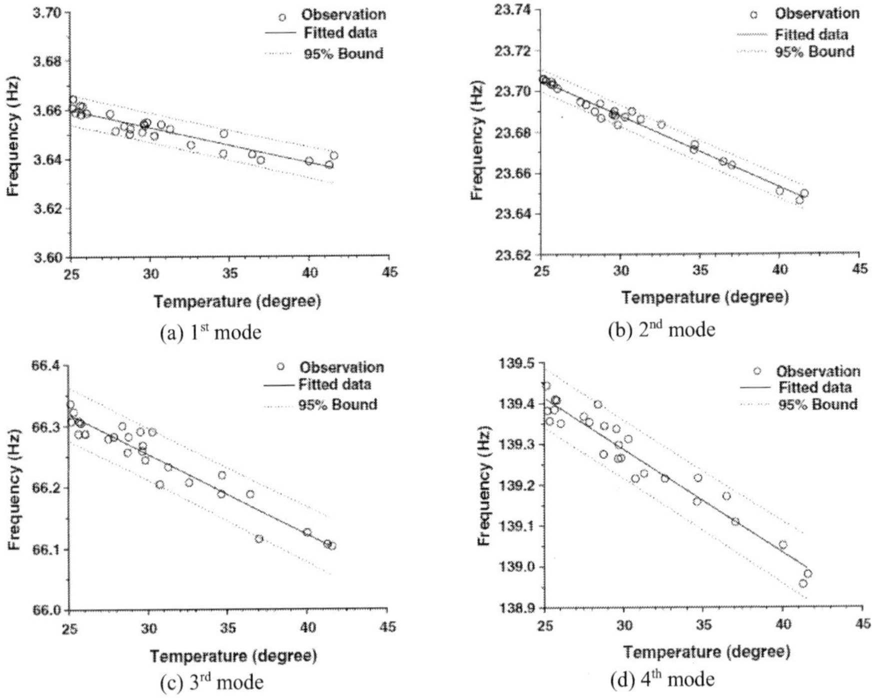

FIGURE 9.6 Relation of natural frequencies to temperature of the steel beam: (a) first mode, (b) second mode, (c) third mode, and (d) fourth mode.

TABLE 9.1
Regression Coefficients of the Models

	Steel Beam			Aluminum Beam			Concrete Slab		
Mode	β_0	β_T	β_T/β_0 ($\times 10^{-4}$)	β_0	β_T	β_T/β_0 ($\times 10^{-4}$)	β_0	β_T	β_T/β_0 ($\times 10^{-4}$)
1	3.69	−0.0014	−3.78	3.28	−0.0024	−7.31	20.65	−0.029	−14.0
2	23.79	−0.0035	−1.47	20.37	−0.0086	−4.22	83.74	−0.089	−10.6
3	66.64	−0.0132	−1.98	60.01	−0.0163	−2.72			
4	140.06	−0.0253	−1.81	117.63	−0.0470	−4.00			

9.5.1.4 Variations in Damping with Respect to Temperature

Variations in the four damping ratios of the steel beam are shown in Figure 9.7. No clear correlation between damping ratios and temperature can be found, indicating that temperature has little effect on damping ratios. Similar conclusions can be drawn from the aluminum beam and the RC slab. The results are not shown here for brevity. Nevertheless, variations in damping ratios are significant.

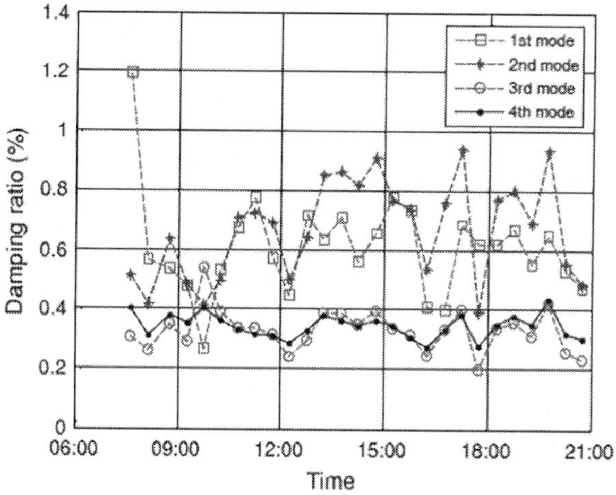

FIGURE 9.7 Relation of damping ratios to temperature of the steel beam.

9.5.2 The Tsing Ma Bridge

The Tsing Ma Bridge is investigated in this section regarding its variations in vibration properties. The configuration of the bridge is shown in Figure 9.8.

The SHM system for the Tsing Ma Bridge comprises about 280 sensors, including 13 accelerometers and 110 temperature sensors (Wong, 2004) as introduced in Chapter 4. Here, only three uniaxial accelerometers installed on the deck (AS-TFN-01, AS-TIN-01, and AS-TJN-01) for measuring vertical acceleration are employed. Hourly acceleration time histories are recorded on tapes, with a sampling frequency of 51.2 Hz. P2-TNN/S includes 28 platinum resistance thermometers installed in one deck section, which are used to calculate the effective temperature of the section. The effective temperature is the average temperature of the cross-section weighted by the areas around the sensors.

The first four vertical frequencies of the structure in different seasons are studied here. The frequencies are identified from the ambient excited acceleration data at the vertical direction measured by sensors AS-TIN-01 and AS-TJN-01 using the

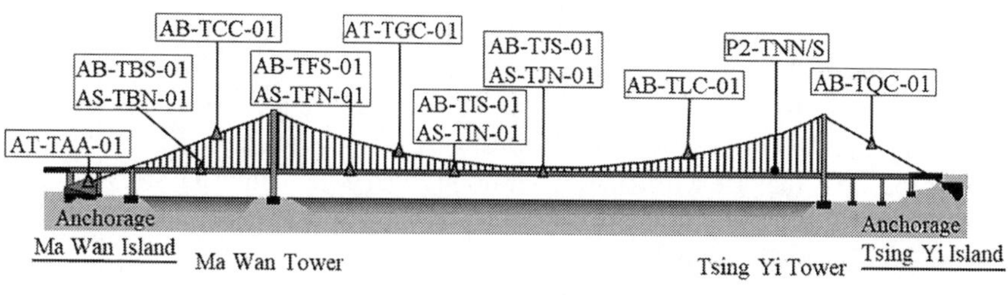

FIGURE 9.8 Accelerometers and thermometers on the Tsing Ma Bridge.

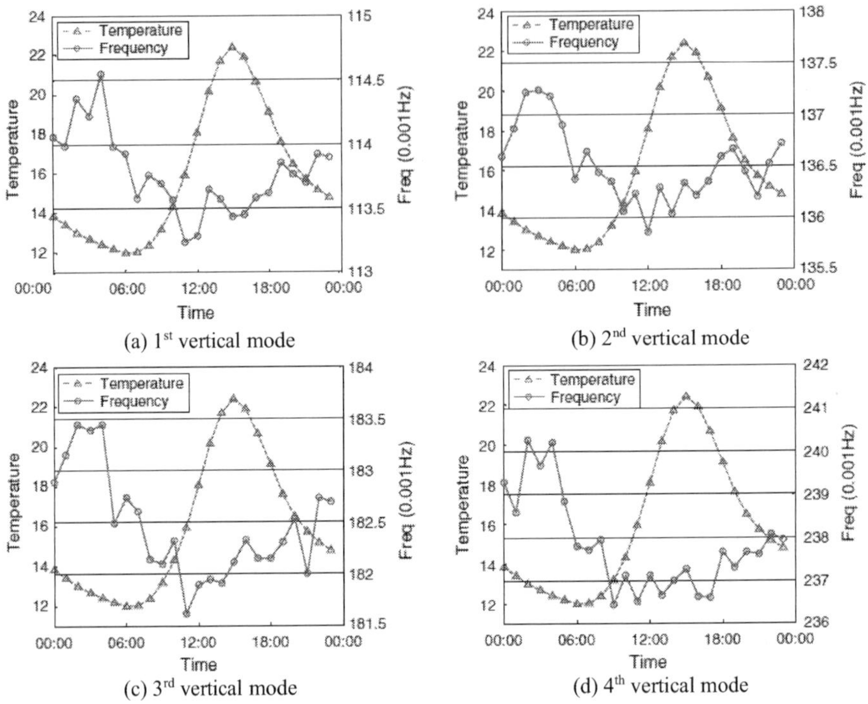

FIGURE 9.9 Variations in frequencies with respect to the temperature on January 17, 2005: (a) first vertical mode, (b) second vertical mode, (c) third vertical mode, and (d) fourth vertical mode.

stochastic subspace identification method (Van Overschee and De Moor, 2012). The vertical acceleration data at AS-TFN-01 are employed as the reference.

Figure 9.9 shows the variations in the first four vertical frequencies and temperature on January 17, 2005. In the figure, the frequency data at each hour are obtained from the acceleration data recorded during that hour, and the temperature is the average effective temperature of the deck section over the hour. All frequencies generally decreased when the temperature went up before noon, whereas they increased as the temperature dropped in the afternoon, although the variations were small. The minimum frequencies and the maximum temperatures did not occur at the same time, and the time difference was about 3 hours. This is because the temperature was non-uniformly distributed across the bridge, and the frequencies are global properties associated with the temperature distribution of the entire structure. Different components have different contributions to the global frequencies, and using one temperature may not represent the variations of the whole structure (Xia et al., 2011). Another reason is that the temperature variations also caused internal stresses of the components, for example, the cable tension forces. The frequency changes due to cable tensions have not been studied here.

The variation percentage of the frequencies with respect to effective temperature is shown in Figure 9.10, where the frequencies are divided by the maximum values

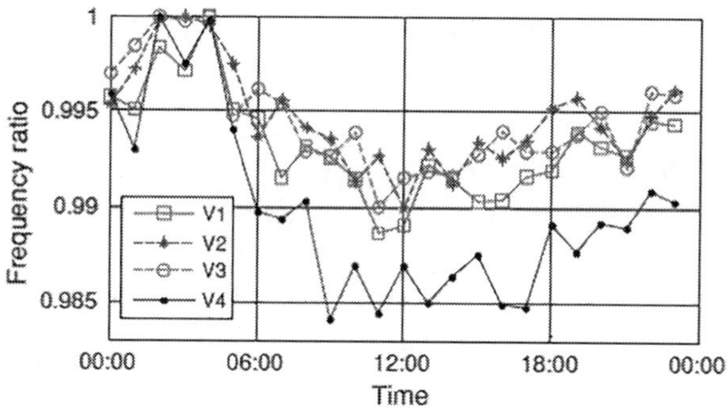

FIGURE 9.10 Variation percentage of frequencies on January 17, 2005.

of the mode. The variation trends of the vertical modal frequencies are very similar, and the maximum variations are about 1%–1.5%.

The frequencies in three other seasons (April 18, July 17, and October 26, 2005) are also studied. Similar results have been observed but are not shown here for brevity. A linear regression model is applied to the first four vertical modal frequencies in 4 days, as shown in Figure 9.11. A good linear correlation can be observed

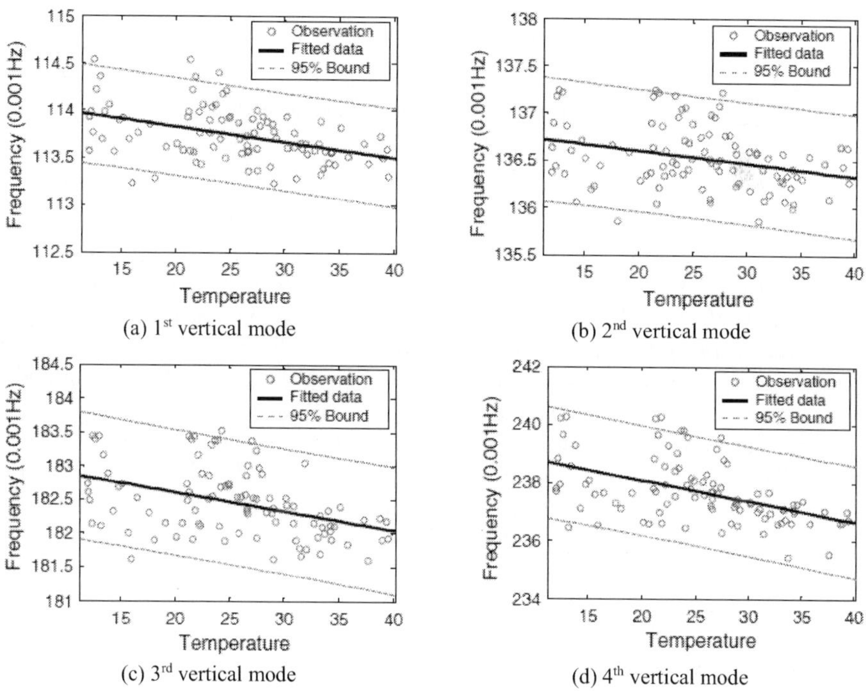

FIGURE 9.11 Relation between natural frequencies and temperature of the Tsing Ma Bridge: (a) first vertical mode, (b) second vertical mode, (c) third vertical mode, and (d) fourth vertical mode.

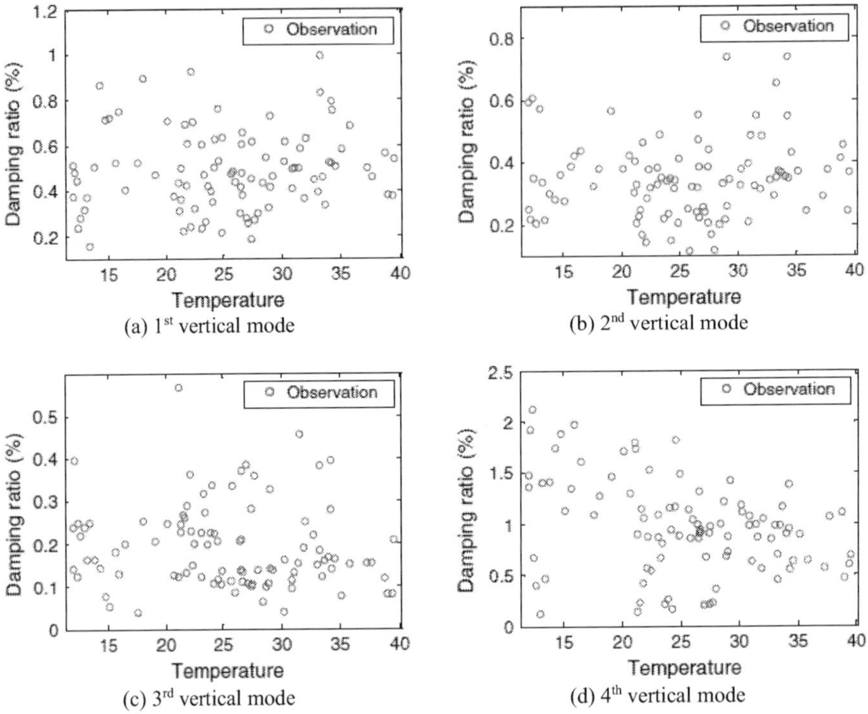

FIGURE 9.12 Relation between damping ratios and temperature of the Tsing Ma Bridge: (a) first vertical mode, (b) second vertical mode, (c) third vertical mode, and (d) fourth vertical mode.

from the figure. The quantity β_T/β_0 for the four modes is calculated as -1.42×10^{-4}, -1.00×10^{-4}, -1.52×10^{-4}, and -2.94×10^{-4}, respectively. The averaged value of β_T/β_0 is -1.72×10^{-4}, very close to that of the laboratory-tested steel beam (0.018%) and also to half of $\theta_E=-3.6\times10^{-4}$°C^{-1}, as described in Equation (9.4). The results imply that variations in the bending frequencies of the structures are mainly caused by the change in the modulus of the material used, although the two structures are quite different in size, type, and boundary conditions.

Damping ratios of the first four modes versus effective temperatures are shown in Figure 9.12. No clear correlation between damping ratios and temperature can be observed. Applying the linear regression model to the damping data shows that their correlation coefficients are 0.10, 0.07, −0.11, and −0.33, respectively.

9.6 CHAPTER SUMMARY

Many studies have shown that varying temperatures affect the structural vibration properties. A general conclusion is that frequencies decrease with increasing temperature, although different magnitudes have been observed. One possible reason is that different temperatures (e.g., air temperature, structural temperature at the surface, and effective structural temperature) are adopted in the correlation analysis. Mode shapes and damping rations do not exhibit an obvious correlation with temperatures.

A quantitative analysis has been derived. It shows that the variation in the modulus of the structural materials under different temperatures is the main factor affecting the structural frequencies. For beam structures, the variation percentage of natural frequencies is approximately half of the thermal coefficients of modulus, regardless of modes, spans, and boundary conditions. Laboratory and field studies verify the results.

REFERENCES

Adams, R. D., Cawley, P., Pye, C. J. and Stone, B. J., 1978, A vibration technique for non-destructively assessing the integrity of structures. *Journal of Mechanical Engineering Science* **20**: 93–100.

Askegaard, V. and Mossing, P., 1988, Long term observation of RC-bridge using changes in natural frequency. *Nordic Concrete Federation* **7**: 20–27.

Balmès, É., Basseville, M., Bourquin, F., Mevel, L., Nasser, H. and Treyssède, F., 2008, Merging sensor data from multiple temperature scenarios for vibration monitoring of civil structures. *Structural Health Monitoring* **7(2)**: 129–142.

Bird, J. O. and Ross, C. T. F., 2015, *Mechanical Engineering Principles*. Abingdon, Oxon, Routledge.

Blevins, R. D., 1979, *Formulas for Natural Frequency and Mode Shape*, New York, Van Nostrand Reinhold.

Brockenbrough, R. L. and Merritt, F. S., 2020, *Structural Steel Designer's Handbook*. New York, McGraw Hill.

CEB- FIP, 1993, *CEB-FIP Model Code 1990: Design Code*. London, T. Telford.

Cornwell, P., Farrar, C. R., Doebling, S. W. and Sohn, H., 1999, Environmental variability of modal properties. *Experimental Techniques* **23(6)**: 45–48.

Deng, Y., Ding, Y. L. and Li, A. Q., 2010. Structural condition assessment of long-span suspension bridges using long-term monitoring data. *Earthquake Engineering and Engineering Vibration* **9(1)**: 123–131.

Desjardins, S. L., Londoño, N. A., Lau, D. T. and Khoo, H., 2006, Real-time data processing, analysis and visualization for structural monitoring of the confederation bridge. *Advances in Structural Engineering* **9(1)**: 141–157.

Ewins, D. J., 2009, *Modal Testing: Theory, Practice and Application*. New York, John Wiley & Sons.

Fu, Y. D. and DeWolf, J. T., 2001, Monitoring and analysis of a bridge with partially restrained bearings. *Journal of Bridge Engineering* **6(1)**: 23–29.

Giraldo, D. F., Dyke, S. J. and Caicedo, J. M., 2006, Damage detection accommodating varying environmental conditions. *Structural Health Monitoring* **5(2)**: 155–172.

Hou, R. and Xia, Y., 2021, Review on the new development of vibration-based damage identification for civil engineering structures: 2010–2019. *Journal of Sound and Vibration* **491**: 115741.

Kottegoda, N. T. and Rosso, R., 1997, *Statistics, Probability and Reliability for Civil and* Environmental Engineers. London, Mc-Graw-Hill.

Li, H., Li, S., Ou, J. and Li, H., 2010, Modal identification of bridges under varying environmental conditions: temperature and wind effects. *Structural Control and Health Monitoring* **17(5)**: 495–512.

Liu, C. and DeWolf, J. T., 2007, Effect of temperature on modal variability of a curved concrete bridge under ambient loads. *Journal of Structural Engineering* **133(12)**: 1742–1751.

Macdonald, J. H. G. and Daniell, W. E., 2005, Variation of modal parameters of a cable-stayed bridge identified from ambient vibration measurements and FE modelling. *Engineering Structures* **27(13)**: 1916–1930.

Ni, Y. Q., Fan, K. Q., Zheng, G., and Ko, J. M., 2005, Automatic modal identification and variability in measured modal vectors of a cable-stayed bridge. *Structural Engineering and Mechanics* **19(2)**: 123–139.

Ni, Y. Q., Hua, X. G., Fan, K. Q. and Ko, J. M., 2005, Correlating modal properties with temperature using long-term monitoring data and support vector machine technique. *Engineering Structures* **27(12)**: 1762–1773.

Peeters, B. and De Roeck, G., 2001, One-year monitoring of the Z24-Bridge: environmental effects versus damage events. *Earthquake Engineering and Structural Dynamics* **30(2)**: 149–171.

Salawu, O. S., 1997, Detection of structural damage through changes in frequency: a review. *Engineering Structures* **19(9)**: 718–723.

Song, W. and Dyke, S., 2006, Ambient vibration based modal identification of the Emerson bridge considering temperature effects. In *The 4th World Conference on Structural Control and Monitoring.****

Van Overschee, P. and De Moor, B., 2012, *Subspace Identification for Linear Systems: Theory—Implementation—Applications*. New York, Springer Science & Business Media.

Wong, K. Y., 2004, Instrumentation and health monitoring of cable-supported bridges. *Structural Control and Health Monitoring* **11(2)**: 91–124.

Xia, Y., Chen, B., Weng, S., Ni, Y.-Q. and Xu, Y.-L., 2012, Temperature effect on vibration properties of civil structures: a literature review and case studies. *Journal of Civil Structural Health Monitoring* **2**: 29–46.

Xia, Y., Hao, H., Zanardo, G. and Deeks, A., 2006, Long term vibration monitoring of an RC slab: Temperature and humidity effect. *Engineering Structures* **28(3)**: 441–452.

Xia, Y., Xu, Y.-L., Wei, Z.-L., Zhu, H.-P. and Zhou, X.-Q., 2011, Variation of structural vibration characteristics versus non-uniform temperature distribution. *Engineering Structures* **33(1)**: 146–153.

Xu, Y. L., Chen, B., Ng, C. L., Wong, K. Y. and Chan, W. Y., 2010, Monitoring temperature effect on a long suspension bridge. *Structural Control and Health Monitoring* **17(6)**: 632–653.

Xu, Z. D. and Wu, Z. S., 2007, Simulation of the effect of temperature variation on damage detection in a long-span cable-stayed bridge. *Structural Health Monitoring* **6(3)**: 177–189.

Zhou, H. F., Ni, Y. Q. and Ko, J. M., 2010, Constructing input to neural networks for modeling temperature-caused modal variability: Mean temperatures, effective temperatures, and principal components of temperatures. *Engineering Structures* **32(6)**: 1747–1759.

10 Epilogue: Conclusions and Challenges

10.1 CONCLUSIONS

Bridges are subjected to daily, seasonal, and yearly environmental temperature variations induced by solar radiation and varying ambient air temperature. Numerous studies have investigated temperature behaviors of bridges. Many of them have found that varying temperatures have an important effect on bridges' responses, even more than other load actions. Moreover, bridges are simultaneously subjected to multiple loads, such as wind and traffic. Separating these loading actions is necessary to assess each loading effect accurately. Finally, varying temperatures may cause more significant changes in structural vibration properties than damage. Therefore, an accurate investigation of the temperature behavior is crucial to the safety and serviceability of bridges.

Although the thermal behavior of bridges has been extensively investigated through field monitoring, the majority of the studies focus on regression analyses of the relation between the temperature and bridge responses. However, the limited and inadequate sensors in a SHM system fail to obtain the temperature of the entire bridge and the most adverse temperature effect. Field monitoring data must be integrated with numerical analyses.

However, many numerical heat-transfer and structural analyses adopt a "divide-and-conquer" approach in practice. That is, the heat-transfer analysis is performed on the deck/towers with a 2D or local 3D model. The calculated temperatures of the components are then input into another global bridge model to calculate temperature responses via the structural analysis. This process is inefficient because it requires considerable manual intervention. The results are also inaccurate because the simplified 2D/3D models fail to simulate the real temperature distribution of the entire bridge. An integrated heat-transfer and structural analysis has been developed in this book using a unified FE model.

Although field monitoring and numerical analyses can be integrated together to study the temperature behavior more accurately, the study lacks generalization. Theoretical analyses of temperature behavior have been conducted based on the structural mechanics. General and simple formulas are derived, providing a convenient means to study the temperature mechanisms and offering engineers straightforward tools for design and analysis. Such formulas are rare in the literature.

Four types of bridges, namely, suspension bridges, cable-stayed bridges, arch bridges, and beam bridges, have been studied comprehensively via field monitoring, numerical simulation, and theoretical analysis. Multiple real bridges have been used as the testbeds to verify the accuracy of the proposed analyses and formulas.

DOI: 10.1201/9781003480136-10

10.2 CHALLENGES

Although a considerable amount of contents have been studied, some issues are not well solved. This section raises some problems in the current studies and proposes possible solutions.

10.2.1 Heat-Transfer Analysis

Thermal boundary conditions play a critical role in the heat-transfer analysis. Air temperature and solar radiation contribute most to the external thermal boundary conditions. However, almost all SHM systems do not install pyranometers to measure the solar radiation around the bridge. Most studies rely on empirical formulas or the meteorological data from weather stations around. This is one main source of the temperature simulation errors. Therefore, more pyranometers should be installed in the existing SHM systems, and new SHM systems should include the sensors.

The second source of errors is the internal thermal boundary conditions. The air temperatures along the vertical and transverse directions within a box girder are non-uniform because of the airflow inside the girder. For example, the air temperature close to the upper deck is higher than that close to the bottom deck during the daytime; the air temperature in the south may be higher than that in the north. However, almost all existing studies have treated the interior bridge temperatures as uniform and ignored the non-uniformity due to a limited number of sensors. Computational fluid dynamics may be used to simulate the airflow inside the box girder and calculate non-uniform air temperatures (Zhang et al., 2024).

Material thermal properties (such as specific heat capacity, thermal conductivity, absorption coefficient, etc.) significantly affect the simulation results. Most of current studies use the data from literatures, which may differ from the real parameters. These thermal properties need to be measured in the laboratory or on site. For example, spectrophotometers can be used to measure the absorptivity of solar spectrum, and Fourier transform infrared spectrometer can be used to measure the emissivity of the middle/far infrared spectrum.

10.2.2 Machine Learning Techniques

Existing studies of the thermal behaviors of bridges can be categorized into physics-based (also known as white-box) or data-driven (black-box) approaches. Heat-transfer analyses belong to the former. As explained before, this approach has the following practical difficulties. First, most numerical analyses adopt a divide-and-conquer strategy; namely, the heat-transfer and structural analyses are conducted on different FE models. Second, considerable effort in FE modeling is needed, and the calculation is time-consuming. Third, the thermal boundary conditions are usually not available or incomplete, causing the simulation results to be inaccurate. Lastly, the analysis is structure-dependent, and the results cannot be directly applied to other bridges.

Most data-based studies adopt linear regression analysis to explore the relationship between temperature and bridge responses at limited measurement points. However, this approach cannot capture the behavior of the entire bridge. Moreover, data-based machine learning models often lack physical interpretability, leading to overfitting the available data and limiting their applicability to other structures. They lack adaptability and generalization.

Grey-box machine learning-based techniques have been developed in the past few years by integrating the strengths of the physics-based and data-driven methods. However, grey-box models for the temperature behavior of bridges are lacking. The grey-box model should utilize the advantages of the white-box and black-box but avoid their weakness. Moreover, transfer learning-based neural networks can be developed to transfer the grey-box models from one bridge to another. Therefore, the neural networks should have a generalization ability to obtain temperature behaviors with similar geometric dimensions. This may be worth of further studies.

10.2.3 Global Warming

Global warming is a major global threat. Most studies on global warming focus on the change of air temperature and its consequences to the environment and ecology of inhabitants, including the migration and extinction of animals, the rise of the global sea level, extreme weather conditions, and the melting of glaciers. Regarding the safety of civil engineering structures (bridges, buildings, dams, etc.) in the context of global warming, researchers have focused on issues such as the carbonation of concrete, steel corrosion, groundwater intrusion, extreme waves and storms, scour and flood risks to bridges, and accelerated deterioration of pavement materials due to increased temperatures (Talukdar et al., 2012). However, studies on the global warming-induced structural temperature of bridges and its direct consequence to bridge safety are few. How much structural temperature has changed and how much temperature-induced responses occurred when the mean air temperature rose during the past few decades are unclear.

The effects of global warming on bridge safety cannot be ignored for the following reasons: (1) The bridge surface temperature is much higher than the air temperature, and the former directly affects the structural behavior. (2) The maximum temperature change is more significant than the average temperature change, and structural safety is governed by the maximum values (or extreme values). (3) The design service life of bridges is generally 100 years or longer, whereas the current design standards are based on historical climatic data but do not consider the increasing global warming issue. Thus, the temperature behavior of bridges during their long-term design service life must be investigated.

One practical challenge is that the long-term actual data of bridges is lacking. The Tsing Ma Bridge in Hong Kong is equipped with a long-term real-time SHM system. Being established in 1997, this system is one of the earliest ones. Other earliest SHM systems include the Akashi Kaikyo Bridge in Japan since 1998 (Kitagawa, 2004) and the Confederation Bridge in Canada since 1997 (Desjardins et al., 2006). These rare and precious data may provide the actual temperature and behavior of the bridges during the past decades. They can be used as testbeds to study the global warming effects on bridges' temperature behaviors.

Moreover, to predict the future temperature behaviors of the bridges in their service lives, the local air temperature and solar radiation in the future are required. The global climate models (GCMs) are the primary tools and sources for predictions of the future potential impacts of different systems under climate change. GCMs from the Coupled Model Intercomparison Project Phase 6 effectively reproduce large-scale atmospheric circulations. By modulating the intensity of radiative forcing, GCMs can produce future projections under five Shared Socioeconomic Pathways (i.e., SSP1-1.9, SSP1-2.6, SSP2-4.5, SSP3-7.0, and SSP5-8.5). However, limited by a coarse resolution, GCMs are not expected to provide detailed information on small-scale local climate dynamics, which rely on resolving local features, such as complex terrain topology, heterogeneous land covers, and land uses. With trustworthy observations for validation, the small-scale climate details can be estimated from large-scale circulations using downscaling techniques, such as statistical downscaling, dynamical downscaling, and hybrid statistical-dynamical downscaling (Fowler et al., 2007). The conventional statistical downscaling technique aims at building an empirical mathematical relationship between large-scale predictors and fine-scale predictands directly, while the dynamical downscaling approach captures the complex local physics and climate responses, which makes it extremely computationally expensive. To balance computational efficiency, output accuracy, and physical interpretability, the hybrid statistical-dynamical technique is a promising solution for constructing solid and efficient predictions of finer-resolution environment details derived from GCMs (Walton et al., 2015).

By using the historical data of the bridge and GCMs during the past few decades, the statistical-dynamical downscaling framework can be developed. A multiscale downscaling approach that spans global, regional, and local scales may be sought to bridge the scale gap between the GCM and the local bridge. Based on these, the future local environmental conditions, namely, wind, air temperature, and solar radiation, can be projected. The integrated heat-transfer and structural analysis will be conducted for each emission scenario to forecast bridge temperature distribution and responses. The safety condition of the bridges will then be evaluated. Extreme value models of temperature will be developed. Current bridge design standards for the temperature load will be evaluated. Recommendations about structural analysis, design, and field monitoring will be offered.

REFERENCES

Desjardins, S. L., Londono, N. A., Lau, D. T. and Khoo, H., 2006, Real-time data processing, analysis and visualization for structural monitoring of the confederation bridge. *Advances in Structural Engineering* **9(1)**: 141–157.

Fowler, H. J., Blenkinsop, S. and Tebaldi, C., 2007, Linking climate c hange modelling to impacts studies: Recent advances in downscaling techniques for hydrological modelling. *International Journal of Climatology* **27(12)**: 1547–1578.

Kitagawa, M., 2004, Technology of the Akashi Kaikyo Bridge. *Structural Control and Health Monitoring* **11(2)**: 75–90.

Talukdar, S., Banthia, N. and Grace, J., 2012, Carbonation in concrete infrastructure in the context of global climate change–Part 1: Experimental results and model development. *Cement and Concrete Composites* **34(8)**: 924–930.

Walton, D. B., Sun, F., Hall, A. and Capps, S., 2015, A hybrid dynamical–statistical downscaling technique. Part I: Development and validation of the technique. *Journal of Climate* **28(12)**: 4597–4617.

Zhang, L., Shan, Y., Li, L., Wang, F. and Xia, Y., 2024, Thermal boundary conditions for heat transfer analysis of bridges considering non-uniform distribution of internal air temperature by computational fluid dynamics. *Journal of Civil Structural Health Monitoring* **14(5)**: 1295–1310.

Index

Note: **Bold** page numbers refer to tables, *italic* page numbers refer to figures.

FSC
www.fsc.org
MIX
Papier | Fördert gute Waldnutzung
FSC® C083411

Kolibri 360 GmbH
Ferdinand-Jühlke-Straße 7
99095 Erfurt, Deutschland
produktsicherheit@kolibri360.de